Theory of Many-Particle Systems

Translation Series

Theory of Many-Particle Systems

I. P. Bazarov
P. N. Nikolaev
Moscow State University

Translated by
George Adashko

Library of Congress Cataloging-in-Publication Data

Bazarov, I. P.
Theory of many-particle systems.

Translation of: Teorii͡a sistem mnogikh chastit͡s.
Bibliography: p.
1. Many-body problem. 2. Statistical physics.
I. Nikolaev, P. N. (Pavel Nikolaevich). II. Title. III. Series.
QC174.17.P7B3913 1989 530.1'44 88-35109
ISBN 978-0-88318-601-5

Contents

Chapter 3. Cluster form of the equation of state of a real gas

Chapter 4. Integral equations in the theory of the liquid state

Chapter 5. Numerical methods in statistical physics

Chapter 6. Principles of the theory of the crystalline state

Chapter 7. Correlation theory of crystals

Foreword

Many-particle systems are investigated by thermodynamic and statistical methods. This book deals with the statistical theory of macroscopic objects. Formulation of the fundamental problems is followed by a systematic treatment of the gaseous, liquid, and crystalline states of matter. The free energy of a system is calculated anew, starting from expressions that are asymptotic in the number of particles. This leads to a number of original results. The theoretical conclusions are compared in all cases with the experimental data or with data obtained by the molecular-dynamics or by the Monte Carlo method. In view of the rapid advances in computational techniques, these methods have recently led to substantial progress in the theory of many-particle systems. This approach to the exposition makes for a better understanding of the ability of the various theoretical methods to solve specific problems. Our aim is to meet the needs of both theoreticians and experimentalists for a treatment of the theory of many-particle systems.

Foreword

Chapter 1
Initial premises and basic equations of the theory of many-particle systems

1. Fundamental problems of the theory of many-particle systems and their probabilistic description

The task of the theory of many-particle systems, or statistical physics, is to explain the properties of macroscopic systems on the basis of our knowledge of the forces involved in the interaction between particles.

Statistical physics encompasses the theory of equilibrium states (statistical thermodynamics) and the theory of nonequilibrium processes (physical kinetics) in various systems.

Owing to the large number of particles in macroscopic systems ($N \sim N_A = 6022 \times 10^{23}$ mol^{-1}), their disordered motion acquires new properties. We have here an example of a "dialectic" transformation of quantity into quality, wherein an increase in the number of mechanically moving particles in a body generates a qualitatively new form of motion, viz., thermal motion.

Statistical physics and thermodynamics deal with the thermal form of the motion of matter and with its variety of laws. Thermodynamics describes the properties of equilibrium systems from first principles (the laws of thermodynamics), and invokes an aggregate of other experimental results (equations of state, etc.), without resorting to any notions concerning the molecular structure of matter. Statistical physics, on the other hand, is based on some of a number of molecular models of the systems, from which it derives by appropriate methods the equations of state and all other properties.

It is advantageous, in view of the different methods used to investigate many-particle systems in thermodynamics and in statistical physics, to describe separately these branches of theoretical physics, although it is impossible at present to draw a clear-cut boundary between them.

Thermodynamic systems are described by macroscopic parameters, such as density, elasticity, temperature, entropy, and others. These parameters, which define adequately the state of such a system, are uniquely connected by various relations, so that thermodynamics, just as mechanics, is a dynamic theory.

In those cases when the thermodynamic relations do not satisfy our requirements and we wish to establish the properties of a many-particle system by starting from the properties of the particles themselves, we must turn to the molecular theory of these systems, i.e., to statistical physics.

Let us consider initially, for simplicity, a classical molecular model of a body comprising a system of material points that move in accordance with the laws of classical mechanics.*

2. Principles of the mechanics of many-particle systems

The solution of the fundamental problem of statistical physics begins with the definition of the microstate of a macroscopic system, i.e., the mechanical state of an aggregate of a large number of particles. We describe, therefore, first those principles of classical and quantum mechanics which will be needed later on.

2.1 Classical mechanics

In classical mechanics we start out from the Hamiltonian or canonical equations. For a system of N particles with holonomic ideal constraints, these equations take in the absence of dissipative forces the form

$$\dot{q}_i = \frac{\partial H}{\partial p_i}, \quad \dot{p}_i = -\frac{\partial H}{\partial q_i} \quad \begin{pmatrix} i = 1,2,\dots,n, \\ n = 3N \end{pmatrix}, \tag{1.1}$$

where $H = H(t, q_1, q_2,\dots, q_n, p_1, p_2,\dots, p_n)$ is the Hamiltonian of the system; q_i and p_i are the generalized coordinates and momenta of the particles, and t is the time.[3]

The canonical Hamiltonian equation can be solved with the aid of an action function S that satisfies a first-order partial differential equation (the Hamilton–Jacobi equation)

$$\frac{\partial S}{\partial t} + H\left(q_1,\dots,q_n,\frac{\partial S}{\partial q_1},\dots,\frac{\partial S}{\partial q_n},t\right) = 0, \tag{1.2}$$

where

$$H\left(q, \frac{\partial S}{\partial q}, t\right)$$

*The laws of classical mechanics are the approximate laws of atomic physics, so that classical statistical mechanics is the limiting case of quantum statistical physics. This limiting approach explains better the main ideas of statistical physics, and is in fact the basis of our treatment. Furthermore, to calculate the macroscopic parameters of many-particle systems it is necessary, in contrast to classical statistical mechanics, to average twice, since quantum mechanics is itself a statistical theory.

is the system Hamiltonian with the generalized momenta replaced by the partial derivatives of S with respect to the corresponding coordinates. The total integral of the Hamilton–Jacobi equation is

$$S = S(t, q_1,...,q_n, \alpha_1,...,\alpha_n) + \alpha_{n+1} , \tag{1.3}$$

where α_i are arbitrary constants.

According to the Jacobi theorem, the equations

$$p_i = \frac{\partial S}{\partial q_i}, \quad \beta_i = \frac{\partial S}{\partial \alpha_i} \quad (i = 1,2,...,n) \tag{1.4}$$

determine the solution of the canonical set of equations (β_i are arbitrary constants).

In the case of a conservative system, the Hamiltonian does not depend explicitly on the time, is equal to the sum of the kinetic and potential energies of the system, and is an integral of the motion, i.e., a function of q_i and p_i that remains constant as the system moves. In fact,

$$\frac{dH}{dt} = \frac{\partial H}{\partial t} + \sum_{i=1}^{n} \left(\frac{\partial H}{\partial q_i}\dot{q}_i + \frac{\partial H}{\partial p_i}\dot{p}_i \right)$$

and according to Eq. (1.1) we have under the condition $\partial H/\partial t = 0$,

$$H(q_1,...,q_n, p_1,...,p_n) = \text{const.}$$

A conservative closed system with n degrees of freedom has altogether $(2n - 1)$ integrals of motion. This can be easily verified by dividing the remaining $(2n - 1)$ canonical equations (1.1) by $\dot{p}_1 = -\partial H/\partial q_1$. We obtain then the $(2n - 1)$ equations

$$\frac{dq_1}{dp_1} = -\frac{\dfrac{\partial H}{\partial p_1}}{\dfrac{\partial H}{\partial q_1}}, ..., \frac{dp_n}{dp_1} = \frac{\dfrac{\partial H}{\partial q_n}}{\dfrac{\partial H}{\partial p_1}} , \tag{1.5}$$

whose integration leads to $(2n - 1)$ integrals of the motion.

The most important role in physics is played by seven integrals of motion: the energy, the three momentum components, and the three angular-momentum components of the system. These integrals describe the main properties of space and time, viz., the law of energy conservation (homogeneity of time), the law of momentum conservation (homogeneity of space), and the law of angular-momentum conservation (isotropy of space).

Regarding q_i and p_i as rectangular coordinates of the particles in $2n$-dimensional space, we define the state of a system as a point in $2n$-dimensional space called phase space. The phase spaces of an individual particle and of an aggregate of particles are usually referred to as the μ and Γ spaces, respectively.[4]

A change of the state of a system is described by the motion of a phase point along a phase trajectory.

The phase space of a Hamiltonian system has two important properties: (a) only one phase trajectory passes through each point of phase space, so that a phase trajectory cannot intersect itself*; (b) the volume occupied by a continuous aggregate of phase points remains unchanged by motion of these points. The first of these properties follows from the uniqueness of the solutions of the set (1.1) of canonical equations. Let us prove the second property, which represents the Liouville theorem, that a given volume in phase space is invariant. We consider, following Gibbs,[1] not one but a very large (in the limit-infinite) number N_0 of identical systems in different states—the Gibbs phase ensemble. At the instant of time t_0 the ensemble occupies a region $\mathscr{G}_0$ of phase space with volume

$$\Gamma_0 = \int \cdots \int_{\mathscr{G}_0} dq_{10}\, dq_{20} \cdots dq_{n0}\, dp_{10}\, dp_{20} \cdots dp_{n0} \,. \tag{1.6}$$

At the instant t, the ensemble will occupy a region $\mathscr{G}$ having a volume

$$\Gamma = \int \cdots \int_{\mathscr{G}} dq_1\, q_2 \cdots dq_n\, dp_1\, dp_2 \cdots dp_n \,. \tag{1.7}$$

To determine the connection between Γ_0 and Γ we rewrite Eq. (1.7) in the form

$$\Gamma = \int \cdots \int_{\mathscr{G}_0} \left| \frac{\partial(q_1, q_2, \ldots, q_n, p_1, p_2, \ldots, p_n)}{\partial(q_{10}, q_{20}, \ldots, q_{n0}, p_{10}, p_{20}, \ldots, p_{n0})} \right| dq_{10}\, dq_{20} \cdots dq_{n0}\, dp_{10}\, dp_{20} \cdots dp_{n0} \,, \tag{1.8}$$

where

$$\mathscr{D} = \frac{\partial(q_1, q_2, \ldots, q_n, p_1, p_2, \ldots, p_n)}{\partial(q_{10}, q_{20}, \ldots, q_{n0}, p_{10}, p_{20}, \ldots, p_{n0})} \tag{1.9}$$

is the Jacobian of the transformation from the variables $q_1, q_2, \ldots, q_n, p_1, p_2, \ldots, p_n$ to the variables $q_{10}, q_{20}, \ldots, q_{n0}, p_{10}, p_{20}, \ldots, p_{n0}$.

It is easily seen that

$$\mathscr{D}(t_0) = 1 \,. \tag{1.10}$$

In accord with the properties of Jacobians we have

$$\frac{\partial(q_1, \ldots, q_n, p_1, \ldots, p_n)}{\partial(q_{10}, \ldots, q_{n0}, p_{10}, \ldots, p_{n0})} = \frac{\partial(q_1, \ldots, q_n, p_1, \ldots, p_n)}{\partial(q_i', \ldots, q_n', p_1', \ldots, p_n')} \frac{\partial(q_1', \ldots, q_n', p_1', \ldots, p_n')}{\partial(q_{10}, \ldots, q_{n0}, p_{10}, \ldots, p_{n0})} \,, \tag{1.11}$$

where q_i' and p_i' are the values of the coordinates and of the momenta at an arbitrary instant t'. Differentiating the identity (1.11) with respect to t, we have

*Configuration space (the space of the generalized coordinates) does not have this property.

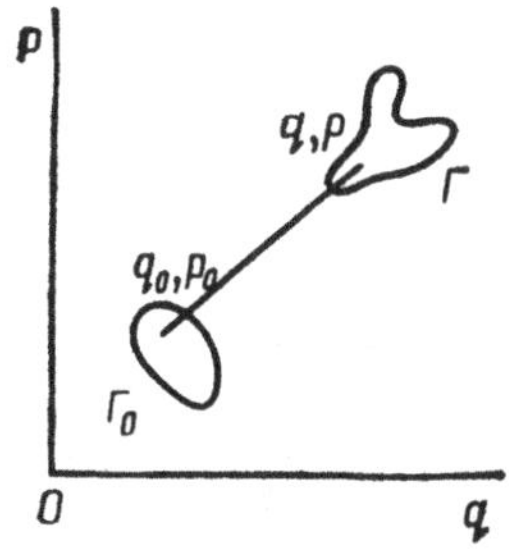

Figure 1.

$$\frac{d\mathscr{D}}{dt} = \frac{\partial(q'_1,\dots,q'_n,p'_1,\dots,p'_n)}{\partial(q_{10},\dots,q_{n0},p_{10},\dots,p_{n0})} \sum_{i=1}^{n} \left[\frac{\partial(q_1,q_2,\dots,\dot{q}_i,\dots,q_n,p_1,\dots,p_n)}{\partial(q'_1,\dots,q'_n,p'_1,\dots,p'_n)} \right.$$
$$\left. + \frac{\partial(q_1,\dots,q_n,p_1,\dots,\dot{p}_i,\dots,p_n)}{\partial(q'_1,\dots,q'_n,p'_1,\dots,p'_n)} \right]$$
$$= \frac{\partial(q'_1,\dots,q'_n,p'_1,\dots,p'_n)}{\partial(q_{10},\dots,q_{n0},p_{10},\dots,p_{n0})} \sum_{i=1}^{n} \left[\frac{\partial \dot{q}_i}{\partial q'_i} + \frac{\partial \dot{p}_i}{\partial p'_i} \right]. \tag{1.12}$$

Since t' is arbitrary we get, by putting $t' = t$ in Eq. (1.12) and taking Eq. (1.1) into account,

$$\frac{d\mathscr{D}}{dt} = 0, \tag{1.13}$$

i.e., the Jacobian (1.9) is independent of time, and hence by virtue of Eq. (1.10),

$$\mathscr{D}(t) = 1. \tag{1.14}$$

Consequently,

$$\Gamma = \Gamma_0, \tag{1.15}$$

i.e., the phase-space volume occupied continuously by an aggregate of phase points is not changed by the motion of these points (see Fig. 1)—this is the Liouville theorem.

In other words, the motion of the phase points of Hamiltonian systems is similar to that of an incompressible fluid.

We conclude by finding the total time derivative of an arbitrary function $\varphi(t,q,p)$ of the phase-point coordinates $q = \{q_i\}$ and momenta $p = \{p_i\}$ at the instant of time t:

$$\frac{d\varphi}{dt} = \frac{\partial\varphi}{\partial t} + \sum_{i=1}^{n} \left[\frac{\partial\varphi}{\partial q}\dot{q}_i + \frac{\partial\varphi}{\partial p_i}\dot{p}_i \right]. \tag{1.16}$$

Substituting in this equation $\dot{q}_i$ and $\dot{p}_i$ from Eq. (1.1), we get

$$\frac{d\varphi}{dt} = \frac{\partial\varphi}{\partial t} + [\varphi, H], \tag{1.17}$$

where

$$[\varphi, H] = \sum_{i=1}^{n} \left[\frac{\partial \varphi}{\partial q_i}\frac{\partial H}{\partial p_i} - \frac{\partial H}{\partial q_i}\cdot\frac{\partial \varphi}{\partial p_i}\right] \tag{1.18}$$

are Poisson brackets.

Introducing the Liouville operator

$$\Pi = \sum_{i=1}^{n} \left[\frac{\partial H}{\partial p_i}\frac{\partial}{\partial q_i} - \frac{\partial H}{\partial q_i}\frac{\partial}{\partial p_i}\right], \tag{1.19}$$

we can rewrite Eq. (1.16) in the form

$$\frac{d\varphi}{dt} = \frac{\partial \varphi}{\partial t} + \Pi\varphi . \tag{1.20}$$

2.2 Quantum mechanics

In classical mechanics, the state of a system is defined by specifying all the coordinates and momenta of the system, and the change of the state is defined by the canonical equations (1.1). No such definition of the state of a system is possible in quantum mechanics, since the coordinates and momenta of the system cannot be measured simultaneously. The state of a system at a fixed instant of time is defined in quantum mechanics by an aggregate of simultaneously measurable independent physical quantities—a complete set of variables analogous to the aggregate of the canonical variables of classical mechanics. The number of variables in the complete set is equal to the number of degrees of freedom of the system and is less than the aggregate of the classical variables. From the standpoint of classical mechanics, the quantum-mechanical description is therefore incomplete. The complete set of variables of a quantum system is usually chosen to be the set of coordinates (description of the state in the coordinate representation).

The state of a quantum system is described by a wave function $\psi(t, q_1, q_2,...,q_n)$ in the sense that all the probabilities of the results of the measurements of the system can be expressed in terms of this wave function.[5–7]

Thus

$$\psi(t, q_1, q_2,...,q_n)\psi^*(t, q_1, q_2,...,q_n)dq_1\, dq_2 \cdots dq_n \tag{1.21}$$

is the probability of observing the system at the instant of time t in a configuration-space element $q_1, q_1 + dq_1; q_2, q_2 + dq_2; ...; q_n + dq_n$.

The wave function satisfied the normalization condition

$$\int \psi(t, q_1,...,q_n)\psi^*(t, q_1,...,q_n)dq_1 \cdots dq_n = 1 . \tag{1.22}$$

The integral in Eq. (1.22) diverges in some cases. The ratio of $\psi\psi^*$ for two neighboring points determines then the probability of observing the system at the corresponding coordinates.

According to the superposition principle, all the equations satisfied by the wave functions must be linear in ψ.

The wave function depends not only on the complete set of variables, but also on the time as a parameter. The explicit time dependence of the wave function is written out only when the system varies significantly in the course of time. If, however, only the state at a fixed instant of time is of interest, the time dependence of ψ is omitted.

To find the wave function that describes a particular state in which a given quantity (or a set of quantities) has a specified value, each physical quantity is set in correspondence in quantum mechanics with a separate operator such that its eigenvalues yield the possible values of this quantity. The eigenfunctions of this operator describe then the state in question.

Thus, if $\hat{f}$ is the operator of the quantity f, the spectrum of its possible values f_s in the corresponding states ψ_s is obtained from the equation

$$\hat{f}\psi_s = f_s\psi_s (s = 1,2,...) . \tag{1.23}$$

This means that f has in the state ψ_s a definite value f_s. If, on the other hand, the state ψ of the system is unequal to any of the ψ_s, different measurements yield for f different values ranging over a spectrum $f_1, f_2, ..., f_s, ...$. The expectation value of f in the state ψ is then

$$\bar{f} = \int \psi^* \hat{f} \psi \, dq_1 \cdots dq_n . \tag{1.24}$$

Assume n particles without internal degrees of freedom. The principal elements are then $2n = 6N$ Hermitian (self-adjoint) operators

$$\hat{q}_i, \hat{p}_i \quad (i = 1,2,...,n) , \tag{1.25}$$

that meet the condition

$$[\hat{q}_i, \hat{p}_i] = \hat{q}_i\hat{p}_i - \hat{p}_i\hat{q}_i = i\hbar\hat{I} , \tag{1.26}$$

where $\hat{I}$ is the unit operator.

The time variation of a wave function is defined by the Schrödinger equation

$$i\hbar\frac{\partial\psi}{\partial t} = \hat{H}\psi , \tag{1.27}$$

where $\hat{H}$ is the Hamiltonian operator. The Schrödinger equation for stationary states is

$$\hat{H}\psi(q_1,...,q_n) = E\psi(q_1,...,q_n) . \tag{1.28}$$

A wave function can be expanded in a complete set of functions.

Let us find for the operator $\hat{f}$ the time derivative $\hat{\dot{f}}$ defined as the quantity whose mean value is equal to the time derivative of the expectation value $\bar{f}$:

$$\bar{\dot{f}} = \dot{\bar{f}} . \tag{1.29}$$

Differentiating Eq. (1.24) for $\bar{f}$ with respect to time, we get

$$\dot{\bar{f}} = \frac{d\bar{f}}{dt} = \int (\psi^* \dot{\hat{f}}\psi + \dot{\psi}^* \hat{f}\psi + \psi^* \hat{f}\dot{\psi}) dq \, .$$

Taking into account Eq. (1.27) and its complex conjugate

$$-i\hbar \frac{\partial \psi^*}{\partial t} = \hat{H}^* \psi^* \, ,$$

we get

$$\bar{f} = \int \left(\psi^* \dot{\hat{f}}\psi + \frac{i}{\hbar} \hat{H}^* \psi^* \hat{f}\psi - \frac{i}{\hbar} \psi^* \hat{f}\hat{H}\psi \right) dq \, .$$

Since $\hat{H}$ is Hermitian, we obtain ultimately

$$\dot{\bar{f}} = \int \psi^* \left(\dot{\hat{f}} + \frac{i}{\hbar} \hat{H}\hat{f} - \frac{i}{\hbar} \hat{f}\hat{H} \right) \psi \, dq = \bar{\dot{f}} = \int \psi^* \dot{\hat{f}} \psi \, dq \, ,$$

whence

$$\dot{\hat{f}} = \frac{\partial \hat{f}}{\partial t} + [\hat{f}, \hat{H}] \, , \tag{1.30}$$

where

$$[\hat{f}, \hat{H}] = (\hat{f}\hat{H} - \hat{H}\hat{f}) / i\hbar \, . \tag{1.31}$$

In classical mechanics, the particles of a system can be numbered, each of their trajectories determined, and the motion of an individual particle tracked. In quantum mechanics this is impossible by virtue of the particle-wave duality, meaning that quantum particles are indistinguishable.

A particle system is described in quantum mechanics either by symmetric wave functions (a system of bosons) or by antisymmetric ones (a system of fermions subject to Pauli exclusion). Let ψ be the wave function of a particle system. Following permutation of two particles* we have

$$\psi(\ldots, \mathbf{q}_k, \ldots, \mathbf{q}_l, \ldots) = \delta \psi(\ldots, \mathbf{q}_l, \ldots, \mathbf{q}_k, \ldots) \, , \tag{1.32}$$

where

$$\delta = \begin{cases} +1 & \text{for bosons} \\ -1 & \text{for fermions} \end{cases} \tag{1.33}$$

and in both cases

$$|\psi(\ldots, \mathbf{q}_k, \ldots, \mathbf{q}_l, \ldots)|^2 = |\psi(\ldots, \mathbf{q}_l, \ldots, \mathbf{q}_k, \ldots)|^2 \, .$$

Consider a system of noninteracting particles. Let $\{\psi p_i(\mathbf{q}_i)\} (i = \overline{1, N})$ be the wave functions of the various possible states of the particles. The normalized wave function for bosons takes the form

$$\psi = \left(\frac{N_1! N_2! \ldots}{N!} \right)^{1/2} \sum_p \prod_{i=1}^{N} \psi p_i(q_i) \tag{1.34}$$

(the summation is over all different permutations of the indices $p_1, p_2, \ldots, p_N$,

*Here and elsewhere $\mathbf{g}_i$ denotes the radius vector of the ith particle.

$\sum_i N_i = N$, N_i, and N_i is equal to the number of subscripts having identical values of i).

The wave function for a system of fermions can be written in the form

$$\psi = \frac{1}{\sqrt{N!}} \begin{vmatrix} \psi p_1(q_1) & \psi p_1(q_2) & \cdots & \psi p_1(q_N) \\ \psi p_2(q_1) & \psi p_2(q_2) & \cdots & \psi p_2(q_N) \\ \vdots & \vdots & \cdots & \vdots \\ \psi p_N(q_1) & \psi q_N(q_2) & \cdots & \psi p_N(q_N) \end{vmatrix} . \tag{1.35}$$

We see thus that two particles cannot be in the same state, for then the determinant (1.35) would have two equal rows and the wave function would vanish.

Quantum-system states describable by a wave function are called *pure states*. They contain the maximum possible information on the system, i.e., are defined by a complete set of physical quantities. If, however, the system conditions are such that it has no complete set of physical values, its state cannot be described by a definite wave function, i.e., it has no wave function. States that have no wave functions are called *mixed states*.

A system is in a mixed state if it is a part of a closed system. Indeed, let the closed system itself be in a certain state described by a wave function $\psi(x, q)$, where x is the aggregate of the coordinates of our (mixed) system, and q the remaining coordinates of the closed system. This wave function cannot, generally speaking, be a product of functions of x only and of q only, so that the mixed system does not have a wave function of its own. In fact, the probability density of observing a closed system with coordinates x and q is equal to

$$W(x, q) = |\psi(x, q)|^2 ,$$

and the probability density of observing a system with coordinates x (at arbitrary values of q) is

$$W(x) = \int W(x, q)dq = \int |\psi(x, q)|^2 dq .$$

Were the considered system to have a wave function $\psi(x)$, this function would obviously be

$$\psi(x) = \int \psi(x, q)dq$$

and would satisfy the equality

$$|\psi(x)|^2 = W(x)$$

or

$$\left| \int \psi(x, q)dq \right|^2 = \int |\psi(x, q)|^2 dq ,$$

which is impossible, except in the case of noninteracting parts, when

$$\psi(x, q) = \psi(x)\varphi(q) .$$

Thus any closed-system part that interacts with other parts has no wave function. The expectation value $\langle f \rangle$ of any quantity f in such a system can be determined from the wave function $\psi(x, q)$ of the closed system, and is equal, in analogy with Eq. (1.24), to

$$\langle f \rangle = \int \psi^*(x, q)\hat{f}\psi(x, q)dx\, dq\,, \tag{1.36}$$

where the operator $\hat{f}$ of the quantity f of our system acts only on the x coordinates.

Expression (1.36) for the expectation value of a physical quantity f of our system can be written in the form

$$\langle f \rangle = \int [\,\hat{f}\rho(x, x')]_{x=x'}\, dx\,, \tag{1.37}$$

where

$$\rho(x, x') = \int \psi(x, q)\psi^*(x', q)dq\,, \tag{1.38}$$

and after calculating the action of the operator $\hat{f}$ on x only, it is necessary to put $x' = x$ in the function $\rho(x, x')$.

The function $\rho(x, x')$ is called the density matrix, and the operator $\hat{\rho}$ corresponding to this matrix is called the statistical operator or the density operator.

From the definition (1.38) of the density matrix it is seen that it is self-adjoint, or Hermitian:

$$\rho^*(x, x') = \rho(x', x)\,. \tag{1.39}$$

Knowing the density matrix, one can find the expectation values of the physical quantities and the probabilities of their different values. The state of a system that has no wave function can thus be described by a density matrix.

The most general quantum-mechanical description of a system is with the aid of a density matrix. On the other hand, a description using a wave function $\psi(x)$ is a particular case corresponding to a density matrix

$$\rho(x, x') = \psi(x)\psi^*(x')\,. \tag{1.40}$$

3. Liouville equation. Bogolyubov method

The mechanical (or microscopic) state of a classical many-particle system is specified by the particle coordinates and momenta, or by the phase point (q, p). The thermodynamic (and, in general, macroscopic) state of this system is determined by much fewer macroscopic parameters (volume, density, etc.).[8] Each macrostate of the system is made up of a large number of its different microstates or phase points. In the general case, some microstates are more frequently realized, others less so. It is evident therefore that, from the macroscopic viewpoint, the macrostate of a system is characterized not by a specified phase point (i.e., by the canonical variables of the system), but by the relative density of these states in phase space, or by the phase probability density, i.e., by the quantity

$$\rho(t, q, p) = \frac{dN}{d\Gamma}\Big/ \Gamma_0 . \tag{1.41}$$

Here N_0 is the total number of the phase points ($N_0 \rightarrow \infty$ if the microstates vary continuously), and dN is their number in a phase-space volume element $d\Gamma$.

The aggregate of identical macroscopic systems in different microscopic states (or the aggregate of the phase points that represent these system states) is called the Gibbs phase ensemble. If, however, each state of a phase ensemble is characterized by a definite probability, such a phase ensemble is called a statistical ensemble.

A Gibbs statistical ensemble is thus specified by the probability density of the microstate of the system, or by the phase distribution function (1.41), which is obviously normalized to unity:

$$\int \rho\,(t, \mathbf{q}_1, \ldots, \mathbf{q}_N,\ \mathbf{p}_1, \ldots, \mathbf{p}_N)\, d\mathbf{q}_1 \cdots d\mathbf{p}_N = 1 . \tag{1.42}$$

The determination of this function and its use to obtain the macroscopic parameters of a system is the main subject of the statistical theory of many-particle systems.[9]

The equation for the phase distribution function follows directly from the Liouville theorem (1.15) for Hamiltonian systems. In fact, when the phase points move in phase space their number does not change, so that all the phase points located at an instant t in a volume element $d\Gamma$ go over at the instant t' into an element $d\Gamma'$. Consequently,

$$\rho(t, q, p) d\Gamma = \rho(t', q', p') d\Gamma' ,$$

and since $d\Gamma = d\Gamma'$ by the Liouville theorem, we have

$$\rho(t, q, p) = \rho(t', q', p') \tag{1.43}$$

or

$$d\rho(t, q, p)/dt = 0 ,$$

i.e., $\rho(t, q, p)$ is constant along the phase trajectories.

The invariance of the phase distribution function along the phase trajectories (1.43) can be regarded as one of the formulations of the Liouville theorem.

From this invariance and from expression (1.20) for the total time derivative of the wave function we obtain the Liouville equation

$$\frac{\partial \rho}{\partial t} = [H, \rho] . \tag{1.44}$$

This function is the basis of the statistical investigation of classical many-particle systems and is therefore the fundamental equation of the statistical mechanics of these systems.

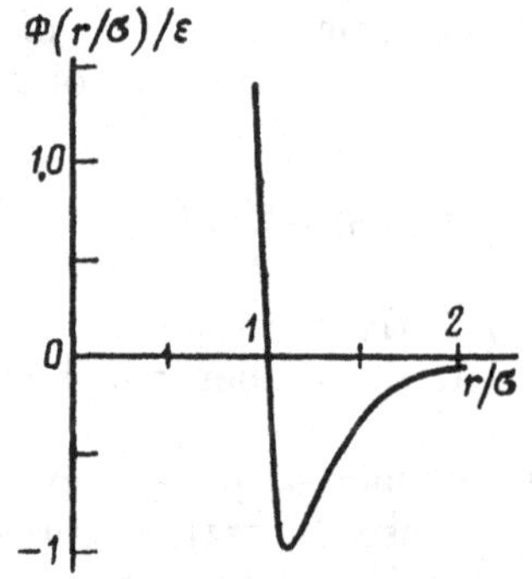

Figure 2.

Equations (1.44) and (1.1) are equivalent, and the statistics stems from the uncertainty of the initial dynamic state of the system.[2] The solution of the Liouville equation for a function of $6N+1$ variables is just as complicated a task as the solution of the dynamic system (1.1). It does, however, lead to simpler equations for the probabilities of one or several particles of the system in elements of the corresponding phase space. The investigation of the properties of these partial distribution functions is the gist of the Bogolyubov method.[2]

Following Bogolyubov, we introduce distribution functions $F_s(t, x_1,...,x_s)$ that are symmetric functions of $x_1,...,x_s$ $[x_i = (\mathbf{p}_i, \mathbf{q}_i)]$ and are so normalized that

$$\frac{1}{V^s} F_s(t, x_1,...,x_s)dx_1 \cdots dx_s \tag{1.45}$$

is the probability that the dynamic states of a given group of s molecules are located at the instant t in infinitely small volumes $dx_1,...,dx_s$ near the points $q_1,...,q_s$. Hence

$$F_s(t, x_1,...,x_s) = V^s \int_{\Omega_V} \cdots \int_{\Omega_V} \mathscr{D}(t, x_1,...,x_N)dx_{s+1} \cdots dx_N , \tag{1.46}$$

where Ω_V is the phase space.

Let the Hamiltonian H of the system be

$$H = \sum_{1 \leqslant i \leqslant N} H(x_i) + \sum_{1 \leqslant i < j \leqslant N} \Phi(|q_i - q_j|), \tag{1.47}$$

where $\Phi(|q_i - q_j|)$ is a two-particle (say, Lennard-Jones) potential (see Fig. 2).

In the absence of an external field, $H(x)$ is equal to the kinetic energy

$$T(p) = \sum_{1 \leqslant \alpha \leqslant 3} \frac{(p^\alpha)^2}{2m}. \tag{1.48}$$

To take the finite volume into account, we must introduce a supplementary potential function $U_V(q)$ which is constant inside V and increases rapidly to infinity as q approaches the boundary surface (see Fig. 3). Therefore,

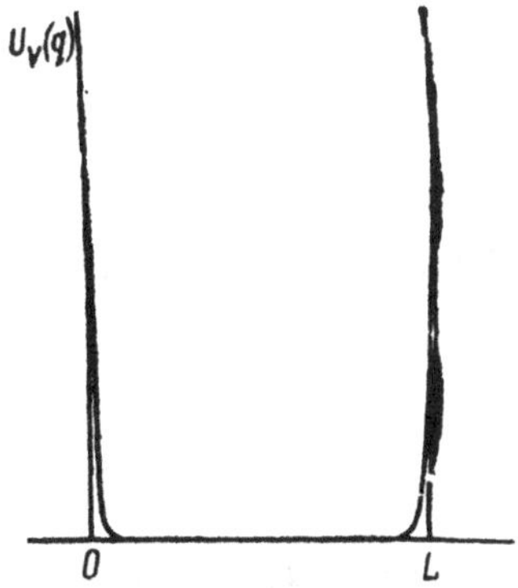

Figure 3.

$$H(x) = T(p) + U_V(q)\,. \tag{1.49}$$

With allowance for Eq. (1.47), we rewrite Eq. (1.44) in the form

$$\frac{\partial \rho}{\partial t} = \sum_{1 \leqslant i \leqslant N} [H(x_i), \rho] + \sum_{1 \leqslant i < j \leqslant N} [\Phi(|q_i - q_j|), \rho]\,. \tag{1.50}$$

We multiply both sides of Eq. (1.50) by V^s and integrate with respect to the variables $x_{s+1},...,x_N$ over all of the phase space Ω_V. By virtue of the definition (1.46) we have

$$\frac{\partial F_s}{\partial t} = \sum_{1 \leqslant i \leqslant N} V^s \int_{\Omega_V} \cdots \int_{\Omega_V} [H(x_i), \rho] dx_{s+1} \cdots dx_N + \sum_{1 \leqslant i < j \leqslant N} V^s \int_{\Omega_V} \cdots \int_{\Omega_V} [\Phi(|q_i - q_j|), \rho] dx_{s+1} \cdots dx_N \tag{1.51}$$

or

$$\begin{aligned} \frac{\partial F_s}{\partial t} = & \sum_{1 \leqslant i \leqslant s} V^s \int_{\Omega_V} \cdots \int_{\Omega_V} [H(x_i), \rho] dx_{s+1} \cdots dx_N \\ & + \sum_{s+1 \leqslant i \leqslant N} V^s \int_{\Omega_V} \cdots \int_{\Omega_V} [H(x_i), \rho] dx_{s+1} \cdots dx_N \\ & + \sum_{1 \leqslant i < j \leqslant s} V^s \int_{\Omega_V} \cdots \int_{\Omega_V} [\Phi(|q_i - q_j|), \rho] dx_{s+1} \cdots dx_N \\ & + \sum_{\substack{1 \leqslant i \leqslant s \\ s+1 \leqslant j \leqslant N}} V^s \int_{\Omega_V} \cdots \int_{\Omega_V} [\Phi(|q_i - q_j|), \rho] dx_{s+1} \cdots dx_N \\ & + \sum_{s+1 \leqslant i < j \leqslant N} V^s \int_{\Omega_V} \cdots \int_{\Omega_V} [\Phi(|q_i - q_j|), \rho] dx_{s+1} \cdots dx_N\,. \end{aligned} \tag{1.52}$$

From the definition of Poisson brackets,

$$\int_{\Omega_V} [H(x_i), \rho] dx_i = 0 \,, \tag{1.53}$$

$$\int_{\Omega_V} \cdots \int_{\Omega_V} [\Phi(|q_i - q_j|), \rho] dx_{s+1} \cdots dx_N = 0 \,, \tag{1.54}$$

hence

$$\sum_{s+1 \leqslant i \leqslant N} V^s \int_{\Omega_V} \cdots \int_{\Omega_V} [H(x_i), \rho] dx_{s+1} \cdots dx_N = 0 \,, \tag{1.55}$$

$$\sum_{s+1 \leqslant i < j \leqslant N} V^s \int_{\Omega_V} \cdots \int_{\Omega_V} [\Phi(|q_i - q_j|), \rho] dx_i dx_j = 0 \,. \tag{1.56}$$

We transform the expression

$$\begin{aligned}
&\sum_{1 \leqslant i \leqslant s} V^s \int_{\Omega_V} \cdots \int_{\Omega_V} [H(x_i), \rho] dx_{s+1} \cdots dx_N \\
&\quad + \sum_{1 \leqslant i < j \leqslant s} V^s \int_{\Omega_V} \cdots \int_{\Omega_V} [\Phi(|q_i - q_j|), \rho] dx_{s+1} \cdots dx_N \\
&= \sum_{1 \leqslant i \leqslant s} \left[H(x_i), V^s \int_{\Omega_V} \cdots \int_{\Omega_V} \rho \, dx_{s+1} \cdots dx_N \right] \\
&\quad + \sum_{1 \leqslant i < j \leqslant s} \left[\Phi(|q_i - q_j|), V^s \int_{\Omega_V} \cdots \int_{\Omega_V} \rho \, dx_{s+1} \cdots dx_N \right] \\
&= \left[\sum_{1 \leqslant i \leqslant s} H(x_i) + \sum_{1 \leqslant i < j \leqslant s} \Phi(|q_i - q_j|), F_s \right] .
\end{aligned} \tag{1.57}$$

In view of the symmetry of the function ρ with respect to the variables $x_1, \ldots, x_N$, we write

$$\begin{aligned}
&\sum_{\substack{1 \leqslant i \leqslant s \\ s+1 \leqslant j \leqslant N}} V^s \int_{\Omega_V} \cdots \int_{\Omega_V} [\Phi(|q_i - q_j|), \rho] dx_{s+1} \cdots dx_N \\
&= (N - s) \sum_{1 \leqslant i \leqslant s} V^s \int_{\Omega_V} \cdots \int_{\Omega_V} [\Phi(|q_i - q_{s+1}|), \rho] dx_{s+1} \cdots dx_N \\
&= \frac{N-s}{V} \int_{\Omega_V} \left[\sum_{1 \leqslant i \leqslant s} \Phi(|q_i - q_{s+1}|), F_{s+1} \right] dx_{s+1} \,.
\end{aligned} \tag{1.58}$$

From Eq. (1.52) we have, taking Eqs. (1.55)–(1.58) into account,

$$\begin{aligned}
\frac{\partial F_s}{\partial t} &= \left[\sum_{1 \leqslant i \leqslant s} H(x_i) + \sum_{1 \leqslant i < j \leqslant s} \Phi(|q_i - q_j|), F_s \right] \\
&\quad + \frac{N-s}{V} \int_{\Omega_V} \left[\sum_{1 \leqslant i \leqslant s} \Phi(|q_i - q_{s+1}|), F_{s+1} \right] dx_{s+1} \,.
\end{aligned} \tag{1.59}$$

We introduce the notation

$$H_s = H_s(x_1,\dots,x_s) = \sum_{i=1}^{s} T(p_i) + \sum_{1\leqslant i<j\leqslant s} \Phi(|q_i - q_j|),$$
$$v = V/N. \tag{1.60}$$

In the new notation, the equation takes the form

$$\frac{\partial F_s}{\partial t} = \sum_{i=1}^{s} [u(q_i), F_s] + [H_s, F_s] + \frac{1-s/N}{v}\int_{\Omega_V}\left[\sum_{i=1}^{s} \Phi(|q_i - q_{s+1}|), F_{s+1}\right] dq_{s+1}\,. \tag{1.61}$$

To investigate the volume properties we take in Eq. (1.61) the limit $N\to\infty$ under the condition that

$$V/N = v = \text{const}\,,$$

and the boundary surface goes off to infinity.* The sought asymptotic expressions for the distribution functions satisfy the equations

$$\frac{\partial F_s}{\partial t} = [H_s, F_s] + \frac{1}{v}\int_{\Omega}\left[\sum_{i=1}^{s} \Phi(|q_i - q_{s+1}|), F_{s+1}\right] dx_{s+1}\,, \tag{1.62}$$

where Ω is the unbounded space of the points $x(q,p)$.

4. von Neumann equation. Bogolyubov method in quantum statistics

The state of a closed system is described by a wave function and is called pure, and the aggregate of the results of identical measurements performed on identical systems in one and the same quantum state constitutes a pure ensemble.

Statistical physics deals not with individual (closed) systems, but mainly with systems that are small parts of a large system (systems in a heat bath) and are described not by a wave function but by a density matrix. Knowing this matrix we can determine the expectation values of various physical quantities.

Let us derive an equation for the density matrix. We use for this purpose a statistical ensemble of a type more general than a pure one, namely, a mixed ensemble (or "mixture").

Consider N identical noninteracting systems, of which N_1 are in the state $\psi_1(x)$, N_2 in the state $\psi_2(x)$, etc. This aggregate of systems forms then a mixed

*This statement is valid, strictly speaking, if it is possible to inscribe a sphere of maximum radius R_m in the volume occupied by the system, and $R_m\to\infty$ as $N\to\infty$.

ensemble described by the wave functions $\psi_1, \psi_2, \ldots$, and by the corresponding probabilities $w_1 = N_1/N$, $w_2 = N_2/N, \ldots$ which add up to unity:

$$w_1 + w_2 + \cdots = 1. \qquad (1.63)$$

Obviously, the expectation value f of a physical quantity (dynamic variable) in a system of this ensemble is defined as*

$$\langle f \rangle = \sum_k w_k \bar{f}_k = \sum_k w_k \int \psi_k^* \hat{f} \psi_k \, dx = \int [\hat{f} \rho(x, x')]_{x'=x} \, dx \,, \qquad (1.64)$$

where

$$\rho(x, x') = \sum_k w_k \psi_k(x) \psi_k^*(x') \qquad (1.65)$$

is the density matrix or the density operator in the matrix coordinate representation, and meets the normalization condition

$$\mathrm{Tr}\, \hat{\rho} = 1, \qquad (1.66)$$

since the normalization conditions for the wave functions and for the probabilities lead to

$$\mathrm{Tr}\, \hat{\rho} = \int \rho\,(x, x)\, dx = \sum_k w_k \int \psi_k^* \psi_k \, dx = 1 \,.$$

Writing the operator $\hat{f}$ in the matrix coordinate representation with the aid of its matrix elements $f(x, x')$, we have

$$\hat{f} \psi_k(x) = \int f(x, x') \psi_k(x') dx' \,.$$

Substituting this expression in Eq. (1.64) we get

$$\langle f \rangle = \sum_k w_k \int \psi_k^* f(x, x') \psi_k(x') dx\, dx' = \int f(x, x') \rho\,(x, x') dx\, dx' \qquad (1.67)$$

or

$$\langle f \rangle = \mathrm{Tr}\, \hat{f} \hat{\rho} \,. \qquad (1.68)$$

Thus, knowing the density operator $\hat{\rho}$ we can calculate the expectation values of various physical quantities for systems in a mixed ensemble; this operator permits, thus, a full description of the considered statistical ensemble.

We have considered so far a statistical ensemble and the corresponding operator $\hat{\rho}$ for some fixed instant of time. Let us determine now their evolution in time.

Since the dynamic systems of the ensemble do not interact with one another, we can regard the evolution of each of them separately. Furthermore, if a relative number w_k of systems is in a state ψ_k at the initial instant, the same number will be in a state $\psi_k(x, t)$ at an instant t. The temporal variation of the state $\psi_k(x, t)$ is determined by the Schrödinger equation

*A physical quantity averaged over a pure ensemble is marked by a overbar ($\bar{f}$) and an average over a mixed ensemble by angle brackets ($\langle f \rangle$).

$$i\hbar\frac{\partial\psi_k(x, t)}{\partial t} = \hat{H}\psi_k(x, t) ,$$

or, in matrix form,

$$i\hbar\frac{\partial\psi_k(x, t)}{\partial t} = \int H(x, x')\psi_k(x, t)\, dx' . \tag{1.69}$$

Differentiating with respect to time the density matrix

$$\rho(x, x') = \sum_k w_k \psi_k(x, t)\psi^*(x', t)$$

and using the wave equation (1.69) and the Hermitian character of the Hamiltonian, $\hat{H}(x, x') = \hat{H}(x', x)$, we obtain the equation of motion of the statistical operator in matrix form:

$$\begin{aligned} i\hbar\frac{\partial\rho(x, x', t)}{\partial t} &= \int \sum_k \{H(x, x'')\, w_k \psi_k(x'', t)\psi_k^*(x', t) \\ &\quad - w_k \psi_k(x, t)\psi_k^*(x'', t)\, H(x'', x)\} dx'' \\ &= \int \{H(x, x'')\,\rho(x'', x', t) - \rho(x, x'', t)\, H(x'', x')\} dx'' \end{aligned} \tag{1.70}$$

or in operator form:

$$\begin{aligned} i\hbar\frac{\partial\hat{\rho}}{\partial t} &= \hat{H}\hat{\rho} - \hat{\rho}\hat{H} , \\ \frac{\partial\rho}{\partial t} &= [\hat{H}, \hat{\rho}] , \end{aligned} \tag{1.71}$$

where

$$[\hat{H}, \hat{\rho}] = \frac{1}{i\hbar}(\hat{H}\hat{\rho} - \hat{\rho}\hat{H}) \tag{1.72}$$

are quantum Poisson brackets.

Equation (1.71) for the density matrix is called the von Neumann equation and is the basic equation of statistical physics of quantum systems. This equation is similar to the Liouville equation (1.44) for the phase distribution density $\rho(q, p, t)$.

Note that the density operator, just as the classical phase density, is symmetric with respect to permutation of particles of any type (bosons or fermions).

To investigate dynamic systems it is usually necessary to know not the total density matrix of the system, but simpler statistical operators that depend on the variables of one, two,...,s particles.

We introduce, following Bogolyubov,[2] statistical operators of particle complexes and set up a chain of equations for these operators.

Consider a system consisting of N identical particles. We denote by $x_1, x_2,...,x_N$ the variables of the first, second,...,Nth particle. The wave functions of the entire system are then functions of these variables and of the time:

$$\psi_n = \psi_n(t, x_1,...,x_N),$$

and the operators acting on these functions are matrices of the form

$$A = A(x_1,...,x_N; x'_1,...,x'_N).$$

The density matrix of the system is equal to

$$\rho(t, x_1,...,x_N; x'_1,...,x'_N) = \sum_n \psi_n(t, x_1,...,x_N)\psi_n^*(t, x'_1,...,x'_N). \tag{1.73}$$

We define the operators of the particle complexes (or of the partial density matrices) in terms of the partial convolutions of the density operator $\hat{\rho}(1, 2,..., N)$,* i.e., in terms of the traces of $\hat{\rho}$ over part of the variables:

$$\hat{\rho}_1(1) = \operatorname*{Tr}_{2,...,N} \hat{\rho},$$

$$\hat{\rho}_2(1, 2) = \operatorname*{Tr}_{3,...,N} \hat{\rho},$$

$$\vdots$$

$$\hat{\rho}_s(1, 2,...,s) = \operatorname*{Tr}_{s+1,...,N} \hat{\rho}, \tag{1.74}$$

or, in greater detail, in the matrix x representation

$$\rho_1(t, x_1, x'_1) = \sum_{x_2,...,x_N} \rho(t, x_1, x_2,...,x_N; x'_1, x_2,...,x_N),$$

$$\rho_2(t, x_1, x_2; x'_1, x'_2) = \sum_{x_s,...,x_N} \rho(t, x_1, x_2,...,x_N; x'_1, x'_2,...,x_N),$$

$$\vdots$$

$$\rho_s(t, x_1,...,x_s; x'_1,...,x'_s) = \sum_{x_{s+1},...,x_N} \rho(t, x_1,...,x_N; x'_1,...,x_N).$$

It can be seen from these equations that

$$\operatorname{Tr}\hat{\rho}_1(1) = \operatorname{Tr}\hat{\rho}_2(1, 2) = \cdots = \operatorname{Tr}\hat{\rho}_s(1, 2,...,s) = \cdots = 1 \tag{1.75}$$

and

$$\rho_s(1, 2,...,s) = \operatorname*{Tr}_{s+1} \hat{\rho}_{s+1}(1, 2,...,s+1). \tag{1.76}$$

It is also easy to verify that ρ_s is a self-adjoint operator and that the symmetry of the density operator with respect to the particle permutations leads to the following symmetry properties of particle-complex operators:

*The operator $\hat{\rho}$ is a function of the operators $(\hat{q}_1, \hat{p}_1,...,\hat{q}_N, \hat{p}_N)$, which we shall designate as 1, 2,...,N.

for Bose particles (bosons)

$$\hat{P}\hat{\rho}_s = \hat{\rho}_s\hat{P}_s = \rho_s , \tag{1.77}$$

for Fermi particles (fermions)

$$\hat{P}\hat{\rho}_s = \hat{\rho}_s\hat{P} = (-1)^P\hat{\rho} , \tag{1.78}$$

where $\hat{P}$ is a permutation operator acting on the wave functions $\psi(x_1,...,x_s)$ of the particles 1, 2,...,s.

We show now that the mean values of additive dynamic quantities of the form

$$A = \sum_{1\leqslant r\leqslant N} A_1(r)$$

can be calculated with the aid of the operator $\hat{\rho}_1$, and the expectation values of binary quantities

$$A = \sum_{1\leqslant r<s\leqslant N} A_2(r, s)$$

are expressed in terms of the operator $\hat{\rho}_2$.

In fact, in the former case

$$\langle A\rangle = \sum_{1\leqslant r\leqslant N} \mathrm{Tr}[\hat{A}_1(r), \hat{\rho}] ,$$

and since the density operator $\hat{\rho}$ is symmetric with respect to any permutation of the particle "numbers," all the traces $\mathrm{Tr}\,[\hat{A}_1(r), \hat{\rho}]$ with different $r = 1, 2,...,N$ are equal. Therefore,

$$\langle A\rangle = N\,\mathrm{Tr}[A_1(1)\hat{\rho}]$$

$$= N \mathop{\mathrm{Tr}}_{1} \mathop{\mathrm{Tr}}_{2,...,N} [\hat{A}_1(1)\hat{\rho}] = N \mathop{\mathrm{Tr}}_{1} A_1(1) \mathop{\mathrm{Tr}}_{2,...,N} \rho$$

or

$$\langle A\rangle = N\,\mathrm{Tr}\,\hat{A}(1)\hat{\rho}_1(1) . \tag{1.79}$$

Similarly, in the case of quantities of binary type,

$$\langle A\rangle = \sum_{1\leqslant r<s\leqslant N} \mathrm{Tr}[A_2(r, s)\rho]$$

and as a result of the symmetry of $\hat{\rho}$ all the terms of this series are equal, so that

$$\langle A\rangle = \frac{N(N-1)}{2} \mathop{\mathrm{Tr}}_{1,2,...,N} [\hat{A}_2(1, 2)\hat{\rho}] = \frac{N(N-1)}{2} \mathop{\mathrm{Tr}}_{1,2} \hat{A}_2(1, 2) \mathop{\mathrm{Tr}}_{3,...,N} \hat{\rho}$$

or

$$\langle A\rangle = \frac{N(N-1)}{2} \mathrm{Tr}[\hat{A}_2(1, 2)\,\hat{\rho}_2(1,2)] , \tag{1.80}$$

i.e., the expectation values of a binary dynamic quantity are expressed in terms of the two-particle operator $\hat{\rho}_2$. It can also be shown that the expectation value of an "s-fold" quantity

$$A = \sum_{1 \leq r_1 < r_2 < \ldots < r_s \leq N} A_s(r_1, r_2, \ldots, r_s)$$

is equal to

$$\langle A \rangle = \frac{N(N-1)\cdots(N-s+1)}{s!} \mathrm{Tr}[\hat{A}_s(1,2,\ldots,s)\hat{\rho}_s(1,2,\ldots,s)] . \tag{1.81}$$

The usual macroscopic dynamic quantities are either additive or binary. For example, the total momentum and the kinetic energy are additive, while the interaction energy is binary. For practical purposes it suffices therefore to find the simplest statistical operators $\hat{\rho}_1(1)$ and $\hat{\rho}_2(1, 2)$, and sometimes also several operators of higher order. Naturally, it becomes necessary to obtain chains of equations for the partial operators $\hat{\rho}_1$ and $\hat{\rho}_2$ without first obtaining the total operator and explicitly calculating its traces (1.74).

To this end, we start from the basic equation (1.71) of the statistical physics of quantum systems:

$$\frac{\partial \hat{\rho}}{\partial t} = [\hat{H}, \hat{\rho}] .$$

Let the Hamiltonian of a system of like particles consist of the operators of the individual particle energies and the pair-interaction energy

$$\hat{H} = \sum_{1 \leq r \leq N} \hat{H}(r) + \sum_{1 \leq r < s \leq N} \Phi(r, s) , \tag{1.82}$$

where $\hat{H}(r)$ is the individual energy operator of the rth particle and $\Phi(r, s)$ is the interaction energy of the rth and sth particles.

For example, for a system of monatomic particles with central interaction forces we have

$$\hat{H}(r) = -\frac{\hbar^2}{2m} \Delta q_r + U(q) ,$$
$$\Phi(r, s) = \Phi(|q_r - q_s|) , \tag{1.83}$$

where $U(q)$ is the external-field potential and $\Phi(r)$ is the potential energy of the interaction of two atoms with distance r between them.

Equation (1.71) for the density operator of a system with a Hamiltonian (1.82) takes the form

$$i\hbar \frac{\partial \rho}{\partial t} = \left\{ \sum_{1 \leq r \leq N} \Phi(r) + \sum_{1 \leq r < s \leq N} \Phi(r, s) \right\} \hat{\rho}$$
$$- \hat{\rho} \left\{ \sum_{1 \leq r \leq N} H(r) + \sum_{1 \leq r < s \leq N} \Phi(r, s) \right\} . \tag{1.84}$$

Applying the operation $\operatorname*{Tr}_{2,\ldots,N}$ to both sides of this equation, we obtain

$$i\hbar\frac{\partial\hat{\rho}_1(1)}{\partial t} = \sum_{1\leq r\leq N}\left\{\operatorname*{Tr}_{2,\ldots,N}\hat{H}(r)\hat{\rho} - \operatorname*{Tr}_{2,\ldots,N}\hat{\rho}\hat{H}(r)\right\}$$

$$+\sum_{1\leq r<s\leq N}\left\{\operatorname*{Tr}_{2,\ldots,N}\Phi(r,s)\hat{\rho} - \operatorname*{Tr}_{2,\ldots,N}\hat{\rho}\Phi(r,s)\right\}. \tag{1.85}$$

On the basis of the property of the trace

$$\operatorname*{Tr}_{r}\hat{H}(r)\hat{\rho} = \operatorname*{Tr}_{r}\hat{\rho}\hat{H}(r)$$

we have

$$\sum_{1\leq r\leq N}\left\{\operatorname*{Tr}_{2,\ldots,N}\hat{H}(r)\hat{\rho} - \operatorname*{Tr}_{2,\ldots,N}\hat{\rho}\hat{H}(r)\right\} = \operatorname*{Tr}_{2,\ldots,N}\hat{H}(1)\hat{\rho} - \operatorname*{Tr}_{2,\ldots,N}\hat{\rho}\hat{H}(1)$$

$$= H(1)\operatorname*{Tr}_{2,\ldots,N}\hat{\rho} - \left(\operatorname*{Tr}_{2,\ldots,N}\hat{\rho}\right)\hat{H}(1) = \hat{H}(1)\hat{\rho}_1(1) - \hat{\rho}_1(1)\hat{H}(1). \tag{1.86}$$

On the basis of the same property of the trace, furthermore,

$$\sum_{1\leq r<s\leq N}\left\{\operatorname*{Tr}_{2,\ldots,N}\Phi(r,s)\hat{\rho} - \operatorname*{Tr}_{2,\ldots,N}\hat{\rho}\Phi(r,s)\right\}$$

$$= \sum_{2\leq s\leq N}\left\{\operatorname*{Tr}_{2,\ldots,N}\Phi(1,s)\hat{\rho} - \operatorname*{Tr}_{2,\ldots,N}\hat{\rho}\Phi(1,s)\right\}.$$

From symmetry considerations, however, we see that

$$\operatorname*{Tr}_{2,\ldots,N}\Phi(1,s)\hat{\rho} - \operatorname*{Tr}_{2,\ldots,N}\hat{\rho}\Phi(1,s) = \operatorname*{Tr}_{2,\ldots,N}\Phi(1,2)\hat{\rho} - \operatorname*{Tr}_{2,\ldots,N}\hat{\rho}\Phi(1,2).$$

We can therefore write

$$\sum_{1\leq r<s\leq N}\left\{\operatorname*{Tr}_{2,\ldots,N}\Phi(r,s)\hat{\rho} - \operatorname*{Tr}_{2,\ldots,N}\hat{\rho}\Phi(r,s)\right\}$$

$$= (N-1)\left\{\left\{\operatorname*{Tr}_{2}\Phi(1,2)\operatorname*{Tr}_{2,\ldots,N}\hat{\rho}\right\} - \operatorname*{Tr}_{2}\left\{\operatorname*{Tr}_{3,\ldots,N}\hat{\rho}\Phi(1,2)\right\}\right\}$$

$$= (N-1)\left\{\operatorname*{Tr}_{2}\Phi(1,2)\hat{\rho}_2(1,2) - \operatorname*{Tr}_{2}\hat{\rho}_2(1,2)\Phi(1,2)\right\}.$$

We obtain thus from Eq. (1.85), on the basis of Eq. (1.86), the first equation

$$i\hbar\frac{\partial\hat{\rho}_1(1)}{\partial t} = \hat{H}(1)\,\hat{\rho}_1(1) - \hat{\rho}_1(1)\hat{H}\,(1)$$

$$+(N-1)\left\{\operatorname*{Tr}_{2}\Phi(1,2)\,\hat{\rho}_2(1,2) - \operatorname*{Tr}_{2}\hat{\rho}_2(1,2)\Phi(1,2)\right\}. \tag{1.87}$$

We apply now the operation $\operatorname*{Tr}_{3,\ldots,N}$ to both sides of Eq. (1.84), and obtain

$$i\hbar\frac{\partial\hat{\rho}_2(1,2)}{\partial t}=\sum_{1\leqslant r\leqslant N}\left\{\operatorname*{Tr}_{3,\ldots,N}\hat{H}(r)\hat{\rho}-\operatorname*{Tr}_{3,\ldots,N}\hat{\rho}H(r)\right\}$$
$$+\sum_{1\leqslant r<s\leqslant N}\left\{\operatorname*{Tr}_{3,\ldots,N}\Phi(r,s)\hat{\rho}-\operatorname*{Tr}_{3,\ldots,N}\hat{\rho}\Phi(r,s)\right\}. \tag{1.88}$$

Reasoning as before, we have

$$\sum_{1\leqslant r\leqslant N}\left\{\operatorname*{Tr}_{3,\ldots,N}\hat{H}(r)\hat{\rho}-\operatorname*{Tr}_{3,\ldots,N}\hat{\rho}\hat{H}(r)\right\}$$
$$=\operatorname*{Tr}_{3,\ldots,N}\hat{H}(1)\hat{\rho}+\operatorname*{Tr}_{3,\ldots,N}\hat{H}(2)\hat{\rho}-\operatorname*{Tr}_{3,\ldots,N}\hat{\rho}\hat{H}(1)-\operatorname*{Tr}_{3,\ldots,N}\hat{\rho}\hat{H}(2)$$
$$=[\hat{H}(1)+H(2)]\hat{\rho}_2(1,2)-\hat{\rho}_2(1,2)[\hat{H}(1)+\hat{H}(2)]\,. \tag{1.89}$$

Next

$$\sum_{1\leqslant r<s\leqslant N}\left\{\operatorname*{Tr}_{3,\ldots,N}\Phi(r,s)\hat{\rho}-\operatorname*{Tr}_{3,\ldots,N}\hat{\rho}\Phi(r,s)\right\}$$
$$=\left\{\operatorname*{Tr}_{3,\ldots,N}\Phi(1,2)\hat{\rho}-\operatorname*{Tr}_{3,\ldots,N}\hat{\rho}\Phi(1,2)\right\}$$
$$+\sum_{3\leqslant s\leqslant N}\left\{\operatorname*{Tr}_{3,\ldots,N}\Phi(1,s)\hat{\rho}+\operatorname*{Tr}_{3,\ldots,N}\Phi(2,s)\hat{\rho}\right.$$
$$\left.-\operatorname*{Tr}_{3,\ldots,N}\hat{\rho}\Phi(1,s)-\operatorname*{Tr}_{3,\ldots,N}\Phi(1,s)\hat{\rho}\right\}=\Phi(1,2)\hat{\rho}_2(1,2)$$
$$-\hat{\rho}_2(1,2)\Phi(1,2)+(N-2)\left\{\operatorname*{Tr}_{3}[\Phi(1,3)+\Phi(2,3)]\hat{\rho}_3(1,2,3)\right.$$
$$\left.-\operatorname*{Tr}_{3}\hat{\rho}_3(1,2,3)[\Phi(1,3)+\Phi(2,3)]\right\}.$$

We obtain thus from Eq. (1.88), on the basis of Eq. (1.89), the second equation

$$i\hbar\frac{\partial\hat{\rho}_2(1,2)}{\partial t}=\hat{H}_2(1,2)\hat{\rho}_2(1,2)-\hat{\rho}_2(1,2)\,\hat{H}_2(1,2)$$
$$+(N-2)\left\{\operatorname*{Tr}_{3}[\Phi(1,3)+\Phi(2,3)]\hat{\rho}_3(1,2,3)\right.$$
$$\left.-\operatorname*{Tr}_{3}\hat{\rho}_3(1,2,3)[\Phi(1,3)+\Phi(2,3)]\right\}, \tag{1.90}$$

where $H_2(1,2)$ is the Hamiltonian of one particle pair:

$$\hat{H}_2(1,2)=\hat{H}(1)+\hat{H}(2)+\Phi(1,2)\,. \tag{1.91}$$

We obtain similarly the general equation

$$i\hbar \frac{\partial \hat{\rho}_s}{\partial t} = \hat{H}_s(1, 2,\ldots,s)\hat{\rho}_s(1, 2,\ldots,s)$$
$$- \hat{\rho}_s(1, 2,\ldots,s)\hat{H}_s(1, 2,\ldots,s)$$
$$+ (N - s) \operatorname*{Tr}_{s+1} \left\{ \sum_{1 \leqslant r \leqslant s} \Phi(r, s+1)\hat{\rho}_{s+1}(1,\ldots,s+1) \right.$$
$$\left. - \hat{\rho}_{s+1}(1,\ldots,s+1) \sum_{1 \leqslant r \leqslant s+1} \Phi(r, s+1) \right\}, \tag{1.92}$$

where $\hat{H}_s(1,\ldots,s)$ is the Hamiltonian of a system of s particles:

$$\hat{H}_s(1,\ldots,s) = \sum_{1 \leqslant r \leqslant s} \hat{H}(r) + \sum_{1 \leqslant r_1 < r_2 \leqslant s} \Phi(r_1, r_2). \tag{1.93}$$

The first term in Eq. (1.92) determines the change of ρ_s due to the motion and interaction of the molecules of the complex in question, while the second term with $(N - s)$Tr... takes into account the influence of all the remaining $N - s$ molecules of the system.

Equations (1.92) for $s = 1, 2,\ldots$ are a chain of expressions for the partial operators $\hat{\rho}_1,\ldots,\hat{\rho}_s$ and make it possible to determine them approximately from the total density operator $\hat{\rho}$.

Two ways of solving this chain can be indicated. The first can be used when the interaction between the molecules is weak. The terms that determine the influence of the other molecules on the given complex can be regarded as small and calculated by perturbation theory. The second possibility is to "close" the chain by expressing, say, $\hat{\rho}_3$ in terms of $\hat{\rho}_1$ and $\hat{\rho}_2$. This possibility is particularly important when dynamic systems in the condensed state are investigated, say liquids, and there are no suitable "small parameters" to permit series expansions.

5. Basic premises of the theory of equilibrium systems. Canonical, grand canonical, and microcanonical distributions

The most advanced branch of statistical physics is the theory of equilibrium states, or statistical thermodynamics, to which the succeeding chapters of this book are mainly devoted.

Consider a system that is either adiabatically isolated or in a heat bath and has definite external parameters $a_1,\ldots,a_n$.

We elucidate first the meaning, from the molecular point of view, of the equilibrium internal parameters of a system.

Let a system in thermodynamic equilibrium be subject to certain external constraints and let the laws of motion of its individual particles be known. The value of any macroscopic internal parameter of the system (density, pressure, etc.) is governed by the states of motion $\{q, p\}$ of its particles and is therefore a rapidly varying function of time. In an experiment, however, one measures the expectation value of some parameter over a definite time interval. Thus, for example, pressure is caused by impacts of individual molecules against a wall. The number of these impacts is different at different instants of time, and the equilibrium pressure measured with a manometer is defined as the time average of this quantity.

Thus, from the molecular standpoint, the equilibrium state f_0 of any internal parameter is an average, over an infinitely long time interval T, of a corresponding (to this parameter) function $f(q, p)$ of the coordinates and momenta of the system particles:

$$f_0 = \bar{\bar{f}} = \lim_{T \to \infty} \frac{1}{T} \int_0^T f(q, p) dt \tag{1.94}$$

(the superior double bar denotes here averaging over time). This shows that to calculate any macroscopic internal parameter of a system in thermodynamic equilibrium it is necessary to know the time dependence of all the q and p. By solving the equations of mechanics we can find this dependence and calculate from Eq. (1.94) the equilibrium parameters f_0.

Although the behavior of a many-particle system is determined by the mechanical motion of its constituent particles, this does not mean that the laws of thermal motion can be reduced to the laws of mechanics, since the laws governing the behavior of a macroscopic system differ qualitatively, in view of the large number of particles, from those of mechanics.*

To determine this different behavior of a system consisting of many particles, the thermodynamic parameters are not calculated in accordance with Eq. (1.94). The basic procedure in statistical mechanics is to replace the time averaged value $\bar{\bar{f}}$ of some quantity in one system by an average over an aggregate of the same quantity in a suitably chosen ensemble of identical systems:

$$\langle f \rangle = \int f(q, p) \rho(q, p) \, d\Gamma, \tag{1.95}$$

where $\rho(q, p)$ is the sought phase density of the probability of the system state.

To find, for any system, a statistical ensemble (or a distribution function) that ensures the equality

$$\bar{\bar{f}} = \langle f \rangle \tag{1.96}$$

is the most important problem of statistical thermodynamics.[11–16]

To solve this problem, we start from the basic equations of statistical physics—the Liouville equation (1.41) for a classical system and the von Neumann equation (1.72) for quantum systems.

*"The discovery that heat is some molecular action marked an epoch in science. Since, however, I can say nothing about heat except that it constitutes a certain displacement of molecules, it is better to remain silent." (Friedrich Engels, *Dialectics of Nature*).

The phase density of a system in thermodynamic equilibrium is independent of time:

$$\frac{\partial \rho(t, q, p)}{\partial t} = 0,$$

therefore, according to Eq. (1.44),

$$[H, \rho] = 0$$

or

$$\sum_i \left\{ \frac{\partial H}{\partial q_i} \frac{\partial \rho}{\partial p_i} - \frac{\partial H}{\partial p_i} \frac{\partial \rho}{\partial q_i} \right\} = 0.$$

The general integral of this first-order linear partial differential equation is an arbitrary function of all $(2n - 1)$ integrals of the system (1.5) (the energy H, the three momentum components J_2, J_3, J_4, the three angular momentum components J_5, J_6, J_7, etc.):

$$\rho(q, p) = \varphi(H, J_2, J_3, \ldots, J_{2n-1}). \tag{1.97}$$

The mean value $\langle f \rangle$ of any state function $f(q, p)$ of a system, recalculated from Eq. (1.95) using the phase probability density (1.97) is, in general, a function of all $(2n - 1)$ integrals of its motion.

In thermodynamics, however, any equilibrium internal parameter f_0 of the system [equal to $\bar{\bar{f}}$ according to Eq. (1.94)], depends, given the external parameters $a_1, \ldots, a_n$, only on the energy H:

$$\bar{\bar{f}} = \psi(H).$$

Therefore, if account is taken of this thermodynamic requirement, Eq. (1.96) is valid if the phase density $\rho(q, p)$ depends only on the energy integral H:

$$\rho\,(q, p) = \varphi(H). \tag{1.98}$$

Following Gibbs, we postulate for a system in a heat bath the distribution

$$\rho\,(q, p) = e^{(F - H)/\theta}, \tag{1.99}$$

where θ and F are constants and θ is positive. It is "apparently the simplest conceivable case, since it possesses the property that when the system consists of parts having different energies the law of distribution over the phases is the same for the individual parts; this property simplifies the investigation to the utmost and serves as a basis for very important thermodynamic relations" (Ref. 1).

Indeed, as a result of the distribution (1.99), the differential equation defined by the first law of thermodynamics has an integrating factor for the heat element δQ, in accord with the second law, and when two bodies of unequal temperature are in contact the entropy of the system increases.

The probability density distribution (1.99) of a microstate of a system in a heat bath is called the Gibbs canonical distribution, and θ is called the modulus of the distribution.

From the normalization condition (1.42) of the phase density (1.99):

$$\int \rho \, d\Gamma = \int e^{(F-H)/\theta} d\Gamma = 1 \tag{1.100}$$

we have

$$e^{-F/\theta} = \int e^{-H/\theta} d\Gamma . \tag{1.101}$$

To explain the physical meaning of the canonical-distribution parameters F and θ we use this distribution to determine average energy of the system:

$$\langle H \rangle = \int H e^{-(F-H)/\theta} d\Gamma = e^{F/\theta} \theta^2 \frac{\partial}{\partial \theta} \int e^{-H/\theta} d\Gamma = \theta^2 e^{F/\theta} \frac{\partial}{\partial \theta} (e^{-F/\theta}) . \tag{1.102}$$

We have obtained the Gibbs–Helmholtz equation for the free energy; consequently, F is the free energy of the system and $\theta = kT$ (T is the thermodynamic temperature and k is the Boltzmann constant).

From Eq. (1.101) we obtain an expression for the free energy of the system:

$$F = -\theta \ln Z , \tag{1.103}$$

where

$$Z = \int \cdots \int e^{-H/\theta} dq_1 \cdots dq_n dp_1 \cdots dp_n = \int e^{-H/\theta} d\Gamma \tag{1.104}$$

is the partition function of the system.

It is known from thermodynamics that all the thermodynamic parameters of a system can be determined if its free energy is known as a function of the external parameters and of temperature. Therefore expressions (1.103) and (1.104) are basic in statistical thermodynamics. The entropy of the system is obviously

$$S = -\frac{\partial F}{\partial T} = -k\frac{\partial F}{\partial \theta} = -k\frac{F - \langle H \rangle}{\theta}$$

$$= -K\langle \ln \rho \rangle = -k \int \rho \ln \rho \, d\Gamma . \tag{1.105}$$

This expression for the entropy is sometimes chosen to be the initial postulate for any system. The gist of the equilibrium condition is that the entropy S must be a maximum.

The Gibbs canonical distribution (1.99) determines the probability density of the microstates of a many-particle system in a heat bath. The normalization (1.100) means that the sum of the probabilities of all the possible macrostates is equal to unity. If all N particles of a system are different, any permutation of the particles corresponds to another microstate of the system. Consequently, the normalization (1.100) takes into account the various (differing from one another) possible microstates of the system. If, on the other hand, the system consists of identical particles, their permutation does not change the microstates and each microstate is represented by $N!$ different phase points. The number of different microstates of a system of identical particles is therefore

$$\frac{1}{N!}\int \rho\, d\Gamma,$$

and consequently, in this case

$$\frac{1}{N!}\int \rho(q,p)d\Gamma = 1. \qquad (1.106)$$

This normalization of the phase density $\rho(q, p)$ of a system of N particles allows for their identity. Only with such a normalization does the system's free energy (1.103) correspond to the thermodynamics of an extensive quantity.

Furthermore, the numerical value of the partition function (1.104) depends on the units used for the coordinate-momentum space. This is manifest in the form of an arbitrary point from which the entropy is reckoned in the final expressions for the thermodynamic quantities. For the normalization of $\rho(q, p)$ to correspond to a limiting transition from quantum statistics (which leads by the third law of thermodynamics to $S = 0$ at $T = 0$ K) to classical thermodynamics, it is necessary to change over in Eq. (1.100) to a dimensionless phase volume:

$$d\Gamma = \frac{\prod_{i=1}^{3N} dq_i dp_i}{h^{3N}}, \qquad (1.107)$$

where $h = 2\pi\hbar$ is Planck's constant.

In accordance with the foregoing, one uses in statistical mechanics of classical systems a canonical Gibbs distribution:

$$\rho(q,p) = \frac{1}{N!} e^{-(F-H)/\theta}, \qquad (1.108)$$

that satisfies the normalization condition

$$\frac{1}{N!\,h^{3N}}\int e^{(F-H)/\theta} \prod_{i=1}^{3N} dq_i dp_i = 1, \qquad (1.109)$$

from which we get

$$F = -\theta \ln Z_N\,, \qquad (1.110)$$

$$Z_N = \frac{1}{N!\,h^{3N}}\int e^{-H/\theta} d\mathbf{q}_1 \cdots d\mathbf{q}_N\, d\mathbf{p}_1 \cdots d\mathbf{p}_N\,. \qquad (1.111)$$

For a system of N identical particles with a Hamiltonian,

$$H = \sum_{\substack{1 \leqslant i \leqslant N \\ 1 \leqslant \alpha \leqslant 3}} \frac{p_{i\alpha}^2}{2m} + U(\mathbf{q}_1,\ldots,\mathbf{q}_N)\,, \qquad (1.112)$$

where $U(q_1,...,q_N)$ is the potential energy of the particle interaction, we have

$$Z_N = \frac{1}{N!\,h^{3N}} \int e^{-H/\theta} d\mathbf{q}_1 \cdots d\mathbf{q}_N \, d\mathbf{p}_1 \cdots d\mathbf{p}_N = \frac{1}{N!\,h^{3N}} Z_p Q_N \,, \quad (1.113)$$

where

$$Z_p = \int_{-\infty}^{+\infty} \cdots \int \exp\left\{ - \sum_{\substack{1 \leqslant i \leqslant N \\ 1 \leqslant \alpha \leqslant 3}} \frac{p_{i\alpha}^2}{2m\theta} \right\} d\mathbf{p}_1 \cdots d\mathbf{p}_N = (2\pi m\theta)^{3N/2} \,, \quad (1.114)$$

$$Q_N = \int_V \cdots \int_V \exp\left\{ - \frac{U(\mathbf{q}_1 \cdots \mathbf{q}_N)}{\theta} \right\} d\mathbf{q}_1 \cdots d\mathbf{q}_N \quad (1.115)$$

is the configuration integral.

According to Eq. (1.110), the free energy of the system is

$$F = 3N\theta \ln \lambda - \theta \ln Q_N / N! , \quad (1.116)$$

where $\lambda = h/\sqrt{2\pi m\theta}$ is the thermal de Broglie wavelength.

It can be seen from Eq. (1.116) that the determination of the free energy of a system, and hence of all its thermodynamic functions, reduces to calculation of the configuration integral (1.115).

Various methods of calculating Q_N of many-particle systems will be described in the next chapter.

In the case of an ideal gas, for which $U(\mathbf{q}_1,...,\mathbf{q}_N) = 0$, the configuration integral can be calculated exactly:

$$Q_N = V^N ,$$

and the free energy is equal to

$$F = 3N\theta \ln \lambda - N\theta \ln \frac{V}{(N!)^{1/N}} .$$

The statistical theory determines the bulk thermodynamic properties of various systems in the statistical limit:

$$N \to \infty, \quad V \to \infty, \quad V/N = v = \text{const.} \quad (1.117)$$

Therefore, using Stirling's formula, we have for large N,

$$F = 3N\theta \ln \lambda - N\theta \ln(ev) \quad (1.117')$$

and

$$f = \frac{F}{N} = 3\theta \ln \lambda - \theta \ln(ev) . \quad (1.118)$$

From expression (1.117) for the free energy of an ideal gas we obtain all its properties:

$$p = -\left(\frac{\partial F}{\partial V}\right)_\theta = \frac{N\theta}{V},$$

$$E = F - \theta\left(\frac{\partial F}{\partial \theta}\right)_V = -\theta^2\left[\frac{\partial}{\partial \theta}\left(\frac{F}{\theta}\right)\right]_V = \frac{3}{2}N\theta,$$

$$C_V = \left(\frac{\partial E}{\partial T}\right)_V = \frac{3}{2}kN,$$

$$S = -k\left(\frac{\partial F}{\partial \theta}\right) = N\ln\left[\frac{(2\pi m\theta)^{3/2} v e^{5/2}}{h^3}\right],$$

etc.

The ideal-gas problem is thus exactly solvable in statistical physics.

The Gibbs canonical distribution (1.108) determines the probability density of the microstates of a system of N particles in a volume V at a temperature T. Let us find the phase probability density of a system having a variable number of particles and placed in a heat bath.

A feature of the thermodynamics of such a system is that their thermodynamic potential $\omega(T, V, \mu)$ is high:

$$\omega = F - G = F - \mu N = -pV,$$

whence

$$F = \omega + \mu N, \tag{1.119}$$

where $G = \mu N$ is the Gibbs energy, μ the chemical potential, and p the pressure.

Substituting Eq. (1.119) in Eq. (1.108) we obtain an expression for the probability densities of the microstates of a system with a variable number of particles, or the Gibbs grand canonical distribution

$$\rho(q, p, N) = \frac{1}{N!} e^{(\omega + \mu N - H_N)/\theta}, \tag{1.120}$$

which satisfies the normalization condition*

$$\sum_{N=0}^{\infty} \frac{1}{N!} \int e^{(\omega + \mu N - H_N)/\theta} d\Gamma_N = 1. \tag{1.121}$$

From Eq. (1.121) we obtain an expression for the grand thermodynamic potential

$$\omega(V, T, \mu) = -\theta \ln Z_G, \tag{1.122}$$

where

$$Z_G = \sum_{N=0}^{\infty} \frac{Z^N}{N!} \int e^{-H_N/\theta} d\Gamma_N \tag{1.123}$$

is the grand partition function and $Z = e^{\mu/\theta}$ is the activity.

Since $\omega = -pV$, we obtain from Eqs. (1.122) and (1.123) the equation of state of the system

*The subscript N labels quantities pertaining to systems of N particles.

$$pV = \theta \ln \sum_{N=0}^{\infty} \frac{Z^N}{N!} \int e^{-H_N/\theta} d\Gamma_N \,. \tag{1.124}$$

An important statistical-equilibrium state occurs when all the systems of an ensemble have equal energies. Such an ensemble is called microcanonical, and the distribution of the systems over the states is called the Gibbs microcanonical distribution. We can find this distribution as the limiting one in a layer between two close boundary energies E and $E + \Delta E$ ($\Delta E \to 0$), with the phase density having an arbitrary energy dependence inside the layer and equal to zero outside the layer. This distribution, which is obtained from the part of a canonical ensemble between two energy boundaries, is consequently independent of the modulus when the difference between its boundary values becomes infinitely small, but is completely determined by the energy and is identical with the distribution if the density is uniform between boundary energies that approach the same limiting value.

The probability density of states of an isolated system takes thus the form

$$\rho(q,p) = \frac{1}{\omega(E)} \delta[H(q,p) - E] \,, \tag{1.125}$$

where $\delta(x)$ is the Dirac delta function and $1/\omega(E)$ is a normalization factor; $\omega(E)dE$ has the meaning of the phase-space volume of an infinitely thin layer between the energy hypersurfaces E and $E + \Delta E$.

All three considered distributions are, except in a few cases, thermodynamically equivalent, i.e., the canonical distribution can, for example, be used also to describe energywise isolated systems, since the relative fluctuations of macroscopic quantities are small.

We turn now to quantum statistics.

In a state of statistical equilibrium, when the density operator $\hat{\rho}$ and the Hamiltonian $\hat{H}$ of the system do not depend explicitly on the time, the fundamental equation of statistical physics of quantum systems [the von Neumann equation (1.72)] takes the form

$$[\hat{H}, \hat{\rho}] = 0 \,, \tag{1.126}$$

i.e., the operator $\hat{\rho}$ commutes in this case with the Hamiltonian and is consequently an integral of the motion. Therefore, in analogy with the Gibbs expression (1.99) for the phase density of a classical equilibrium system in a heat bath, we choose the density operator of a quantum equilibrum system in a heat bath in the form

$$\hat{\rho} = e^{(F - \hat{H})/\theta} \,, \tag{1.126'}$$

or, in the coordinate representation,

$$\rho(x,x') = \sum_n e^{(F - E_n)/\theta} \psi_n(x) \psi_n^*(x') \,, \tag{1.127}$$

where $\psi_n(x)$ are the eigenfunctions of the Hamiltonian

$$\hat{H}\psi_n = E_n \psi_n \,.$$

Expression (1.127) for $\rho(x, x')$ shows that the probability of a system state defined by a wave function ψ_n with energy E_n is

$$W_n = e^{(F - E_n)/\theta}. \tag{1.128}$$

If the energy levels are multiple, at a multiplicity ω_n of the level energy E_n the probability of a state with such an energy is

$$W_n = e^{(F - E_n)/\theta}\omega_n. \tag{1.128'}$$

A distribution (1.128) or (1.128′) is called a quantum canonical distribution.

We determine the physical meaning of the parameters F and θ of the canonical distribution (1.128) by calculating with its aid the system equilibrium (average) energy and pressure (or another force A related to an external parameter a) under the normalization condition

$$\sum_{n=0}^{\infty} W_n = e^{F/\theta} \sum_{n=0}^{\infty} e^{-E_n/\theta} = 1. \tag{1.129}$$

The average system energy is

$$\langle E \rangle = \sum_{n=0}^{\infty} E_n W_n = \sum_{n=0}^{\infty} E_n e^{(F - E_n)/\theta} = e^{F/\theta}\theta^2 \frac{\partial}{\partial\theta} \sum_{n=0}^{\infty} e^{-E_n/\theta},$$

and since

$$\sum_{n=0}^{\infty} e^{-E_n/\theta} = e^{-F/\theta},$$

according to Eq. (1.129), we have

$$\langle E \rangle = F - \theta \frac{\partial F}{\partial \theta}. \tag{1.130}$$

This equation coincides with the Gibbs–Helmholtz equation for the free energy. Consequently, F is the free energy of the system, and $\theta = kT$.

From the normalization condition (1.129) we find that the free energy is equal to

$$F = -\theta \ln Z, \tag{1.131}$$

where

$$Z = \sum_{n=0}^{\infty} e^{-E_n/\theta} \tag{1.132}$$

is the partition function.

To calculate the free energy of a quantum system it is thus first necessary to solve the dynamic problem of determining its energy spectrum, find the partition function (1.132), and then F itself from Eq. (1.131).

This problem was solved exactly for only a few simple cases.

By way of example of an exactly solvable statistical system, we model a solid by an aggregate of identical three-dimensional independent oscillators (the Einstein model). Since they are independent, we have

$$\hat{H} = \sum_{i=1}^{N} \hat{H}_i,$$

where

$$\hat{H}_i = -\frac{\hbar^2}{2m}\nabla_i^2 + \frac{m\omega_i^2}{2}(x_i^2 + y_i^2 + z_i^2),$$

and x_i, y_i, and z_i are the Cartesian coordinates of the ith oscillator.

From the solution of the dynamic problem

$$\hat{H}_i\psi_n^i = E_n^i\psi_n^i$$

we obtain the spectrum of the possible oscillator energies

$$E_{n_1n_2n_3}^i = \hbar\omega_i(n_1 + n_2 + n_3 + \tfrac{3}{2}).$$

The partition function of the system is

$$Z = \prod_{i-1}^{N} Z_i,$$

where

$$Z_i = \sum_{n_1,n_2,n_3} e^{-\hbar\omega_i(n_1+n_2+n_3+\frac{3}{2})/\theta} = \frac{e^{-3\hbar\omega_i/2\theta}}{(1-e^{-\hbar\omega_i/\theta})^3}.$$

The free energy of the system is

$$F = \sum_{i=1}^{N} F_i,$$

the free energy of the ith oscillator being

$$F_i = -\theta\ln Z_i = \frac{3\hbar\omega_i}{2} + 3\theta\ln(1-e^{-\hbar\omega_i/\theta}).$$

The average energy of each oscillator at thermal equilibrium is

$$\langle E_i\rangle = F_i - \theta\frac{\partial F_i}{\partial\theta} = -\theta^2\frac{\partial}{\partial\theta}\left(\frac{F_i}{\theta}\right) = \frac{3\hbar\omega_i}{2} + \frac{3\hbar\omega_i}{e^{\hbar\omega_i/\theta}-1},$$

and the average system energy is

$$\langle E\rangle = \sum_{i=1}^{N}\langle E_i\rangle.$$

The frequencies ω_i of identical cubic-crystal oscillators are obviously equal; therefore,

$$F_1 = F_2 = \cdots = F_N = F_0 = \frac{3\hbar\omega}{2} + 3\theta\ln(1-e^{-\hbar\omega/\theta}),$$

$$F = NF_0,\quad E = \langle E\rangle = NE_0,$$

$$E_0 = \frac{3\hbar\omega}{2} + \frac{3\hbar\omega}{e^{\hbar\omega/\theta}-1}.$$

The heat capacity C_V of the system is

$$C_V = k\frac{\partial E}{\partial\theta} = 3kN\left(\frac{\hbar\omega}{\theta}\right)^2\frac{e^{\hbar\omega/\theta}}{(e^{\hbar\omega/\theta}-1)^2}.$$

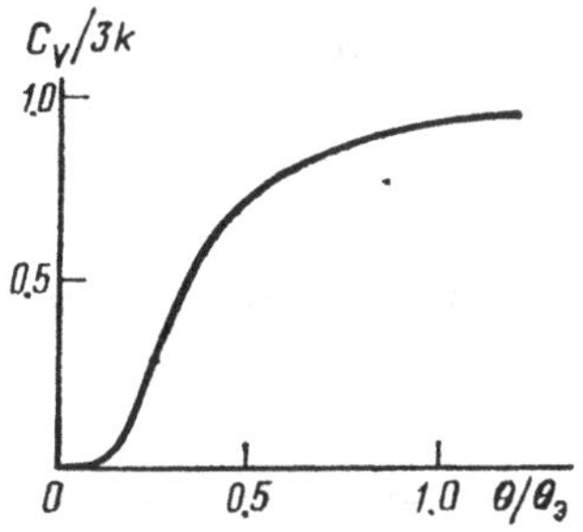

Figure 4.

The temperature dependence of C_V is shown in Fig. 4. At $\theta/\hbar\omega \gg 1$ (high temperatures)

$$C_V = 3kN,$$

i.e., we get the classical relation for the heat capacity of a solid—the Dulong and Petit law. At $\theta/\hbar\omega \ll 1$ (low temperatures)

$$C_V = 3kN\left(\frac{\hbar\omega}{\theta}\right)^2 e^{-\hbar\omega/\theta},$$

which agrees only approximately with experiment, since no account is taken of the collective character of the particle motion in a real crystal. At $\theta/\hbar\omega \ll 1$ we actually have

$$C_V = a\theta^3 \quad \text{(the Debye law)}.$$

Analysis of exactly solvable problems in statistical physics is important for the understanding of the most distinctive features of systems.[10] An exactly solvable system is frequently chosen as a base, and the better it describes the main properties of the investigated system the faster the convergence of the perturbation-theory series used to determine the properties of this system.

6. Bogolyubov equation chain for equilibrium systems

The Gibbs canonical distribution (as well as his other distributions) yields in principle the mean values of all physical quantities. In many cases, however, these quantities can be obtained not from a function of all the coordinates, but from the probability density for one, two, or three particles. We introduce these partial distribution functions for classical equilibrium systems, starting from the configurational Gibbs canonical distribution $\mathscr{D}(q_1,\ldots,q_N)$, which determines the probability density of a given configuration for all N particles, and is equal to

$$D(q_1,...,q_N) = \int \rho\,(q_1,...,q_N, p_1,...,p_N)\, dp_1 \cdots dp_N$$
$$= \frac{1}{Q} e^{-U(q_1,...,q_N)/\theta} \tag{1.133}$$

(here q_i and p_i are vectors; their Cartesian components will be designated q_i^α, p_i^α; $\alpha = 1, 2, 3$).

Let

$$F_s(q_1,...,q_s) \quad (s = 1, 2,...)$$

be symmetric functions of the arguments $q_i,...,q_s$, so normalized that the expression

$$\frac{1}{V^s} F_s(q_1,...,q_s)\, dq_1 \cdots dq_s$$

represents the probability of finding a given group of molecules in infinitely small volumes $dq_1,...,dq_s$ about the points $q_1,...,q_s$. By virtue of this definition we obviously have

$$F_s(q_1,...,q_s) = V^s \int D\,(q_1,...,q_s)\, dq_{s+1} \cdots dq_N\,. \tag{1.134}$$

It is easy to find the equations for these partial distribution functions. In fact, differentiating Eq. (1.133) with respect to q_i^α we obtain

$$\frac{\partial D}{\partial q_1^\alpha} = -\frac{1}{\theta}\frac{\partial U}{\partial q_1^\alpha} D. \tag{1.135}$$

If $\Phi(r) = \Phi(|q_i - q_j|)$ is the interaction potential of two particles, and the potential energy of the system is

$$U = \sum_{1 \leqslant i < j \leqslant N} \Phi(|q_i - q_j|)\,,$$

then, representing this energy in the form

$$U = \sum_{1 \leqslant i < j \leqslant s} \Phi(|q_i - q_j|) + \sum_{\substack{1 \leqslant i \leqslant s \\ s+1 \leqslant j \leqslant N}} \Phi(|q_i - q_j|)$$
$$+ \sum_{s+1 \leqslant i < j \leqslant N} \Phi(|q_i - q_j|),$$

we get

$$\frac{\partial U}{\partial q_1^\alpha} = \frac{\partial U_s}{\partial q_1^\alpha} + \sum_{s+1 \leqslant j \leqslant N} \frac{\partial \Phi(|q_1 - q_j|)}{\partial q_1^\alpha}$$

and

$$\frac{\partial D}{\partial q_1^\alpha} = -\frac{1}{\theta}\frac{\partial U_s}{\partial q_1^\alpha} D - \frac{1}{\theta}\sum_{s+1 \leqslant j \leqslant N} \frac{\partial \Phi(|q_1 - q_j|)}{\partial q_1^\alpha} D, \tag{1.136}$$

where

$$U_s = \sum_{1 \leqslant i < j \leqslant s} \Phi(|q_i - q_j|)$$

is the potential energy of s particles.

Multiplying Eq. (1.136) by V^s, integrating with respect to the variables $q_{s+1} \cdots q_N$, and recognizing that

$$V^s \sum_{s+1 \leqslant j \leqslant N} \int_V \cdots \int_V \frac{\partial \Phi(|q_1 - q_j|)}{\partial q_1^\alpha} D dq_{s+1} \cdots dq_N$$

$$= V^s (N-s) \int_V \cdots \int_V \frac{\partial \Phi(|q_1 - q_{s+1}|)}{\partial q_1^\alpha} D dq_{s+1} \cdots dq_N$$

$$= \frac{N-s}{V} \int_V \frac{\partial \Phi(|q_1 - q_{s+1}|)}{\partial q_1^\alpha} F_{s+1} dq_{s+1} ,$$

we get

$$\frac{\partial F_s}{\partial q_1^\alpha} + \frac{1}{\theta} \frac{\partial U_s}{\partial q_1^\alpha} F_s + \frac{1 - s/N}{v\theta} \int_V \frac{\partial \Phi(|q_1 - q_{s+1}|)}{\partial q_1^\alpha} F_{s+1} dq_{s+1} = 0 ,$$

where

$$v = V/N; \quad \alpha = 1, 2, 3; \quad s = 1, 2, \ldots .$$

In the statistical limit ($N \to \infty$, $V \to \infty$, $V/N =$ const) the functions F_s satisfy the chain of Bogolyubov equations:

$$\frac{\partial F_s}{\partial q_1^\alpha} + \frac{1}{\theta} \frac{\partial U_s}{\partial q_1^\alpha} F_s + \frac{1}{\theta v} \int \frac{\partial \Phi(|q_1 - q_{s+1}|)}{\partial q_1^\alpha} F_{s+1} dq_{s+1} = 0 , \quad (1.137)$$

where the integration extends over all of three-dimensional space.

The equation chain (1.137) is solved using the normalization conditions

$$\lim_{V \to \infty} \frac{1}{V} \int F_1(q) dq = 1, \quad (1.138)$$

$$\lim_{V \to \infty} \frac{1}{V} \int F_{s+1}(q_1, \ldots, q_{s+1}) dq_{s+1} = F_s(q_1, \ldots, q_s) , \quad (1.139)$$

and with boundary conditions requiring that the correlations between the particle positions become weaker with increase of the distance between them:

$$F_s(q_1, \ldots, q_s) \to \prod_{1 \leqslant i \leqslant s} F_1(q_i) , \quad (1.140)$$

when all $|q_i - q_j| \to \infty$.

The functions F_s defined by Eqs. (1.137)–(1.140) are the sought asymptotic distribution functions for real systems with a finite but quite large number of particles ($N \sim 10^{23}$) contained in a finite macroscopic volume V.

More convenient in some cases are the "parental" distribution functions $\rho_s(q_1, \ldots, q_s)$ (Ref. 33).

The expression

$$\rho_s(q_1,\ldots,q_s)\,dq_1\cdots dq_s$$

determines the probability of finding, in a system of N particles, some particle in a volume element dq_1 near q_1, another particle in dq_2 near $q_2,\ldots,$ and finally the last particle in dq_s near q_s. It is easy to obtain

$$\rho_s(q_1,\ldots,q_s)=\frac{N!}{(N-s)!}\cdot\frac{1}{V^s}F_s(q_1,\ldots,q_s)\,. \tag{1.141}$$

Actually there are N possibilities of finding one particle in the volume dq_1, $N-1$ possibilities of finding a second particle in $dq_2,\ldots,$ and $N-s+1$ possibilities of observing the last particle in dq_s. The total number of methods of realizing the considered state is thus

$$N(N-1)\cdots(N-s+1)=\frac{N!}{(N-s)!}\,,$$

and therefore

$$\int_V\cdots\int_V\rho_s(q_1,\ldots,q_s)\,dq_1\cdots dq_s=\frac{N!}{(N-s)!}\,. \tag{1.142}$$

In the statistical limit (since $s\ll N$) we have

$$\rho_s(q_1,\ldots,q_s)=\frac{1}{v^s}F_s(q_1,\ldots,q_s)\,. \tag{1.143}$$

Taking Eq. (1.143) into account, we obtain from Eqs. (1.137) an equation chain for the new introduced functions:

$$\frac{\partial\rho_s}{\partial q_1^\alpha}+\frac{1}{\theta}\frac{\partial U_s}{\partial q_1^\alpha}\rho_s+\frac{1}{\theta}\int\frac{\partial\Phi(|q_1-q_{s+1}|)}{\partial q_1^\alpha}\rho_{s+1}\,dq_{s+1}=0 \tag{1.144}$$

with the normalization condition

$$\frac{1}{N^s}\int_V\cdots\int_V\rho_s(q_1,\ldots,q_s)\,dq_1\cdots dq_s=1 \tag{1.145}$$

$$(s=1,2,\ldots)\,.$$

From among all the partial equilibrium distribution functions, of particular importance is the binary function $F_2(q_1,q_2)$ [or $\rho_2(q_1,q_2)$], for in terms of them we can express the thermal and caloric equations of state as well as other thermal functions of the investigated systems. Thus, in the Bogolyubov method the investigation of equilibrium systems reduces not to a calculation of the configuration integral, but to a solution of equations for partial distribution functions.

Quantum generalizations of the classical partial distribution functions are the partial statistical operators $\hat{\rho}_s(1,\ldots,s)$ introduced in Sec. 4. These operators can be used to determine the expectation values of dynamic quantities of additive, binary, and generally "s-fold" type from Eqs. (1.79), (1.80), and (1.81), respectively.

For the equilibrium statistical operators ($\partial\hat{\rho}_s/\partial t = 0$), the Bogolyubov equation chain takes the form (1.92):

$$[H_s(1,\ldots,s), \hat{\rho}_s(1,\ldots,s)]$$

$$+ (N-s) \operatorname*{Tr}_{s+1} \sum_{1 \leqslant i \leqslant s} [\Phi(i, s+1), \hat{\rho}_{s+1}(1,\ldots,s+1)] = 0\,,$$

where

$$\hat{H}(1,\ldots,s) = \sum_{1 \leqslant i \leqslant s} \hat{H}(i) + \sum_{1 \leqslant i_1 < i_2 \leqslant s} \Phi(i_1, i_2)$$

is the Hamiltonian of an isolated system of s particles with paired interaction.

Chapter 2
Methods of statistical thermodynamics

As noted in Chap. 1, the properties of equilibrium systems can be investigated on the basis of the fundamental equations of statistical physics by two methods: the configuration-integral (partition function) method and the particle-complex distribution-function method. In the former case the problem reduces to calculation of the partition function, in the latter to a determination of the partial distribution functions.

We describe in the present chapter the solution of these problems.

7. Group expansion

7.1 Classical systems

One of the methods of calculating the configuration integral

$$Q_N = \int_V \cdots \int_V e^{-U(q_1,\ldots,q_N)/\theta} dq_1 \cdots dq_N \tag{2.1}$$

is that of the group or cluster expansion. It was developed by Mayer in 1937 for a many-particle system with a pair-interaction potential that falls off rapidly with distance.

The potential energy of such a system is obviously

$$U = \sum_{1 \leqslant i < j \leqslant N} \Phi(|q_i - q_j|) . \tag{2.2}$$

Therefore the integrand in Eq. (2.1),

$$\exp[-\beta U] = \prod_{1 \leqslant i < j \leqslant N}\prod \exp[-\beta\Phi(|q_i - q_j|)] \quad (\beta = 1/\theta)$$

is transformed with the aid of the Mayer function

$$f_{ij} = f(r_{ij}) = \exp[-\beta\Phi(|q_i - q_j|)] - 1 \tag{2.3}$$

into the sum

$$\exp[-\beta U] = \prod_{1 \leqslant i < j \leqslant N}\prod (1 + f_{ij}) = 1 + \sum_{1 \leqslant i < j \leqslant N} f_{ij} + \sum f_{ij} f_{kl} + \cdots , \tag{2.4}$$

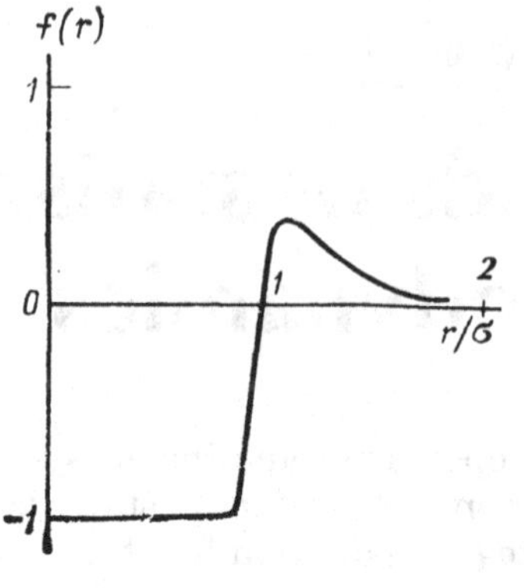

Figure 5.

in which each term corresponds to a subdivision of the particles into definite groups.

Substituting Eq. (2.4) in Eq. (2.1), we obtain a group expansion of the configuration integral Q_N. The Mayer function for a Lennard-Jones interaction potential is shown in Fig. 5. At $r<\sigma$ the function is close to -1, and at $r=r_0=2^{1/6}\sigma$, when the potential reaches a minimum, the function has a maximum and then decreases rapidly and tends to zero.

A general term of the series (2.4) is a sum over all the coupled products of one and the same order n, i.e., the summation is over all n-particle topologically different diagrams. The expansion (2.4) itself can be represented in terms of irreducible diagrams or graphs, in which each point is f coupled to at least two other points (an exception is the case of two points). The number of different irreducible diagrams increases rapidly with increase of n. There exist thus 3 four-particle, 9 five-particle, and 56 six-particle irreducible diagrams.[17]

Integrating Eq. (2.4) over the configurations of the entire system, we obtain for the configuration integral the expression[18]

$$Q_N = V^N + \frac{N(N-1)}{2!} V^{N-1} \int f(|q|)dq + \cdots . \tag{2.5}$$

There are various methods of finding the thermodynamic functions with the aid of the cluster expansion (2.4). The most widely used is a method based on calculation of the grand partition function (1.123):

$$Z_G = \sum_{N=0}^{\infty} \frac{Z^N}{N!} \int e^{-H_N/\theta} d\Gamma_N .$$

This calculation yields the following virial expansions[48] for the logarithms of the grand partition function and for the activity of the series in powers of the density $n = 1/v$:

$$\ln Z_G = N\left(1 + \sum_{i=2}^{\infty} B_i / v^{i-1}\right), \tag{2.6}$$

$$\mu/\theta = \ln Z = \ln(\lambda^3/v) + \sum_{i=2}^{\infty} \frac{i}{i-1} \frac{B_i}{v^{i-1}}, \tag{2.7}$$

where

$$B_i = -\frac{i-1}{i}\beta_i \tag{2.8}$$

are the virial coefficients;

$$\beta_i = \frac{1}{(i-1)!} \int \cdots \int \left(\sum_j \Pi f_{ij} \right) dq_2 \cdots dq_i \tag{2.9}$$

(the products are summed over the irreducible diagrams[17]); λ is the thermal de Broglie wavelength. Using Eqs. (2.6) and (2.7) one can find a virial expansion for any thermodynamic quantity. We obtain thus from Eq. (1.24) the virial equation of state

$$pV = N\theta\left(1 + \sum_{i=2}^{\infty} B_i / v^{i-1}\right), \tag{2.10}$$

and the system free energy is

$$F = G - pV = \mu N - pV$$

$$= -N\theta \ln(ve) + 3N\theta \ln \lambda + N\theta \sum_{i=2}^{\infty} \frac{1}{i-1} \frac{B_i}{v^{i-1}}. \tag{2.11}$$

These virial expansions for the pressure and for the free energy can be obtained also from the canonical distribution, by expressing the expansion (2.5) in the form of a contour integral evaluated by the saddle-point method.[19]

According to the canonical distribution, the free energy (1.116) of a system of n particles in a heat bath is equal to

$$F = 3N\theta \ln \lambda - \theta \ln \frac{Q_N}{N!}. \tag{2.12}$$

Bulk properties of macroscopic systems are investigated in statistical mechanics by going to the statistical limit ($N \to \infty$, $V \to \infty$, $V/N = v = \text{const}$) so as to transform the free energy and other thermodynamic potentials into extensive quantities.

Starting from Eq. (2.12), we calculate now the free energy of the system, taking into account from the very outset its asymptotic form, i.e., representing F as an extensive quantity:

$$F = 3N\theta \ln \lambda - N\theta \ln(eq/N), \tag{2.13}$$

where

$$q(\theta, v) = Q_N^{1/N}. \tag{2.14}$$

Thus in this approach we must find $q(\theta, v)$ in order to determine the free energy of the system and thereby formulate its thermodynamics.

We shall see later on that the new approach to the determination of the free energy from Eq. (2.13) leads to an equation of state that describes real systems better than Eq. (2.10).

We obtain $q(\theta, V)$ in the form of a cluster expansion. To this end, we subject Q_N to a g transformation in which the Mayer function in Eq. (2.5) is assumed proportional to the parameter g:

$$Q_N(g) = V^N + g\frac{N(N-1)}{2!}V^{N-1}\int f(|q|)dq + \cdots .$$

Then, obviously, $Q_N(0)$ corresponds to an ideal gas, while $Q_N(1)$ yields the initial equation (2.5). According to Eq. (2.14),

$$q(g) = \left(V^N + g\frac{N(N-1)}{2!}V^{N-1}\int f(|q|)dq + \cdots\right)^{1/N}. \tag{2.15}$$

We expand $q(g)$ in a Taylor series in g:

$$q(g) = \sum_{n=0}^{\infty}\frac{1}{n!}\left(\frac{d^n q}{dg^n}\right)_{g=0} g^n$$

and put next $g = 1$. The result is

$$q(\theta,v) = N\left(v - \sum_{i=2}^{\infty} b_i/v^{i-2}\right), \tag{2.16}$$

where

$$\begin{aligned} b_2 &= B_2, \\ b_3 &= (B_3 - B_2^2)/2, \\ b_4 &= (2B_4 - 3B_2B_3 - B_2^3)/6, \\ &\vdots \end{aligned} \tag{2.17}$$

Substituting Eq. (2.16) in expression (2.13) for the free energy, we have

$$\begin{aligned} F &= 3N\theta \ln \lambda - N\theta \ln\left[e\left(v - \sum_{i=2}^{\infty} b_i/v^{i-2}\right)\right] \\ &= F_0 - N\theta \ln\left(v - \sum_{i=2}^{\infty} b_i/v^{i-2}\right), \end{aligned} \tag{2.18}$$

where

$$F_0 = N\theta(\ln \lambda^3 - 1).$$

If

$$b = \sum_{i=2}^{\infty} b_i/v^{i-1} \ll v, \tag{2.19}$$

the logarithmic function in Eq. (2.18) can be expanded in a series, and we obtain the usual virial expansion for the free energy (2.11) and the equation of state (2.10).

We have considered so far systems with a pair potential for the interaction between the particles. However, even before Mayer proposed in 1937 his method of expanding the configuration integral in f functions, Ursell[20] introduced in 1927 an expansion of the configuration integral in U functions. Ursell's method is more general: it can be used for particle systems with arbitrary (and not only pair) interaction and can be extended to include quantum-mechanical systems.

Ursell has shown that the Boltzmann factor

$$W_N(q_1,...,q_N) \tag{2.20}$$

that enters in the configuration integral can be represented as a sum of products of U_i functions expressed in terms of the Boltzmann factors $W_1(q_1)$, $W_2(q_1,q_2)$,... of 1, 2,... particles:

$$\begin{aligned}
U_1(q_1) &= W_1(q_1) = 1\,,\\
U_2(q_1,q_2) &= W_2(q_1,q_2) - W_1(q_1)W_2(q_2)\,,\\
U_3(q_1,q_2,q_3) &= W_3(q_1,q_2,q_3) - W_2(q_1,q_2)W_1(q_3)\\
&\quad - W_2(q_1,q_3)W_1(q_2) - W_2(q_2,q_3)W_1(q_1)\\
&\quad + 2W_1(q_1)W_1(q_2)W_1(q_3)\,,\\
&\;\;\vdots
\end{aligned} \tag{2.21}$$

The functions U_i have an important property, viz., they differ from zero only when a group of i particles cannot be broken up into noninteracting groups (we are considering a system of particles with short-range interaction forces).

The Boltzmann factors of 1, 2, 3,... particles can be expressed in terms of the U_i functions of Eq. (2.21).

As a result, the Boltzmann factor takes the form

$$W_N(q_1,...,q_N) = \sum \prod U_e(q_1,...,q_e)\,, \tag{2.22}$$

where the summation must be carried out over all possible distributions of N particles, in m_1 groups of one particle, m_2 of two,... m_e of e particles, with account taken of the condition $\Sigma e m_e = N$.

The configuration integral is then equal to

$$Q_N = N!\sum \prod \frac{(V\bar{b}_e)^{m_e}}{m_e!}\,, \tag{2.23}$$

where b_e are group integrals defined by the expression

$$\bar{b}_e = \frac{1}{Ve!}\int U_e(q_1,...,q_e)dq_1\cdots dq_e\,. \tag{2.24}$$

The integration in Eq. (2.24) is over the volume V. Group integrals can also be written in the form

$$\bar{b}_e = \frac{1}{e!}\int U_e(q_{21},...,q_{e1})dq_{21}\cdots dq_{e1}\,, \tag{2.25}$$

where the integration limits are infinite, and r_{i1} are the particle coordinates measured from the position of particle 1. The first integrals of Eq. (2.24) are of the form

$$\bar{b}_1 = \frac{1}{V}\int dq_1 = 1\,,$$

$$\bar{b}_2 = \frac{1}{2V}\int [W_2(q_1,q_2) - W_1(q_1)W_1(q_2)]dq_1\, dq_2\,. \tag{2.26}$$

$$\vdots$$

The free energy can be calculated on the basis of Eq. (2.23) by various methods, for example, by the saddle point or by the generating function method.[19,21] It is easy to carry out such a calculation with the aid of a g transformation, putting

$$\bar{b}_e \to g^{e-1}\bar{b}_e\,. \tag{2.27}$$

We obtain then an expression for the free energy in the form

$$F = F_0 - N\theta\ln(v - b_1) = F_0 - N\theta\ln\left[v - \sum_{i=2}^{\infty} b_{1i}/v^{i-2}\right], \tag{2.28}$$

where

$$b_{12} = -\bar{b}_2\,,$$

$$b_{13} = -(2\bar{b}_3 - 3\bar{b}_2^{\,2})/2\,, \tag{2.29}$$

$$\vdots$$

In the case when $b_1 \ll v$ we obtain the virial expression for the free energy

$$F = F_0 + N\theta\sum_{i=2}^{\infty}\frac{1}{i-1}\,\frac{b_{2i}}{v^{i-1}}\,, \tag{2.30}$$

where[22]

$$b_{22} = -\bar{b}_2\,,$$

$$b_{23} = -2(\bar{b}_3 - 2\bar{b}_2^{\,2})\,, \tag{2.31}$$

$$\vdots$$

In the additive pair potential approximation we have

$$b_{2i} = B_i,\quad b_{1i} = b_i\quad (i = 2,3,\ldots)\,. \tag{2.32}$$

In the group expansion, the number of topologically different diagrams increases rapidly with increase of e, making it difficult to calculate the virial coefficients for large e. This is precisely why only the first few virial coefficients have been calculated for real potentials.

Ree and Hoover[23] introduced the function

$$\bar{f}_{ij} = e^{-\beta\Phi_{ij}}\,, \tag{2.33}$$

which is connected with the Mayer function f_{ij} by the relation

$$\bar{f}_{ij} - f_{ij} = 1\,.$$

When there is no f coupling of two points on the diagram, an $(\bar{f}_{ij} - f_{ij})$ coupling between them is introduced; it does not make a numerical contribution, but influences strongly the number of diagrams over which the summation is carried out. Thus, in the usual approach with f couplings the virial coefficients B_4, B_5, and B_6 contain, respectively, 3, 10, and 56 topologically different diagrams. In the Ree–Hoover formalism, on the other hand, these coefficients are calculated from 2, 5, and 23 topologically different modified diagrams.

7.2 Quantum systems

As already mentioned, the Ursell method makes a group expansion possible also for the partition function of a quantum system.[24,25] In this case, the Boltzmann factor

$$W(q_1,...,q_N) = \exp[-\beta U_N]$$

is replaced by the Slater sum

$$\overline{W}(1,...,N) = N!\left(\frac{h^2}{2\pi m\theta}\right)^{3N/2} \sum_n \psi_n^*(1,...,N)\exp(-\beta\hat{H})\psi_n(1,...,N), \quad (2.34)$$

where $\hat{H}$ is the Hamiltonian operator of a system of N particles. The functions $\psi_n(1,...,N)$ constitute a complete orthonormalized system of eigenfunctions that are symmetric or antisymmetric, depending on the spin. The Slater sum becomes the Boltzmann factor in the high-temperature limit.[26]

The calculation of virial coefficients for quantum systems is very difficult, for in contrast to classical systems, where the calculation of the virial coefficient calls for finding multiple integrals, it is now necessary to solve first the many-particle quantum-mechanical problem. In this case the exchange degeneracy of the wave functions causes the second virial coefficient to differ from zero even for an ideal quantum gas[17]:

$$B_2 = \pm\frac{1}{2}\left(\frac{\pi\hbar^2}{mkT}\right)^{1/2} \quad (2.35)$$

($+$ and $-$ correspond to boson and fermion systems, respectively).

Since it is difficult to calculate the virial coefficients by purely quantum methods, a quasiclassical method was developed, based on expanding the virial coefficients in powers of h. If the pair-interaction potential has an analytic form $\Phi = \Phi(r)$ and continuous derivatives, the Wigner–Kirkwood method yields the following expression for the second virial coefficient[24]:

$$B_2 = B_{21} + (\hbar^2/2\mu_1)B_{22} + (\hbar^2/2\mu_1)^2 B_{23} + (\hbar^2/2\mu_1)^3 B_{24} + \cdots + B_{20}, \quad (2.36)$$

where B_{21} is the second virial coefficient determined from the classical formula

$$B_{21} = -\tfrac{1}{2}\int f(|q|)dq,$$

and also

$$B_{22} = \frac{\pi\beta^3}{6}\int_0^\infty e^{-\beta\Phi(r)}(\Phi')^2 r^2\, dr\,,$$

$$B_{23} = \frac{\pi\beta^4}{6}\int_0^\infty e^{-\beta\Phi(r)}\left[\frac{(\Phi'')^2}{10} + \frac{(\Phi')^2}{5r^2} + \frac{(\Phi')^3}{9r}\beta - \frac{(\Phi')^4}{72}\beta^3\right] r^2\, dr\,,$$

$$B_{24} = \frac{\pi\beta^5}{6}\int_0^\infty e^{-\beta\Phi(r)}\left[\frac{(\Phi''')^2}{140} + \frac{3(\Phi'')^2}{70r^2} + \frac{(\Phi'')^3}{126}\beta + \frac{\Phi'(\Phi'')^2}{30r}\beta\right.$$

$$\left. + \frac{2(\Phi')^3}{315r^3}\beta - \frac{(\Phi')^2(\Phi'')^2}{120}\beta^2 - \frac{(\Phi')^4}{1080r^2}\beta^2 - \frac{(\Phi')^5}{360r}\beta^3 + \frac{(\Phi')^6}{4320}\beta^4\right] r^2\, dr\,,$$

$$B_{20} = \mp(\hbar^2/2\mu_1)^{3/2}\left[\tfrac{1}{2}(\pi\beta)^{3/2}\right] f_1(\lambda)(2s+1)\,,$$

$$\Phi' = \frac{d\Phi}{dr}\,,\quad \Phi'' = \frac{d^2\Phi}{dr^2}\,,\quad \Phi''' = \frac{d^3\Phi}{dr^3}\,.$$

Here $\mu_1 = m_1 m_2/(m_1 + m_2)$ is the reduced mass of two particles ($\mu_1 = m/2$ for $m_1 = m_2 = m$) and $\lambda = h/(2\pi m\theta)^{1/2}$. The function $f_1(\lambda)$ is equal to unity for large λ and tends rapidly to zero for small λ.

An expansion in powers of $\hbar$ is known also for the third virial coefficient.[24] This approach results in a well converging series at high temperature, but the convergence becomes worse at low temperature.

8. Perturbation theory

The free energy

$$F = -\theta \ln(Z/N!) \tag{2.37}$$

is an extensive (additive) quantity; we write it therefore in the form

$$F = -N\theta \ln[p(\theta,v)/(N!)^{1/N}]\,, \tag{2.38}$$

where $p(\theta,v) = Z^{1/N}$, with Z the statistical integral or the partition function of the system.

The partition function cannot be calculated in the general case of a system of interacting particles, and different perturbation-theory variants are used. We consider here the traditional approach. It was first developed for crystals by Born, Brody, and Schrödinger.[27–29,39] In our procedure we expand $p(\theta,v)$ by a method similar to that used to expand Z in the papers of Leibfried[31] and Nakajima.[32] It is a modification of the g transformation used by us in Sec. 7 to obtain the cluster expansion of $q(\theta,v)$.

Consider a system with a Hamiltonian H. Let H_0 be the Hamiltonian for which the partition function Z_0 can be calculated and which approximates well, in a certain sense, the Hamiltonian H. To simplify the calculations, we introduce $H(g)$ defined as

$$H(g) = H_0 + g(H - H_0)\,,$$

from which it can be seen that $H(0) = H_0$ and $J(1) = H$.

The partition function Z, and hence also $p(\theta,v)$, calculated for a system with a Hamiltonian $H(g)$, are functions of the parameter g, viz., $Z = Z(g)$, $p(\theta,v) = p(\theta,v,g)$. We expand $p(\theta,v,g)$ in a Taylor series in g:

$$p(\theta,v,g) = p(\theta,v,0) + p'(\theta,v,g)\big|_{g=0} g + \frac{1}{2!} p''(\theta,v,g)\Big|_{g=0} g^2 + \frac{1}{3!} p'''(\theta,v,g)\Big|_{g=0} g^3 + \frac{1}{4!} p^{\mathrm{IV}}(\theta,v,g)\Big|_{g=0} g^4 + \cdots \tag{2.39}$$

and put here $g = 1$. We obtain the expansion of $p(\theta,v)$ by perturbation theory, since

$$p(\theta,v,1) = p(\theta,v)\,,$$
$$p(\theta,v,0) = p_0(\theta,v)\,,$$

where $p(\theta,v)$ was calculated for a system with the Hamiltonian H, and $p_0(\theta,v)$ for a system with the Hamiltonian H_0.

We express now the derivatives p^i in terms of the derivatives of the partition function:

$$p' = \frac{1}{N} Z^{1/N-1} Z'\,,$$

$$p'' = \frac{1}{N} Z^{1/N-1} Z'' - \frac{1}{N}\left(1 - \frac{1}{N}\right) Z^{1/N-2} (Z')^2\,,$$

$$p''' = \frac{1}{N} Z^{1/N-1} Z''' - \frac{3}{N}\left(1 - \frac{1}{N}\right) Z^{1/N-2} Z' Z'' + \frac{1}{N}\left(1 - \frac{1}{N}\right)\left(2 - \frac{1}{N}\right) Z^{1/N-3} (Z')^3\,,$$

$$p^{\mathrm{IV}} = \frac{1}{N} Z^{1/N-1} Z^{\mathrm{IV}} - \frac{4}{N}\left(1 - \frac{1}{N}\right) Z^{1/N-2} Z' Z''' - \frac{3}{N}\left(1 - \frac{1}{N}\right) Z^{1/N-2} (Z'')^2 + \frac{6}{N}\left(1 - \frac{1}{N}\right)\left(2 - \frac{1}{N}\right) Z^{1/N-3} (Z')^2 Z'' = -\frac{1}{N}\left(1 - \frac{1}{N}\right)\left(2 - \frac{1}{N}\right)\left(3 - \frac{1}{N}\right) Z^{1/N-4} (Q')^4\,.$$

Substituting the obtained expressions in the expansion (2.39) for $p(\theta,v)$ we get from Eq. (2.38) the free energy of the system

$$F = F_0 - N\theta \ln(1 + \varphi)\,, \tag{2.40}$$

with F_0 the free energy of the system whose Hamiltonian is H_0 and

$$\varphi=\frac{1}{N}(Z'/Z)_0+\frac{1}{2N}\left[(Z''/Z)_0-\left(1-\frac{1}{N}\right)(Z'/Z)_0^2\right]$$
$$+\frac{1}{3!N}\left[(Z'''/Z)_0-3\left(1-\frac{1}{N}\right)(Z''/Z)_0(Z'/Z)_0\right.$$
$$\left.+\left(1-\frac{1}{N}\right)\left(2-\frac{1}{N}\right)(Z'/Z)_0^3\right]$$
$$+\frac{1}{4!N}\left[(Z^{\mathrm{IV}}/Z)_0-4\left(1-\frac{1}{N}\right)(Z'''/Z)_0(Z'/Z)_0\right.$$
$$-3\left(1-\frac{1}{N}\right)(Z''/Z)_0^2+6\left(1-\frac{1}{N}\right)\left(2-\frac{1}{N}\right)(Z''/Z)_0(Z'/Z)_0^2$$
$$\left.-\left(1-\frac{1}{N}\right)\left(2-\frac{1}{N}\right)\left(3-\frac{1}{N}\right)(Z'/Z)_0^4\right]+\cdots. \tag{2.41}$$

In the case when

$$\varphi\ll 1,$$

we obtain the usual expression for the free energy[203]

$$F=F_0-\theta(Z'/Z)_0-\frac{\theta}{2}[(Z''/Z)_0-(Z'/Z)_0^2]$$
$$-\frac{\theta}{3!}[(Z'''/Z)_0-3(Z''/Z)_0(Z'/Z)_0+2(Z'/Z)^3]$$
$$-\frac{\theta}{4!}[(Z^{\mathrm{IV}}/Z)_0-4(Z'''/Z)_0(Z'/Z)_0-3(Z''/Z)^2$$
$$+12(Z''/Z)_0(Z'/Z)_0^2-6(Z'/Z)_0^4]+\cdots. \tag{2.42}$$

Here $(\cdots)_0$ denotes that the derivative in the parentheses is calculated for $g=0$.

Equations (2.40) and (2.41) can be used to calculate the free energy of either a classical or a quantum system.

In the classical case,

$$Z_0=(Z)_0=\int\exp[-\beta H_0]d\Gamma, \tag{2.43}$$

$$(Z_i/Z)_0=-(\beta)^i\int(H-H_0)^i\exp[-\beta H_0]d\Gamma/Z_0$$
$$=-(\beta)^i\langle(H-H_0)^i\rangle_0, \tag{2.44}$$

where $\langle\cdots\rangle_0$ denotes averaging over a distribution with a Hamiltonian H_0.

Substituting Eq. (2.44) in Eq. (2.41) we get

$$\varphi = -\frac{\beta}{N}\langle(H-H_0)\rangle_0 + \frac{\beta^2}{2N}\Big[\langle(H-H_0)^2\rangle_0$$
$$-\Big(1-\frac{1}{N}\Big)(\langle(H-H_0)\rangle)^2\Big] - \frac{\beta^3}{3!N}\Big[\langle(H-H_0)^3\rangle_0$$
$$-3\Big(1-\frac{1}{N}\Big)\langle(H-H_0)^2\rangle_0\langle(H-H_0)\rangle_0$$
$$+\Big(1-\frac{1}{N}\Big)\Big(2-\frac{1}{N}\Big)(\langle(H-H_0)\rangle_0)^3\Big] + \frac{\beta^4}{4!N}\Big[\langle(H-H_0)^4\rangle_0$$
$$-4\Big(1-\frac{1}{N}\Big)\langle(H-H_0)^3\rangle_0\langle(H-H_0)\rangle_0 - 3\Big(1-\frac{1}{N}\Big)(\langle(H-H_0)^2\rangle_0)^2$$
$$+6\Big(1-\frac{1}{N}\Big)\Big(2-\frac{1}{N}\Big)\langle(H-H_0)^2\rangle_0(\langle(H-H_0)\rangle_0)^2$$
$$-\Big(1-\frac{1}{N}\Big)\Big(2-\frac{1}{N}\Big)\Big(3-\frac{1}{N}\Big)(\langle(H-H_0)\rangle_0)^4\Big] - \cdots. \tag{2.45}$$

We obtain next from Eq. (2.40) the free energy of the system.

In the case of quantum systems, we calculate F by using the rule for differentiating the operator $\exp[-\beta\hat{H}(g)]$ with respect to the parameter g (Ref. 33; this rule will be derived in the next section):

$$\frac{d}{dg}\exp[-\beta\hat{H}(g)]$$
$$= -\beta\int_0^1 \exp[-\beta\hat{H}(g)(1-\alpha)]\frac{d\hat{H}(g)}{dg}\times\exp[-\beta\hat{H}(g)\alpha]\,d\alpha$$
$$= -\beta\int_0^1 \exp[-\beta\hat{H}(g)(1-\alpha)](\hat{H}-\hat{H}_0)$$
$$\times\exp[-\beta\hat{H}(g)\alpha]\,d\alpha. \tag{2.46}$$

Consequently, the first and second derivatives of the partition function are equal to

$$Z' = -\beta\,\mathrm{Tr}(\hat{H}-\hat{H}_0)\exp[-\beta\hat{H}(g)], \tag{2.47}$$

$$Z'' = \beta^2\,\mathrm{Tr}(\hat{H}-\hat{H}_0)\int_0^1 \exp[-\beta\hat{H}(g)(1-\alpha)](\hat{H}-\hat{H}_0)$$
$$\times\exp[-\beta\hat{H}(g)\alpha]\,d\alpha. \tag{2.48}$$

Differentiating further, we obtain the higher-order derivatives. We ultimately have

$$Z_0 = \mathrm{Tr}\,\exp[-\beta\hat{H}_0], \tag{2.49}$$

$$(Z'/Z)_0 = -\beta\,\mathrm{Tr}(\hat{H}-\hat{H}_0)\exp[-\beta\hat{H}_0]/\mathrm{Tr}\exp[-\beta\hat{H}_0]$$
$$= -\beta\langle(\hat{H}-\hat{H}_0)\rangle_0\,, \tag{2.50}$$
$$(Z''/Z)_0 = \beta^2\,\mathrm{Tr}(\hat{H}-\hat{H}_0)\int_0^1 \exp[-\beta\hat{H}_0(1-\alpha)](\hat{H}-\hat{H}_0)$$
$$\times\exp[-\beta\hat{H}_0\alpha]\,d\alpha/Z_0\,. \tag{2.51}$$

The expressions obtained permit φ, and with it the free energy of the system, to be calculated from Eq. (2.41).

9. Cell-cluster expansion

Crystal structures are usually considered by starting from the Born theory of the crystal lattice, and the anharmonic corrections are obtained by using the perturbation theory described in the preceding section. Different variants of a self-consistent approximation for the description of crystals have also been developed. We consider here a cell-cluster expansion of the free energy F, in which the zeroth approximation is that of the self-consistent field, and the partition function expansion is of the Ursell–Mayer type (when this expansion is used in nonideal-gas theory the zeroth approximation is chosen to be the ideal gas). In contrast to the known approach,[34] we calculate F by starting from the expression for $q(\theta,v)=Q^{1/N}$, which we obtain by a g transformation, assuming the modified Mayer function is proportional to g.

Consider for simplicity a monatomic crystal with N particles in a macroscopic volume V. We divide the volume into N Wigner–Seitz cells (with centers at the equilibrium positions of the crystal atoms)—primitive cells having the total symmetry of the Bravais lattice and representing a region of space in which each point (other than those on the boundary) lies closer to the center of its own cell than to any other center. An example of a Wigner–Seitz cell for an fcc Bravais lattice ("rhombic dodecahedron") is shown in Fig. 6. The multiple partition function is made up of contributions from cells containing a single particle plus contributions, which we neglect, from cells containing many particles (some of the cells turn out to be empty in this case). The configuration integral thus takes the form

$$Q=\int_{B-3}\exp[-\beta U]\,dq_1\cdots dq_N\,, \tag{2.52}$$

where the integration with respect to q_i is carried out over the volume of the ith Wigner–Seitz cell. We assume that the interaction potential U is a sum of pair interactions:

$$U=\sum_{i<j}\Phi(|q_i-q_j|)\,. \tag{2.53}$$

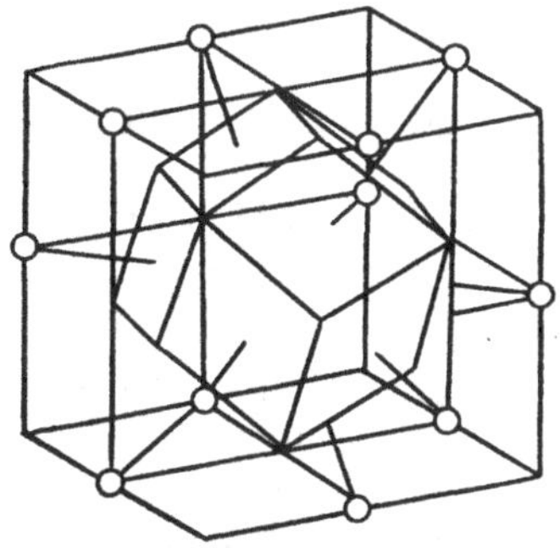

Figure 6.

We introduce the notation

$$U_0 = \tfrac{1}{2} \sum_j \Phi(|q_i^0 - q_j^0|),$$

$$\Phi_1(q_i) = \sum_j [\Phi(|q_i - \bar{q}_j|) - \Phi(|q_i^0 - \bar{q}_j|)], \tag{2.54}$$

$$\Phi_2(q_i, q_j) = \Phi(|q_i - q_j|) + \Phi(|q_i^0 - \bar{q}_j|) + \Phi(|q_j^0 - \bar{q}_i|)$$
$$- \Phi(|q_i - \bar{q}_j|) - \Phi(|q_j - \bar{q}_i|) - \Phi(|q_i^0 - q_j^0|).$$

The potential U then takes the form

$$U = NU_0 + \sum_i \Phi_1(q_i) + \tfrac{1}{2} \sum_{i,j} \Phi_2(q_i, q_j). \tag{2.55}$$

The superior bar in Eq. (2.54) denotes averaging over the single-particle distribution function obtained by solving the self-consistent-field equation. If this function is taken to be a Dirac δ function (it is the solution of the self-consistent field at $\theta = 0$ K), we obtain as the zeroth approximation the Einstein anharmonic model.

On the other hand, it can be assumed that $\Phi(|q_i - \bar{q}_j|) = \varphi(q_i, q_j)$, where $\varphi(q_i, q_j)$ is a certain arbitrary function chosen such that the obtained series converge at the fastest rate.

We introduce a modified Mayer function in the form

$$\exp[-\beta\Phi_2(q_i, q_j)] - 1 \equiv f_{ij} \tag{2.56}$$

and expand the Boltzmann factor in a series

$$\exp[-\beta U] = \exp\Big(-\beta NU_0 - \beta \sum_i \Phi_1(q_i)\Big)\Big(1 + \tfrac{1}{2} \sum_{i,j} f_{ij} + \cdots\Big). \tag{2.57}$$

Substituting this expression in the configuration integral, we get

$$Q = Q_0\Big(1 + \tfrac{1}{2} \sum_{i>j} \langle f_{i,j} \rangle + \cdots\Big), \tag{2.58}$$

where

$$Q_0 = \exp[-\beta N U_0] q_0^N,$$

$$q_0 = \int_{B-3} \exp[-\beta \Phi_i(q_i)] dq_i,$$

$$\langle f_{ij} \rangle = \frac{1}{q_0^2} \iint_{B-3} \exp[-\beta\Phi_1(q_i) - \beta\Phi_1(q_j)] f_{ij} dq_i\, dq_j. \tag{2.59}$$

Here Q_0 is the first-approximation configuration integral. We carry out in Eq. (2.58) a g transformation:

$$Q(g) = Q_0\left(1 + \tfrac{1}{2} g \sum_{i,j} \langle f_{ij} \rangle + \cdots \right). \tag{2.60}$$

Then

$$q(g) = Q_0^{1/N}\left[1 + \frac{g}{2N} \sum_{i,j} \langle f_{ij} \rangle + \cdots \right] \tag{2.61}$$

and we obtain for the free energy the expression

$$F = F_0 - \theta N \ln\left(1 + \frac{1}{2N} \sum_{i,j} \langle f_{ij} \rangle + \cdots \right). \tag{2.62}$$

We consider now a crystal with only nearest-neighbor interaction. Then,

$$\tfrac{1}{2} \sum_{i,j} \langle f_{ij} \rangle \to \tfrac{1}{2} z N q_1, \tag{2.63}$$

where q_1 is equal to $\langle f_{ij} \rangle$ for two neighboring atoms i and j. In the next sum, which contains the product of two functions f_{ij}, it is easy to determine the terms of the form $f_{ij} f_{kl}$ $(k, l \neq i, j)$:

$$\langle f_{ij} f_{kl} \rangle = \langle f_{ij} \rangle \langle f_{kl} \rangle = q_1^2. \tag{2.64}$$

The number of such terms is[34]

$$\tfrac{1}{2}(\tfrac{1}{2} Nz)(\tfrac{1}{2} Nz - 2z + 1). \tag{2.65}$$

We write the remaining terms in the form

$$\langle f_{ij} f_{kl} \rangle = q_{2j}. \tag{2.66}$$

We ultimately obtain for Q,

$$Q = Q_0\Big[1 + (\tfrac{1}{2} Nz) q_1 + \tfrac{1}{2}(\tfrac{1}{2} Nz)(\tfrac{1}{2} Nz - 2z + 1) q_1^2 + (\tfrac{1}{2} Nz) \sum_{j=2}^{z} q_{2j} + O(f^3)\Big]. \tag{2.67}$$

We carry out a g transformation with this expression

$$Q(g) = Q_0\Big[1 + g(\tfrac{1}{2}Nz)q_1 + g^2\tfrac{1}{2}(\tfrac{1}{2}Nz)(\tfrac{1}{2}Nz - 2z + 1)q_1^2 + g^2(\tfrac{1}{2}Nz)\sum_{j=2}^{z} q_{2j} + g^3 o(f^3)\Big]. \tag{2.68}$$

Hence

$$Q'|_{g=0} = Q_0\tfrac{1}{2}Nzq_1,$$

$$Q''|_{g=0} = \Big\{(\tfrac{1}{2}Nz)(\tfrac{1}{2}Nz - 2z + 1)q_1^2 + Nz\sum_{j=2}^{z} q_{2j}\Big\}Q_0, \tag{2.69}$$

etc.

It is now easy to obtain $q(\theta,v)$:

$$\begin{aligned} q(\theta,v) &= Q_0\Big[1 + \frac{1}{2}zq_1 + \frac{1}{2N}\Big[\frac{1}{2}Nz\Big(\frac{1}{2}Nz - 2z + 1\Big)q_1^2 \\ &\quad + Nz\sum_{j=2}^{z} q_{2j} - (1 - 1/N)\frac{1}{4}N^2z^2q_1^2\Big] + \cdots\Big] \\ &= Q_0\Big[1 + \frac{1}{2}zq_1^2 + \frac{z}{4}(1 - 2z)q_1^2 + \frac{z^2}{8}q_1^2 + \frac{z}{2}\sum_{j=2}^{z} q_{2j} + \cdots\Big] \\ &= Q_0\Big[1 + \frac{1}{2}zq_1 + \Big(\frac{z}{4} - \frac{3}{8}z^2\Big)q_1^2 + \frac{z}{2}\sum_{j=2}^{z} q_{2j} + o(f^3)\Big]. \end{aligned} \tag{2.70}$$

This expression allows us to determine the free energy of the system

$$F = F_0 - \theta N\ln\Big[1 + \frac{1}{2}zq_1 + \Big(\frac{z}{4} - \frac{3}{8}z^2\Big)q_1^2 + \frac{z}{2}\sum_{j=2}^{z} q_{2j} + o(f^3)\Big]. \tag{2.71}$$

An approach based on the determination of $q(\theta,v)$ with the aid of a g transformation leads to an explicit expression for F if an interaction takes place not only between nearest neighbors; this could not be done by the method proposed in Ref. 34. Setting apart a form asymptotic not in N but, e.g., in $Nz/2$ (Ref. 34) is inconsistent and can complicate the test of the series for convergence. Thus, for the harmonic correlation contribution in the Einstein model of a solid, the first two terms of the expansion in Ref. 34 agree splendidly with experiment (within 0.5%) when the asymptote in $Nz/2$ is left out of F. The next term, however, makes the agreement with experiment worse and it is concluded on this basis that many terms must be included if one is to come close to the exact value. Turning to Eq. (1.71), we see that if the take the Nmth root of Q rather than the Nth, and choose m appropriately, the first two terms add up to almost the exact value, but additional terms lead to irregular results. In particular, the third term makes the agreement with experiment worse, as is the case in Ref. 34. We emphasize that a series asymptotic in N_i rather than $Nz/2$ or Nm is a physical requirement that must be met by the systems in statistical physics.

10. Variational theorems for statistical systems

An effective tool for the investigation of statistical systems is the method of variational theorems.

We consider first the Bogolyubov statistical variational theorem which is a generalization, to include the temperature, of Fock's quantum-mechanical dynamic theorem for the ground-state energy. This makes possible an estimate of the free energy of a system having a Hamiltonian H, without resorting to expansion in terms of some small parameter, and permits the thermodynamic properties of the system to be investigated in a relatively wide temperature range. A variational theorem that does not involve any expansion in terms of a small parameter obviates the need for perturbation theory. Furthermore, in view of the presence of additional equations (the minimization equations), whose solutions are not infinitely smooth functions of the temperature and of the external parameters, it becomes possible, by starting with microscopic data, to find and describe (albeit in a variational approximation) the first- and second-order phase transitions that can occur in the investigated system.[35] The statistical variational theorem for quantum systems was reported by Bogolyubov in 1956 to a seminar under his direction at the Steklov Mathematics Institute and published in his candidate's dissertation[35] and in a monograph.[36]

Let us prove the Bogolyubov theorem for quantum systems. We represent the Hamiltonian H of the system in the form*

$$H = H_0 + H_1 \,. \tag{2.72}$$

We establish a rule for differentiation of the operator $\exp[H(\lambda)]$ with respect to the parameter λ. It is directly evident that

$$\frac{d}{dt}\exp[H(\lambda)t] = H(\lambda)\exp[H(\lambda)t] \,,$$

$$\frac{d}{dt}\frac{d}{d\lambda}\exp[H(\lambda)t] = H(\lambda)\frac{d}{d\lambda}\exp[H(\lambda)t] + \frac{dH(\lambda)}{dt}\exp[H(\lambda)t] \,,$$

where t is a certain parameter.

We put

$$\frac{d}{d\lambda}\exp[H(\lambda)t] = \exp[H(\lambda)t]\, V/t \,,$$

then

$$\frac{dV(t)}{d\lambda} = \exp[-H(\lambda)t]\frac{dH(\lambda)}{d\lambda}\exp[H(\lambda)t] \,.$$

At $t = 0$ we have for the derivative

*For simplicity, we omit hereafter the carets of the operator symbols.

$$\frac{d}{d\lambda}\exp[H(\lambda)t] = 0\,,$$

therefore,

$$V(0) = 0, \quad V(t) = \int_0^t \exp[-H(\rho)\alpha]\frac{dH(\lambda)}{d\lambda}\exp[H(\lambda)\alpha]d\alpha$$

and consequently

$$\frac{d}{d\lambda}\exp[H(\lambda)] = \exp[H(\lambda)]\int_0^1 \exp[-H(\lambda)\alpha]\frac{dH(\lambda)}{d\lambda}\exp[H(\lambda)\alpha]d\alpha\,. \tag{2.73}$$

We show now that in the particular case

$$H = A + B\lambda\,, \tag{2.74}$$

where A and B are Hermitian operators, the convexity condition

$$\frac{d^2}{d\lambda^2}\operatorname{Tr}\exp(A + B\lambda) \geqslant 0 \tag{2.75}$$

is met. In fact, from Eq. (2.73) we have

$$\begin{aligned}\frac{d}{d\lambda}\operatorname{Tr}\exp(A + B\lambda) &= \operatorname{Tr}\frac{d}{d\lambda}\exp(A + B\lambda)\\ &= \operatorname{Tr}\exp(A + B\lambda)\int_0^1 \exp[-(A + B\lambda)\alpha]B\exp[(A + B\lambda)\alpha]d\alpha\\ &= \operatorname{Tr}[\exp(A + B\lambda)B]\end{aligned}$$

and

$$\begin{aligned}\frac{d^2}{d\lambda^2}\operatorname{Tr}\exp(A + B\lambda) &= \frac{d}{d\lambda}\operatorname{Tr}[\exp(A + B\lambda)B]\\ &= \operatorname{Tr}\Big(\exp(A + B\lambda)\int_0^1 \exp[-(A + B\lambda)\alpha]B\\ &\quad\times\exp[(A + B\lambda)\alpha]d\alpha\, B\Big)\,.\end{aligned}$$

The eigenvalues of the operator H, in a representation in which it is diagonal, will be designated E_n. Then

$$\begin{aligned}\frac{d^2}{d\lambda^2}\operatorname{Tr}\exp(A + B\lambda) &= \sum_{m,n}\exp(E_m)\int_0^1 \exp(-E_m\alpha)B_{mn}\exp(E_n\alpha)d\alpha\, B_{nm}\\ &= \sum_{m,n}|B_{mn}|^2\frac{\exp(E_n) - \exp(E_m)}{E_n - E_m} \geqslant 0\,,\end{aligned}$$

thus proving Eq. (2.75). But if $f(\lambda)$ is a certain continuous differentiable function, then

$$f(\lambda) \geqslant f(0) + \lambda f'(0)$$

if

$$\frac{d^2 f}{d\lambda^2} \geqslant 0\,.$$

Therefore,

$$\mathrm{Tr}\exp(A+B\lambda)\geqslant \mathrm{Tr}\exp(A)+\lambda\mathrm{Tr}\,B\exp(A)\,. \tag{2.76}$$

Putting

$$A=-\frac{H_0}{\theta},\quad B=-\frac{H_1-\langle H_1\rangle}{\theta},\quad \lambda=1,$$

$$\langle H_1\rangle=\mathrm{Tr}[H_1\exp(-H_0/\theta)]/\mathrm{Tr}\exp(-H_0/\theta)\,,$$

we obtain from Eq. (2.76),

$$\mathrm{Tr}\exp(-H/\theta)\geqslant\exp(-\langle H_1\rangle/\theta)\mathrm{Tr}\exp(-H_0/\theta)\,,$$

from which we have for the free energy

$$F(H)\leqslant F(H_0)+\langle H-H_0\rangle_{H_0}\equiv F_{\max}\,, \tag{2.77}$$

which is in fact the statement of the Bogolyubov variational theorem (principle).

The Bogolyubov variational theorem determines the upper bound $F_{\max}$ of the free energy. But it can also be used to determine its lower bound $F_{\min}$.[37] Indeed, applying the Bogolyubov theorem (2.77) to F_0 we get

$$F_0\leqslant F+\langle H-H_0\rangle_H\,, \tag{2.78}$$

hence

$$F\geqslant F+\langle H-H_0\rangle_H\equiv F_{\min}\,. \tag{2.79}$$

Combining inequalities (2.77) and (2.79), we obtain the estimate

$$F_0+\langle H-H_0\rangle_H\leqslant F\leqslant F_0+\langle H-H_0\rangle_{H_0}\,. \tag{2.80}$$

Note that whereas to find $F_{\max}$ we average over H_0, to determine $F_{\min}$ we must average over H. For a system with two-particle interaction this calls for knowledge of the two-particle density matrix, which can be determined only approximately.[38]

Note also that the variational theorem (2.77), while permitting a highly accurate calculation of the free energy of the system, leads to inaccurate derivatives of the free energy with respect to temperature in the entire range of system states. Thus, the heat capacity and entropy calculated from Eq. (2.77),

$$S=-\frac{\partial F_{\max}}{\partial T},\quad C_V=-T\frac{\partial^2 F_{\max}}{\partial T^2}\,, \tag{2.81}$$

the resultant values may be negative for some temperatures. The reason is that the free energy F approximated by $F_{\max}$ does not always have the temperature dependence that the free energy must have, i.e., does not always satisfy the Gibbs–Helmholtz equation.

This shortcoming of the variational principle (2.77) can be eliminated if, following Ref. 41, the temperature T' of the approximating system is introduced as the variational parameter, and if a best fit of F to $F_{\max}$ is obtained by minimizing the latter with respect to T' (or with respect to $\theta'=kT'$). The inequality (2.77) can be written in the form

$$F \leqslant \mathrm{Tr}\, H\rho_0 + \theta\, \mathrm{Tr}\, \rho_0 \ln \rho_0 = \overline{E} - TS_0\,, \tag{2.82}$$

where $\rho_0 = \exp(-H/\theta)/\mathrm{Tr}\exp(-H_0/\theta)$ is the approximating Hamiltonian, $\overline{E} = \mathrm{Tr}\, H\rho_0$ is the approximate internal energy of the system, and $S_0 = -k\, \mathrm{Tr}\, \rho_0 \ln \rho_0$ is the entropy of a system having a free energy $F_0 = -\theta \ln \mathrm{Tr} \exp(-H_0/\theta)$.

The choice of the approximating Hamiltonian H_0 is arbitrary provided that it yields all the thermodynamic functions. We substitute therefore in Eq. (2.82),

$$H_0 \to \frac{\theta}{\theta'} H_0\,.$$

Then

$$\rho_0' = \exp(-H_0/\theta')/\mathrm{Tr}\exp(-H_0/\theta)$$

and

$$F \leqslant \mathrm{Tr}\, H\rho_0' + \theta\, \mathrm{Tr}\, \rho_0' \ln \rho_0' = F_0' + \left\langle H - \frac{\theta}{\theta'} H_0 \right\rangle_0 \equiv F_{\max} \tag{2.83}$$

or

$$F_{\max}(T, T', V) = \overline{E}(T', V) - TS_0(T', V) \geqslant F(T, V)\,. \tag{2.84}$$

The necessary condition that $F_{\max}$ be a minimum with respect to T' leads to

$$\frac{\partial F_{\max}}{\partial T} = 0\,, \tag{2.85}$$

and from Eqs. (2.84) and (2.85) we get

$$\frac{\partial E_0}{\partial T'} = T \frac{\partial S_0}{\partial T'}$$

and

$$\frac{\partial E_0}{\partial S_0} = T\,. \tag{2.86}$$

Thus, if ρ_0 is a function of the temperature T' and of other parameters $\alpha_i (i = \overline{1,n})$, then $F_{\max}$ is a function of T, T', α_i, and V, and the best upper-bound approximation of the free energy is obtained by determining T' and α_i from the condition that $F_{\max}$ be a minimum:

$$\frac{\partial F_{\max}}{\partial T'} = 0, \quad \frac{\partial F_{\max}}{\partial \alpha_i} = 0, \quad \delta^2 F_{\max} \geqslant 0. \tag{2.87}$$

These values of T' and α_i are functions of T:

$$T' = T'(T), \quad \alpha_i = \alpha_i(T)$$

and

$$F_{\max} = F_{\max}[T, T'(T), \alpha_i(T), V]\,.$$

We now obtain

$$S = -\left(\frac{\partial T_{\max}}{\partial T}\right)_V ,$$

$$C_V = -T\left(\frac{\partial^2 F_{\max}}{\partial T^2}\right)_V .$$

It is easily seen that under conditions (2.87) we get

$$S = -\left(\frac{\partial F_{\max}}{\partial T}\right)_V = S_0 , \tag{2.88}$$

$$C_V = T\left(\frac{\partial S_0}{\partial T}\right)_V = \left(\frac{\partial E_0}{\partial T}\right)_V \tag{2.89}$$

and the Gibbs–Helmholtz equation

$$F_{\max} = E_0 + T\left(\frac{\partial F_{\max}}{\partial T}\right)_V \tag{2.90}$$

is valid. In fact,

$$S = -\left(\frac{\partial F_{\max}}{\partial T}\right)_{V,\alpha_i T'} - \left(\frac{\partial F_{\max}}{\partial T'}\right)_{T,\alpha_i,V}\frac{dT'}{dT} - \sum_i\left(\frac{\partial F_{\max}}{\partial \alpha_i}\right)_{V,T,T'}\frac{d\alpha_i}{dT} ,$$

and, taking Eqs. (2.87) and (2.84) into account, we obtain Eq. (2.88).

Substituting Eq. (2.88) in expression (2.84) for $F_{\max}$ we get Eq. (2.90), and by differentiating Eqs. (2.90) and (2.88) with respect to T we get

$$\frac{\partial E_0}{\partial T} = -T\frac{\partial^2 F_{\max}}{\partial T^2} = C_V ,$$

$$T\frac{\partial S}{\partial T} = T\frac{\partial S_0}{\partial T} = C_V ,$$

which proves Eq. (2.89).

Since

$$S_0 = -k\,\mathrm{Tr}\,\rho_0' \ln \rho_0' ,$$

while ρ_0' is a positive-definite operator and $\mathrm{Tr}\,\rho_0' = 1$, the entropy S_0 is positive, so that S is also positive. It can be shown that C_V is also always positive.[41]

11. Statistical thermodynamics according to the Bogolyubov method

One of the most effective statistical treatments of equilibrium and kinetic processes in many-particle systems is by the Bogolyubov distribution-function method.[2]

For a system of N particles interacting in a volume V via central forces with a pair potential $\Phi(|q_1 - q_2|)$, the thermal and caloric equations of state are expressed in terms of a binary distribution function. In fact, the pressure p and the internal energy E of the system are determined by the relations

$$p = -\left(\frac{\partial F}{\partial V}\right)_\theta = \frac{\theta}{Q}\frac{\partial Q}{\partial V}, \tag{2.91}$$

$$E = F - \theta\left(\frac{\partial F}{\partial \theta}\right)_V = -\theta^2\frac{\partial}{\partial \theta}\left(\frac{F}{\theta}\right). \tag{2.92}$$

To find the derivative $\partial Q/\partial V$ we transform in the expression for the configuration integral

$$Q = \int_V \cdots \int_V \exp\left[-\frac{1}{\theta}\sum_{1 \leqslant i < j \leqslant N} \Phi(|q_i - q_j|)\right] dq_1 \cdots dq_N$$

to new variables $x_i^a = V^{-1/3} q_i^a$, so that $x_i = (1/V)\, q_i$ and $dq_i = V\, dx_i$. Then

$$Q = V^N \int_1 \cdots \int_1 \exp\left[-\frac{1}{\theta}\sum_{i<j} \Phi(|X_i - X_j| V^{1/3})\right] dX_1 \cdots dX_N,$$

where the integration is over a unit volume. The integral contains V only as a parameter, and the differentiation with respect to the volume is easy:

$$\frac{\partial Q}{\partial V} = NV^{N-1}\int_1 \cdots \int_1 \exp[-U/\theta]\, dX_1 \cdots dX_N$$
$$- \frac{V^N}{\theta}\int_1 \cdots \int_1 \exp[-U/\theta]\sum_{i<j}\frac{\partial \Phi(|X_i - X_j| V^{1/3})}{\partial V} dX_1 \cdots dX_N.$$

But

$$\frac{\partial \Phi(|X_i - X_j| V^{1/3})}{\partial V} = \Phi'(|X_i - X_j| V^{1/3})\tfrac{1}{3} V^{-2/3}|X_i - X_j|$$
$$= \frac{1}{3V}|q_i - q_j|\Phi'(|q_i - q_j|).$$

Therefore, returning to the old coordinates, we obtain

$$\frac{\partial Q}{\partial V} = \frac{NQ}{V} - \frac{1}{3\theta V}\int_V \cdots \int_V \exp[-U/\theta]\sum_{i<j}|q_i - q_j|\Phi'(|q_i - q_j|)$$
$$= \frac{N\theta}{V} - \frac{N(N-1)Q}{6\theta V^3}\int_V\int_V |q_1 - q_2|\Phi'(|q_1 - q_2|)F_2(q_1,q_2)\, dq_1\, dq_2,$$

and consequently

$$p = \frac{N\theta}{V} - \frac{N(N-1)}{6V^3}\int_V\int_V |q_1 - q_2|\Phi'(|q_1 - q_2|)F_2(q_1,q_2)\, dq_1\, dq_2$$

or

$$\frac{pv}{\theta} = 1 - \frac{(1 - 1/N)}{6\theta v}\frac{1}{V}\int_V\int_V |q_1 - q_2|\Phi'(|q_1 - q_2|)F_2(q_1,q_2)\, dq_1\, dq_2, \tag{2.93}$$

where $v = V/N$.

The internal energy of the system is, according to Eq. (2.92),

$$E=\frac{3}{2}N\theta+\frac{1}{Q}\int U\exp[-U/\theta]dq_1\cdots dq_N$$
$$=\frac{3}{2}N\theta+\frac{N(N-1)}{2V^2}\int_V\int_V\Phi(q_1-q_2|)F_2(q_1,q_2)dq_1\,dq_2\,. \quad (2.94)$$

To determine the heat capacity and the isothermal elastic modulus we need the three-particle and four-particle distribution functions in addition to the two-particle one.[39] Taking the second derivatives of the free energy with respect to temperature and volume, we arrive at a Bogolyubov thermodynamics in which all the thermodynamic relations are satisfied, including the condition that the thermal and caloric equations of state be compatible:

$$\theta\left(\frac{\partial p}{\partial\theta}\right)_V=\left(\frac{\partial E}{\partial V}\right)_\theta+p\,. \quad (2.95)$$

If the three-particle as well as the two-particle interactions are taken into account, the thermal and caloric equations are[38]

$$p=\frac{N\theta}{V}-\frac{N(N-1)}{6V^3}\int_V\int_V|q_1-q_2|\Phi_2'(|q_1-q_2|)F_2(q_1,q_2)dq_1\,dq_2$$
$$-\frac{N(N-1)(N-2)}{18V^4}\int_V\int_V\int_V|q_1-q_2|\Phi_{3r_{12}}'(r_{12},r_{23},r_{31})$$
$$+|q_2-q_3|\Phi_{3r_{23}}'(r_{12},r_{23},r_{31})+|q_3-q_1|\Phi_{3r_{31}}'(r_{12},r_{23},r_{31})$$
$$\times F_3(q_1,q_2,q_3)dq_1\,dq_2\,dq_3\,, \quad (2.96)$$

$$E=\frac{3}{2}N\theta+\frac{N(N-1)}{2V^2}\int_V\int_V\Phi_2(|q_1-q_2|)F_2(q_1,q_2)dq_1\,dq_2$$
$$+\frac{N(N-1)(N-2)}{3!V^3}\int_V\int_V\int_V\Phi_3(r_{12},r_{23},r_{31})F_3(q_1,q_2,q_3)dq_1\,dq_2\,dq_3\,. \quad (2.97)$$

Thus, to formulate the Bogolyubov thermodynamics of a system it is necessary to define its lower-order distribution functions (or lower-order density matrices in the quantum region) as solutions of the chain of equations:

$$\frac{\partial F_s}{\partial q_1^\alpha}+\frac{1}{\theta}\frac{\partial U_s}{\partial q_1^\alpha}F_s+\frac{1}{v\theta}\int\frac{\partial\Phi(|q_1-q_{s+1}|)}{\partial q_1^\alpha}F_{s+1}dq_{s+1}=0\,. \quad (2.98)$$

These equations can be solved by various methods. We consider here a solution for systems with equations in which a typical small parameter (density, plasma parameter, interaction strength) can be separated. The solutions themselves are then sought as expansions of the distribution functions in powers of this parameter. Another approach to the solution of the chain (2.98) will be described in succeeding sections.

11.1 System with short-range interaction forces at low density

In this case, a scaling transformation in Eqs. (2.98) to the dimensionless quantities $q'^{\alpha} = q^{\alpha}/r_0$, where r_0 is the particle effective radius, shows that the integral terms of the resultant equations contain the parameter

$$\varepsilon = r_0{}^3/v\,, \tag{2.99}$$

which is small ($\varepsilon \ll 1$) for a gas under ordinary conditions. The solution of these equations can be sought as series in powers of this small parameter.

Formally, however, Eq. (2.98) can be solved directly by expanding F_s in powers of the density $1/v$:

$$F_s = F_s^0 + \frac{1}{v}F_s^1 + \frac{1}{v^2}F_s^2 + \cdots . \tag{2.100}$$

We substitute in Eq. (2.98) this expansion, the correlation-decay conditions (1.40), and the normalization conditions (1.138) and (1.139). Equating next the coefficients of like powers of the density, we get

$$\begin{gathered}\frac{\partial F_s^0}{\partial q_1^\alpha} + \frac{1}{\theta}\frac{\partial U_s}{\partial q_1^\alpha}F_s^0 = 0\,,\\ \frac{\partial F_s^1}{\partial q_1^\alpha} + \frac{1}{\theta}\frac{\partial U_s}{\partial q_1^\alpha}F_s^1 + \frac{1}{\theta}\int \frac{\partial \Phi(|q_1 - q_2|)}{\partial q_1^\alpha}F_{s+1}^0\, dq_{s+1} = 0\,,\\ \vdots\end{gathered} \tag{2.101}$$

$$\begin{gathered}F_s^0(q_1,\ldots,q_s) - \prod_{1\leqslant i\leqslant s} F_1^0(q_i) \to 0, \quad |q_i - q_j| \to \infty,\\ F_s^1(q_1,\ldots,q_s) - \prod_{1\leqslant i\leqslant s} F_1^1(q_i) \prod_{\substack{1\leqslant j\leqslant s\\ j\neq i}} F_1^0(q_j) \to 0, \quad |q_i - q_j| \to \infty\,,\\ \vdots\end{gathered} \tag{2.102}$$

$$\lim_{V\to\infty}\frac{1}{V}\int_V F_1^0(q)dq = 1, \quad \lim_{V\to\infty}\frac{1}{V}\int_V F_1^1(q)dq = 0, \quad \ldots . \tag{2.103}$$

We put

$$F_s^0 = C_s^0(q_1,\ldots,q_s)\exp[-U_s/\theta]\,.$$

The first equation of Eq. (2.101) shows then that

$$\frac{\partial C_s^0}{\partial q_1^\alpha} = 0,$$

i.e., that C_s^0 does not depend on q_1. From its symmetry relative to $q_1,\ldots,q_s$ we find that C_s^0 is a constant. Taking the boundary conditions (2.102) into account we have $C_s^0 = (C_1^0)^s$, and the normalization conditions (2.103) yield $C_s^1 = 1$. We have ultimately

$$F_s^0 = \exp[-U_s/\theta], \quad s = 2,3,\ldots,$$
$$F_1^0 = 1. \tag{2.104}$$

We consider now the second equation of Eq. (2.101):

$$\frac{\partial F_s^1}{\partial q_1^\alpha} + \frac{1}{\theta}\frac{\partial U_s}{\partial q_1^\alpha}F_s^1 + \frac{1}{\theta}\int \frac{\partial \Phi(|q_1 - q_{s+1}|)}{\partial q_1^\alpha}\exp[-U/\theta]dq_{s+1} = 0.$$

It is easily seen that

$$\frac{1}{\theta}\frac{\partial \Phi(|q_1 - q_{s+1}|)}{\partial q_1^\alpha}\exp[-U_{s+1}/\theta]$$
$$= -\exp[-U_s/\theta]\frac{\partial}{\partial q_1^\alpha}\exp\left[-\sum_{1\leqslant i\leqslant s}\frac{\Phi(|q_i - q_{s+1}|)}{\theta}\right]$$
$$= -\exp[-U_s/\theta]\frac{\partial}{\partial q_1^\alpha}\prod_{1\leqslant i\leqslant s}[1 + f(|q_i - q_{s+1}|)],$$

where $f(r)$ is the Mayer function.

Since the forces are short range, $f(r)$ decreases rapidly with increase of r. The integrals in question converge therefore absolutely, and consequently

$$\int \frac{\partial}{\partial q_1^\alpha}\left\{1 + \sum_{1\leqslant i\leqslant s} f(|q_i - q_{s+1}|)\right\}dq_{s+1} = \int \frac{\partial f(|q_1 - q_{s+1}|)}{\partial q_1^\alpha}dq_{s+1}$$
$$= \frac{\partial}{\partial q_1^\alpha}\int f(|q_1 - q_{s+1}|)dq_{s+1} = \frac{\partial}{\partial q_1^\alpha}\int f(|q|)dq = 0.$$

Thus

$$\frac{\partial F_s^1}{\partial q_1^\alpha} + \frac{F_s^1}{\theta}\frac{\partial U_s}{\partial q_1^\alpha} = \exp[-U_s/\theta]\frac{\partial}{\partial q_1^\alpha}\int\left\{\prod_{1\leqslant i\leqslant s}[1 + f(|q_i - q_{s+1}|)]\right.$$
$$\left. -1 - \sum_{1\leqslant i\leqslant s} f(|q_i - q_{s+1}|)\right\}dq_{s+1}.$$

Putting here

$$F_s^1 = C_s^1(q_1,\ldots,q_s)\exp[-U_s/\theta], \tag{2.105}$$

we get

$$\frac{\partial C_s^1}{\partial q_1^\alpha} = \frac{\partial}{\partial q_1^\alpha}\int\left\{\prod_{1\leqslant i\leqslant s}[1 + f(|q_i - q_{s+1}|)] - 1\right.$$
$$\left. - \sum_{1\leqslant i\leqslant s} f(|q_i - q_{s+1}|)\right\}dq_{s+1}.$$

By virtue of the symmetry of C_s^1 relative to $q_1,\ldots,q_s$ we have

$$C_s^1 = \int\left\{\prod_{1\leqslant i\leqslant s}[1 + f(|q_i - q_{s+1}|)] - 1\right.$$
$$\left. - \sum_{1\leqslant i\leqslant s} f(|q_i - q_{s+1}|)\right\}dq_{s+1} + k_s, \quad k_s = \text{const.}$$

Recognizing that

$$\int \left\{ \prod_{1 \leqslant i \leqslant s} [1 + f(|q_i - q_{s+1}|)] - 1 - \sum_{1 \leqslant i \leqslant s} f(|q_i - q_{s+1}|) \right\} dq_{s+1} \to 0,$$

as all $|q_i - q_j| \to \infty$, and on the basis of Eqs. (2.102), (2.103), and (2.105) we have $k_s = 0$.

Ultimately

$$F_s^1 = \exp[-U_s/\theta] \int \left\{ \prod_{1 \leqslant i \leqslant s} [1 + f(|q_i - q_{s+1}|)] - 1 - \sum_{1 \leqslant i \leqslant s} f(|q_i - q_{s+1}|) \right\} dq_{s+1}. \tag{2.106}$$

The remaining coefficients of the expansions (2.100) can be obtained similarly. Substituting Eqs. (2.104) and (2.106) in Eq. (2.100) we get

$$F_s = \exp[-U_s/\theta] \left\{ 1 + \frac{1}{v} \int \left(\prod_{1 \leqslant i \leqslant s} [1 + f(|q_i - q_{s+1}|)] - 1 - \sum_{1 \leqslant i \leqslant s} f(|q_i - q_{s+1}|) \right) dq_{s+1} \right\}. \tag{2.107}$$

In particular,

$$F_1 = 1, \tag{2.108}$$

$$F_2 = \mu(|q_1 - q_2|),$$

$$\mu(|q|) = \exp\{-\Phi(|q|)/\theta\} \left\{ 1 - \frac{1}{v} \int f(|q - q'|) f(|q'|) dq' + \cdots \right\}. \tag{2.109}$$

Substituting Eq. (2.109) in Eq. (2.93) we get in the statistical limit the thermal equation of state

$$\frac{pv}{\theta} = 1 + B_2/v + B_3/v^2 + \cdots, \tag{2.110}$$

where the virial coefficients are

$$B_2 = -\tfrac{1}{2}\beta_1 = -\tfrac{1}{2} \int_V f(|q|) dq,$$

$$B_3 = -\tfrac{2}{3}\beta_3 = -\tfrac{1}{3} \int_V \int_V f(|q - q'|) f(|q|) f(|q'|) dq\, dq'.$$

Solution of the equations for the distribution functions, by expansion in powers of the density, leads thus, without complicated combinatorics, to the results of the Ursell–Mayer theory described in Sec. 7 of the present chapter.

Note that the coefficients in the expansions of the distribution functions depend substantially on temperature. Figure 7 shows plots for $F_2^0(r)$ of a particle system with Lennard-Jones interaction at $\theta/\varepsilon = 1$ and $\theta/\varepsilon = 3$.

The temperature dependence of the distribution-function expansion coefficients determines also the behavior of the virial coefficients (2.10); this

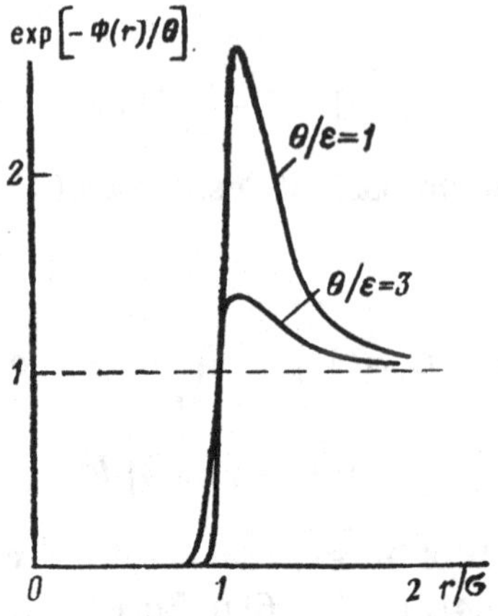

Figure 7.

will be discussed in greater detail below. We note here only that the convergence of the expansion (2.110) depends both on r_0^3/v and on the virial coefficients B_i as functions of temperature.

11.2 System with Coulomb interaction

Consider now a volume V containing a system of N particles of two oppositely charged species, with zero total system charge (a plasma). For simplicity, let each particle have a charge e.

The Coulomb interaction, in contrast to the intramolecular, is long range—it decreases slowly with distance. Therefore, each charged particle interacts at all times with a large aggregate of neighboring particles. The plasma is thus not a gas but a unique system held together by long-range forces (the fourth state of matter).

In view of the long-range character of the Coulomb forces, the previously used expansions (2.100) of the distribution functions in powers of the density will not do, since the coefficients of these powers diverge. The distribution functions F for a plasma must thus be expanded in powers of another small parameter. Finding this parameter calls for a scaling transformation in the equations for F_2 to dimensionless quantities, relative to a suitable unit of length. This length and the parameter itself are determined from the following physical considerations. Each charged particle is surrounded by charges predominantly of opposite sign, so that the field of this charge is screened and extends in fact over a certain length $r_d \equiv D$ called the Debye screening radius and having for a plasma temperature θ the value

$$D = \sqrt{\frac{\theta v}{\sum_k 4\pi n_k e_k^2}} = \sqrt{\frac{\theta v}{4\pi e^2}}, \tag{2.111}$$

where

$$v = V/N, \quad n_k = N_k/N, \quad k = a,b,$$

and N_k is the number of particles of species k.

Thus, at $r > D$ the motion of the particles in their self-consistent field is random ($F_{ij} \approx F_i F_j$).

A sphere of radius D contains many particles. The small parameter for a plasma is therefore the ratio of the average volume v per particle to the volume of the Debye sphere, or

$$\varepsilon = \frac{v}{D^3} \ll 1, \tag{2.112}$$

which is valid if

$$\frac{e^2}{D} \ll \theta, \tag{2.113}$$

i.e., when the interaction energy of two particles separated by on Debye radius is much less than the average kinetic energy of the particles.

Let us find the first approximation of the plasma binary distribution function under these conditions.

By Coulomb's law, the pair-interaction potential of particles i and j is

$$\Phi_{ij}(r) = \frac{e_i e_j}{r}. \tag{2.114}$$

In addition to the Coulomb term, the expression for the mutual potential should include also the potential $\overline{\Phi}_{ij}(r)$ corresponding to the short-range repulsion forces, for example, in the form of a potential barrier

$$\overline{\Phi}_{ij} = \begin{cases} \infty & r \leqslant \sigma \\ 0 & r > \sigma \end{cases},$$

where σ is the effective charged-particle diameter. Such a potential makes the particles impenetrable and leads to vanishing of the binary distribution function $F_{ij}(r)$ at distances $0 \leqslant r \leqslant \sigma$ between the particles.

To determine F_{ij} at $r > \sigma$ in first-order approximation we confine ourselves to the Coulomb interaction (2.114).

The first two Bogolyubov-chain equations for a system of two-particle species are a natural generalization of the corresponding equations (2.98) for a system of particles of the same species, and take the form[2]

$$\frac{\partial F_i(q)}{\partial q^\alpha} + \frac{1}{\theta} \int \sum_j n_j \frac{\partial \Phi_{ij}}{\partial q^\alpha} F_{ij}(q,q') dq' = 0, \tag{2.115}$$

$$\frac{\partial F_{ij}(q,q')}{\partial q^\alpha} + \frac{1}{\theta} \frac{\partial \Phi_{ij}}{\partial q^\alpha} F_{ij}(q,q') + \frac{1}{\theta} \int \sum_k n_k \frac{\partial \Phi_{ik}}{\partial q^\alpha} F_{ijk}(q,q',q'') dq'' = 0. \tag{2.116}$$

The plasma parameter (2.112) was obtained by making Eqs. (2.115) and (2.116) nondimensional. It is also possible, however, to find the distribution functions directly from solutions of these equations, as expansions in powers of v. It is only necessary to assume that $(1/\theta)\Phi_{ij}(r)$ is proportional to v:

$$\frac{1}{\theta}\Phi_{ij}(r) = v\varphi_{ij}(r), \tag{2.117}$$

where

$$\varphi_{ij}(r) = \frac{\bar{e}_i\bar{e}_j}{D^2}\frac{1}{r}, \tag{2.118}$$

$$\bar{e}_i = \frac{e_i}{\sqrt{\sum_i 4\pi n_i e_i^2}}. \tag{2.119}$$

The correlation-weakening conditions for $F_{ij}(q,q')$ are

$$F_{ij}(q,q') \to F_i F_j, \quad |q - q'| \to \infty. \tag{2.120}$$

This asymptotic expression is the principal part of the binary distribution function at finite $|q - q'|$, since the particles move in the strong self-consistent field of all the particles, and the contribution of an individual particle to the field is small. Consequently, the addition F_{ij} in the asymptotic relation (2.120) is proportional to the small parameter v:

$$F_{ij} = F_i F_j + v g_{ij}. \tag{2.121}$$

Similarly, the three-particle distribution function can be represented in the form

$$F_{ijk} = F_i F_j F_k + v g_{ij} F_k + v g_{ik} F_j + v g_{jk} F_i + v^2 g_{ijk}. \tag{2.122}$$

Here q_{ij} and q_{ijk} are, respectively, the two- and three-particle correlation functions.

We seek the solution of the system (2.115) and (2.116) in the form

$$F_i = g_i{}^0 + v g_i{}^1 + \cdots, \quad g_{ij} = g_{ij}{}^0 + v g_{ij}{}^1 + \cdots. \tag{2.123}$$

Substituting Eq. (2.123) and Eqs. (2.121) in Eq. (2.115) we have in zeroth order

$$\frac{\partial g_i^0(q)}{\partial q^\alpha} + g_i^0(q)\int \sum_j \frac{\partial \varphi_{ij}}{\partial q^\alpha} n_j g_j^0(q')dq' = 0. \tag{2.124}$$

By virtue of the neutrality of the system we have

$$\sum_j \varphi_{ij} n_j = 0,$$

therefore

$$\int \sum_j \frac{\partial \varphi_{ij}}{\partial q^\alpha} n_j \, dq' = 0.$$

Equation (2.124) has the obvious solution

$$g_i{}^0 = \text{const}.$$

The normalization condition yields

$$g_i{}^0 = 1. \tag{2.125}$$

This corresponds to a spatially homogeneous distribution.

Consider now Eq. (2.116) in zeroth order:

$$\frac{\partial q_{ij}^0}{\partial q^\alpha} + \frac{\partial \varphi_{ij}}{\partial q^\alpha} q_{ij}^0 + \int \sum_k \frac{\partial \varphi_{ik}}{\partial q^\alpha} n_k \{1 + g_{ij}^0 + g_{ik}^0 + g_{jk}^0\} dq'' = 0. \tag{2.126}$$

Recognizing that the system is electrically neutral and that spatial homogeneity makes g_{ik} and φ_{ik} even functions, we get

$$\int \sum_k \frac{\partial \varphi_{ik}}{\partial q^\alpha} n_k q_{ik}^0 dq'' = 0,$$

and Eq. (2.126) takes the form

$$\frac{\partial}{\partial q^\alpha}\left\{q_{ij}^0 + \varphi_{ij} + \int \sum_k \varphi_{ik} n_k g_{jk}^0 dq''\right\} = 0.$$

We integrate this equation and take it into account that $g_{ij}^0 \to 0$ and $\varphi_{ij} \to 0$ as $|q - q'| = |R| \to \infty$, so that

$$q_{ij}^0(|R|) + \int \sum_k \varphi_{ij}(|R - r|) n_k q_{jk}^0(|r|) dr = -\varphi_{ij}(|R|). \tag{2.127}$$

We seek the solution of this equation by the Fourier-transform method, putting

$$g_{ij}{}^0 = \int C_{ij}(|\nu|) \exp(i\nu R) d\nu, \tag{2.128}$$

where

$$\nu R = \sum_{1 \leqslant \alpha \leqslant 3} \nu^\alpha R^\alpha, \quad d\nu = \prod_{1 \leqslant \alpha \leqslant 3} d\nu^\alpha.$$

Then

$$\varphi_{ij}(|r|) = \frac{1}{4\pi} \int \exp(i\nu r) Y_{ij}(|\nu|) d\nu, \tag{2.129}$$

where

$$Y_{ij}(|\nu|) = \frac{2}{\pi \nu^2} \frac{\bar{e}_i \bar{e}_j}{D^2}. \tag{2.130}$$

Substituting Eqs. (2.128) and (2.129) in Eq. (2.127) we obtain a system of linear algebraic equations:

$$C_{ij}(|\nu|) + 2\pi^2 \sum_k Y_{ik}(|\nu|) n_k C_{jk}(|\nu|) = -\frac{1}{4\pi} Y_{ij}(|\nu|) \exp(-i\nu q). \tag{2.131}$$

We have taken it into account here that

$$\int \exp(-i\nu q'')g_{jk}(|q''-q'|)dq'' = (2\pi)^3 C_{jk}(|q|).$$

The solution of the system (2.131) for C_{ij} is

$$C_{ij} = -\frac{\bar{e}_i\bar{e}_j}{2\pi^2}\frac{1}{|\nu|^2D^2 + 4\pi\sum_k n_k\bar{e}_k^2}$$

or

$$C_{ij} = \frac{\bar{e}_i\bar{e}_j}{2\pi^2}\frac{1}{|\nu|^2D^2+1}. \tag{2.132}$$

We transform to the coordinate representation

$$g_{ij}^0(r) = \frac{4\pi}{r}\int_0^\infty \nu C_{ij}(|\nu|)\sin(\nu r)d\nu$$

$$= -\frac{2\bar{e}_i\bar{e}_j}{\pi r}\int_0^\infty \frac{\nu\sin\nu r}{|\nu|^2D^2+1}d\nu = -\frac{\bar{e}_i\bar{e}_j}{D^2}\frac{\exp(-r/D)}{r}. \tag{2.133}$$

In view of the radial symmetry of the function g_{ij}^0 we have

$$g_i^{\,1}(q) = 0.$$

We can write for the distribution functions F_i and F_{ij} the following expression accurate to second-order terms:

$$F_i(q) = 1,$$

$$F_{ij}(|q-q'|) = 1 - \frac{\bar{e}_i\bar{e}_j}{D^2}v\frac{\exp(-r/D)}{r} = 1 - \varepsilon\bar{e}_i\bar{e}_j\frac{\exp(-r/D)}{r/D} \tag{2.134}$$

or

$$F_{ij}(r) = 1 - \frac{e_ie_j}{r\theta}\exp(-\varkappa r), \tag{2.135}$$

where $\varkappa = 1/D$.

This equation is valid for $r > e^2/\theta$; for small r of order εD it is no longer useful, and for $e_ie_j > 0$ it leads as $r \to 0$ to $F_{ij} = -\infty$, contradicting the definition of the distribution function (see Fig. 8). It is seen fr m Eq. (2.135) that the probability of finding in a plasma two oppositely charged particles separated by a distance r is higher than the probability of equally spaced two particles of like sign. As a result, attraction forces predominate in a plasma.

With the aid of expression (2.135) obtained for the binary function F_{ij} we find the internal energy of the system, using the equation

$$E = \frac{3}{2}N\theta + \frac{1}{2V^2}\sum_{i,j} N_iN_j\int_V\int_V \Phi_{ij}F_{ij}dq_i\,dq_j\,. \tag{2.136}$$

Since the plasma is neutral, the term 1 in F_{ij} makes no contribution to the energy, therefore,

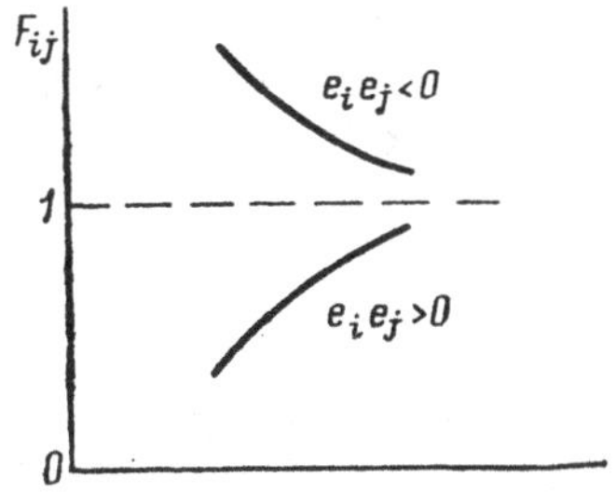

Figure 8.

$$E = \frac{3}{2} N\theta + \frac{1}{2V} \sum_{i,j} N_i N_j \int_0^\infty \frac{e_i e_j}{r} \frac{e_i e_j}{\theta r} \exp(-\varkappa r) 4\pi r^2 dr$$
$$= \frac{3}{2} N\theta - Ne^3 \sqrt{\frac{\pi}{\theta v}}. \quad (2.137)$$

Integrating the Gibbs–Helmholtz equation

$$\frac{\partial}{\partial\theta}\left(\frac{F}{\theta}\right)_V = -\frac{E}{\theta^2},$$

we obtain the free energy of the plasma

$$F = -\theta \int \frac{E}{\theta^2} d\theta + \theta C(V) = -\frac{3}{2} N\theta \ln\theta - \frac{2}{3} Ne^3 \sqrt{\frac{\pi}{\theta v}} + \theta C(V)$$
$$= F_0 - \frac{2}{3} Ne^3 \sqrt{\frac{\pi}{\theta v}}, \quad (2.138)$$

since a very low density plasma ($V \to \infty$, $N = \text{const}$, $\theta = \text{const}$) becomes an ideal gas, and $-\frac{3}{2} N\theta \ln\theta + \theta C(V)$ is its free energy F_0.

This yields for a low-density plasma the thermal equation of state

$$p = -\frac{\partial F}{\partial V} \frac{\theta}{v} - \frac{e^3}{3v} \sqrt{\frac{\pi}{\theta v}}, \quad (2.139)$$

its entropy

$$S = -k\frac{\partial F}{\partial\theta} = S_0 - \frac{Ne^3 k}{3\theta} \sqrt{\frac{\pi}{\theta v}}, \quad (2.140)$$

and its heat capacity

$$C_V = \theta\frac{\partial S}{\partial\theta} = C_{0V} + \frac{Ne^3 k}{3\theta} \sqrt{\frac{\pi}{\theta v}}, \quad (2.141)$$

where S_0 and C_{0V} are the entropy and heat capacity of an ideal gas.

The plasma pressure and entropy are lower, and the heat capacity higher, than those of an ideal gas, since attraction forces predominate in the plasma.

Tyablikov and Tolmachev[40] proposed a method that yields for F_{ij} an interpolation formula that describes correctly its behavior at short and long distances, by transforming from the functions $F_{i_1 i_2 \ldots}$ to the functions $C_{i_1 i_2} \ldots$, using the relations

$$F_{i_1 i_2 \cdots} = C_{i_1 i_2 \cdots} \exp\left(- \frac{1}{\theta} \sum_{2<e} \Phi_{i_k i_e} \right),$$

i.e., the principal "small-scale" part is separated from $F_{i_1 i_2 \ldots}$. The plasma expansion, on the other hand, is used for $C_{i_1 i_2 \ldots}$. A graphic expansion for a plasma similar to the Feynman diagram method of quantum electrodynamics was developed in Ref. 42.

12. Integral equations

In the preceding section we have considered the solution of a chain of Bogolyubov equations, using expansions of the distribution functions in powers of a small parameter. We describe here another method of solving this system, based on "termination" of the chain when additional physical considerations applied to some system make it possible to approximate the leading distribution function F_s $(s \geqslant 2)$ by an expression that includes lower-order functions F_k $(k < s)$. This yields one or a set of integral equations that are closed with respect to the functions $F_1, \ldots, F_{s-1}$. Solving these integral equations, we can find the equilibrium functions of the investigated system.

We consider now some integral equations for the distribution functions.

12.1 Self-consistent-field equation

The first equation of the Bogolyubov chain (1.144) for the distribution functions $\rho_s(q_1, \ldots, q_s)$ is of the form

$$\frac{\partial \rho_1(q)}{\partial q^\alpha} + \frac{1}{\theta} \int_V \frac{\partial \Phi(|q - q'|)}{\partial q^\alpha} \rho_2(q, q') dq' = 0. \qquad (2.142)$$

In crystals, the single-particle function $\rho_1(q)$ is periodic with sharp peaks at the lattice sites, i.e., it can be written in the form

$$\rho_1(q) = \sum_i \rho_1^i(q),$$

where $\rho_1^i(q)$ differs from zero only near the ith crystal-lattice site and is equal to zero elsewhere. Regions containing various sites with nonzero $\rho_1^i(1)$ do not overlap. Moreover, their linear dimensions b are small compared with the distance a between nearest neighbors. The motion of each crystal particle at any instant of time is therefore determined by its interaction with an assembly of nearest neighbors that are on a par with it, and also with particles from other coordination spheres. All this makes it possible to use in the statistical investigation of the crystal properties a factorized approximation of the binary distribution function[33]:

$$\rho_2(q,q') = \rho_1(q)\rho_1(q'), \tag{2.143}$$

since

$$|q - q'| \gg b. \tag{2.144}$$

Substituting Eq. (1.143) in Eq. (1.142) we obtain a closed nonlinear integral self-consistent-field equation for the single-particle distribution function

$$\theta \ln \lambda\rho_1(q) + \int_V \Phi(|q - q'|)\rho_1(q')dq' = 0, \tag{2.145}$$

where λ is the normalization constant.

12.2 Integral equation in the superposition approximation

Consider a spatially homogeneous system with

$$F_1(q) = 1$$

and a binary distribution function

$$F_2(q,q') = \mu(|q - q'|).$$

The second equation of the Bogolyubov chain (2.98) is of the form

$$\frac{\partial F^2}{\partial q^\alpha} + \frac{1}{\theta}\frac{\partial \Phi(|q - q'|)}{\partial q^\alpha}F_2 + \frac{1}{\theta v}\int \frac{\partial \Phi(|q - q''|)}{\partial q^\alpha}F_3\, dq'' = 0 \tag{2.146}$$

as $|q - q'| \to \infty$, we have

$$F_2(q,q') \to F_1(q)F_1(q') = 1.$$

Similarly, a three-particle distribution function $F_3(q, q', q'')$ in which one of the particles goes off to infinity tends to the two-particle function of the remaining two particles. It is therefore natural to use for $F_3(q, q', q'')$ the approximation

$$F_3(q,q',q'') = F_2(q,q')F_2(q',q'')F_2(q'',q), \tag{2.147}$$

called the superposition approximation.

Substituting Eq. (2.147) in Eq. (2.146) we obtain[2]

$$\frac{\partial}{\partial q^\alpha}F_2(q,q') + \frac{1}{\theta}\frac{\partial \Phi(|q - q'|)}{\partial q^\alpha}F_2(q,q')$$
$$+ \frac{1}{\theta v}F_2(q,q')\int \frac{\partial \Phi(|q - q''|)}{\partial q^\alpha}F_2(q,q'')F_2(q',q'')dq'' = 0. \tag{2.148}$$

By virtue of the radial symmetry of the two-particle distribution function, we have

$$\int \frac{\partial \Phi(|q-q''|)}{\partial q^\alpha} \mu(|q-q''|) dq'' = 0,$$

meaning that

$$\begin{aligned} &\int \frac{\partial \Phi(|q-q''|)}{\partial q^\alpha} \mu(|q-q''|)\mu(|q''-q'|) dq'' \\ &= \int \frac{\partial \Phi(|q-q''|)}{\partial q^\alpha} \mu(|q-q''|)\{\mu(|q''-q'|)-1\} dq'' \\ &= \int \frac{\partial \Phi(|q-q'-q_1|)}{\partial q^\alpha} \mu(|q-q'-q_1|)\{\mu(|q_1|)-1\} dq_1. \end{aligned}$$

Therefore,

$$\begin{aligned} &\int \frac{\partial \Phi(|q-q''|)}{\partial q^\alpha} \mu(|q-q''|)\mu(|q''-q'|) dq'' \\ &= \frac{\partial}{\partial q^\alpha} \int \{\mu(|q_1|)-1\} \left\{ \int_\infty^{|q-q'-q_1|} \mu(r) \frac{\partial \Phi(r)}{\partial r} dr \right\} dq_1. \end{aligned} \tag{2.149}$$

We introduce the notation

$$A(|q|) = \frac{1}{v} \int \{\mu(|q_1|)-1\} \left\{ \int_\infty^{|q-q_1|} \mu(r) \frac{\partial \Phi(r)}{\partial r} dr \right\} dq_1. \tag{2.150}$$

Substituting Eq. (2.149) in Eq. (2.148) and letting $q-q' \to q$, we get

$$\frac{\partial}{\partial q^\alpha} \mu(|q|) + \frac{1}{\theta} \mu(|q|) \frac{\partial}{\partial q^\alpha} \{A(|q|) + \Phi(|q|)\} = 0. \tag{2.151}$$

Since

$$\left.\begin{aligned} \mu(|q|) &\to 1 \\ \Phi(|q|) &\to 0 \\ A(|q|) &\to 0 \end{aligned}\right\} \quad \text{as } |q| \to \infty,$$

it follows from Eq. (2.151) that

$$\mu(|q|) = \exp\left\{ -\frac{A(|q|) + \Phi(|q|)}{\theta} \right\}.$$

For the binary distribution function we obtain ultimately the nonlinear integral equation

$$\begin{aligned} W(|q|) &= \Phi(|q|) + A(|q|) = \Phi(|q|) \\ &\quad + \frac{1}{v} \int \{\mu(|q_1|)-1\} \left\{ \int_\infty^{|q-q_1|} \mu(r) \frac{d\Phi(r)}{dr} dr \right\} dq_1, \\ \mu(|q|) &= \exp\left\{ -\frac{1}{\theta} W(|q|) \right\}. \end{aligned} \tag{2.152}$$

If an expansion in powers of the density $1/v$ is used to solve Eq. (2.151), the first two terms in the series for μ coincide with those of the series obtained for μ by solving the chain of the Bogolyubov equations.

To simplify Eq. (2.152) we reduce the three-dimensional integrals to one-dimensional ones:

$$-\theta \ln \mu(r) = \Phi(r) + \frac{2\pi}{vr}\int_0^\infty \rho\left\{\int_{|r-\rho|}^{r+\rho} E(t)t\,dt\right\}\{\mu(\rho)-1\}d\rho, \quad (2.153)$$

$$E(t) = \int_\infty^t \mu(t)\frac{d\Phi(t)}{dt}\,dt.$$

The function

$$h(r) = \mu(r) - 1 \quad (2.154)$$

is called the pair correlation function.

We define the structure factor $S(k)$ as a Fourier transform of $h(r)$, using the relation

$$S(k) - 1 = \frac{1}{v}\int h(r)\exp(i\mathbf{kr})d\mathbf{r}. \quad (2.155)$$

The cross sections for single coherent scattering of x rays, electrons, or slow neutrons are expressed in terms of $S(k)$. Therefore, after determining $S(k)$ from experiment, we can calculate $h(r)$ and $\mu(r)$ by taking inverse Fourier transforms. The expression obtained for $\mu(r)$ by solving the integral equation (2.153) can thus be compared with experiment directly, rather than using the expressions obtained for the pressure

$$p = \frac{\theta}{v} - \frac{2\pi}{3v^2}\int_0^\infty r\Phi'(r)\mu(r)r^2dr \quad (2.156)$$

and for the energy

$$E = \frac{3}{2}\theta N + \frac{2\pi N}{v}\int_0^\infty \Phi(r)\mu(r)r^2dr. \quad (2.157)$$

12.3 The Percus–Yevick equation

We introduce a direct correlation function with the aid of the integral equation

$$h(r) = C(r) + \frac{1}{v}\int C(|r-r'|)h(r')dr'. \quad (2.158)$$

We choose now $C(r)$ in the form

$$C(r) = [1 + h(r)]\,[1 - \exp(\Phi(r)/\theta)] \quad (2.159)$$

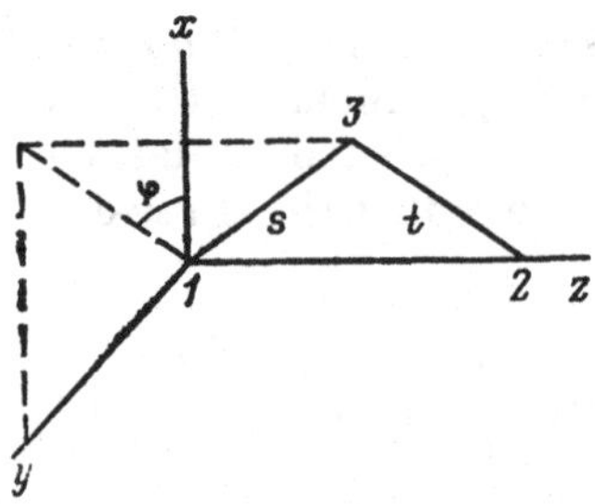

Figure 9.

and substitute in Eq. (2.158), transforming to the function $\mu(r)$:

$$\mu(r)\exp[\Phi(r)/\theta]$$
$$=1+\frac{1}{v}\int[\mu(|r-r'|)-1]\mu(r')\{1-\exp[\Phi(r')/\theta]\}dr'. \tag{2.160}$$

This is known as the Percus–Yevick equation.[43–45] It has a remarkable property—it can be solved analytically for a system of hard spheres.[46,47]

The interaction potential of a hard-sphere system is of the form

$$\Phi(r)=\begin{cases}0 & (r>\sigma)\\ \infty & (r<\sigma)\end{cases}. \tag{2.161}$$

Hence

$$\exp[-\Phi(r)/\theta]=H(r-\sigma), \tag{2.162}$$

where H is the Heaviside unit step function

$$H(x)=\begin{cases}0 & (x<0)\\ 1 & (x>0)\end{cases}. \tag{2.163}$$

We introduce the function

$$\tau(r)=r\mu(r)\exp[\Phi(r)/\theta]. \tag{2.164}$$

For the chosen potential we have

$$\tau(r)=\begin{cases}-rC(r) & (r<\sigma)\\ r\mu(r) & (r>\sigma)\end{cases}. \tag{2.165}$$

The newly introduced function allows us to rewrite Eq. (2.160) as

$$\tau(r)=r-\frac{2\pi}{v}\int_0^{\tau}ds\,\tau(s)\int_{|r-s|}^{r+s}dt\,H(t-\sigma)\tau(t)$$
$$+2\pi\int_0^{\sigma}ds\,\tau(s)\int_{|r-s|}^{r+s}dt\,t. \tag{2.166}$$

We have used here a bipolar coordinate frame (see Fig. 9); the location of the third particle is determined by the distances s and t from the first and second particles, respectively, and by the azimuthal angle φ.

Introducing the notation

$$A = 1 + \frac{4\pi}{v}\int_0^\sigma ds\, \tau(s) \tag{2.167}$$

we can rewrite Eq. (2.166) in the form

$$\tau(r) = Ar - \frac{2\pi}{v}\int_0^\sigma ds\, \tau(s)\int_{|r-s|}^{r+s} dt\, H(t-\sigma)\tau(t). \tag{2.168}$$

Laplace transforms of both sides of Eq. (2.168) yield

$$\int_0^\infty dr\, \exp(izr)\tau(r) = (-iz)^{-2}A - \frac{2\pi}{v}\int_0^\infty dr\, \exp(izr)$$
$$\times \int_0^\sigma ds\, \tau(s)\int_{|r-s|}^{r+s} dt\, H(t-\sigma)\tau(t). \tag{2.169}$$

We transform the triple integral in the right-hand side of this equation[49]:

$$K = \int_0^\sigma ds\, \tau(s)\int_0^\infty dt\, H(t-\sigma)\tau(t)\int_{|t-s|}^{t+s} dr\, \exp(izr)$$
$$= (iz)^{-1}\left\{\int_0^\sigma ds\, \tau(s)\exp(izs)\int_\sigma^\infty dt\, \tau(t)\exp(izt)\right.$$
$$\left. - \int_0^\sigma ds\, \tau(s)\exp(-izt)\int_\sigma^\infty dt\, \tau(t)\exp(izt)\right\} \tag{2.170}$$

and introduce the notation

$$F(z) = \int_0^\sigma ds\, \tau(s)\exp(izs) = -\int_0^\infty ds\, sC(s)\exp(izs), \tag{2.171}$$

$$G(z) = \int_\sigma^\infty ds\, \tau(s)\exp(izs) = \int_0^\infty ds\, s\mu(s)\exp(izs). \tag{2.172}$$

In this notation, Eq. (2.169) becomes

$$F(z) + G(z) = (iz)^{-2}A - \frac{2\pi}{v}(iz)^{-1}G(z)[F(z) - F(-z)]. \tag{2.173}$$

From this we determine $G(z)$:

$$G(z) = \frac{(iz)^{-2}A - F(z)}{1 + (2\pi/v)(iz)^{-1}[F(z) - F(-z)]}. \tag{2.174}$$

$F(z)$ is an entire function, since it is the Laplace transform of a regular function in a finite region $(0, \sigma)$. At the same time, $G(z)$ is the Laplace transform of a regular function in the semiinfinite (σ, ∞) plane, and is therefore regular in the upper half-plane and can have poles only either on the real axis or in the lower half-plane. According to Eq. (2.174), $G(z)$ has a second-order pole at the point $z = 0$.[47]

We introduce a new function

$$L(z) = (iz)^4 G(z)\{(iz)^{-2}A - F(-z)\}. \tag{2.175}$$

It is an entire function. It can be completely determined by investigating its behavior at infinity.

Substituting Eq. (2.174) in Eq. (2.175) we get

$$L(z)=(iz)^4\frac{(iz)^{-4}A^2-(iz)^{-2}A[F(z)+F(-z)]+F(z)F(-z)}{1+2\pi/v[F(z)-F(-z)](iz)^{-1}}. \qquad (2.176)$$

The upper bound of $F(z)$ is estimated at

$$F(z)\ll\bar{\tau}\int_0^\sigma ds\,\exp(izs)=\frac{\bar{\tau}[\exp(i\sigma z)-1]}{iz}$$

[inasmuch as if τ is a continuous function, it is bounded in the region $(0,\sigma)$]. As $z\to\infty$ we have therefore in the lower half-plane

$$F(z)\sim\frac{\exp(i\sigma z)}{z}, \qquad (2.177)$$

i.e., $F(z)$ increases exponentially.

Integrating the expression for $F(-z)$ by parts

$$F(-z)=-\int_0^\sigma ds\,sC(s)\exp(-isz)=\frac{\sigma C(\sigma)\exp(-i\sigma z)}{iz}$$
$$+\frac{[\sigma C'(\sigma)+C(\sigma)]\exp(-i\sigma z)-C(0)}{(iz)^2}+O(z^{-3})$$

we obtain as $z\to\infty$ (in the lower half-plane)

$$F(-z)\sim-\frac{C(0)}{(iz)^2}=\frac{A}{(iz)^2} \qquad (2.178)$$

[the equality $-C(0)=A$ follows from the Percus–Yevick equation].

We can thus determine now the asymptotic form of $L(z)$:

$$\begin{aligned}L(z)&\sim(iz)^4\frac{\{-(iz)^{-2}A+F(-z)\}F(z)}{(2\pi/v)(iz)^{-1}F(y)}\\&\sim(iz)^5\frac{-(iz)^{-2}A+(iz)^{-2}A+O(z^{-3})}{(2\pi/v)}=O(z^2).\end{aligned} \qquad (2.179)$$

According to the Rouché theorem

$$L(z)=\lambda_1+\lambda_2z^2, \qquad (2.180)$$

where λ_1 and λ_2 are constants.

Comparing Eq. (2.180) with Eq. (2.176) (we expand the latter in a series about $z=0$) we have

$$\begin{aligned}\lambda_1&=A,\\ \lambda_2&=-2F(0)+\left(\frac{2\pi}{3iv}\right)F'''(0).\end{aligned} \qquad (2.181)$$

Comparing Eqs. (2.180) and (2.175) we get

$$G(z)\{(iz)^{-2}A-F(-z)\}=\lambda_1(iz)^{-4}+\lambda_2(iz)^{-2}. \qquad (2.182)$$

This relation yields the term $G(z)F(-z)$, which we substitute in Eq. (2.173):

$$F(z) = \lambda_1(iz)^{-2} - \frac{2\pi}{v}\lambda_2(iz)^{-3} - \frac{2\pi}{v}\lambda_1(iz)^{-5}$$
$$+\frac{2\pi}{v}G(z)\{-(iz)^{-1}F(z) + (iz)^3 A\}\,. \qquad (2.183)$$

The inverse Laplace transform of this equation is

$$(2\pi)^{-1}\int_C dz\,\exp(izr)F(z)$$
$$= (2\pi)^{-1}\int_C dz\,\exp(-izr)\left\{\lambda_1(iz)^{-2} - \frac{2\pi}{v}[\lambda_2(iz)^{-3} + \lambda_1(iz)^{-5}]\right\}$$
$$+\frac{1}{v}\int_C dz\,\exp(-izr)G(z)\{-(iz)^{-1}F(z) + (iz)^{-3}A\}\,. \qquad (2.184)$$

Consider the case $r<\sigma$. The second term on the right-hand side of Eq. (2.184) is zero by virtue of the regularity of the integrand in the upper half-plane. The calculation of the remaining terms is straightforward:

$$(2\pi)^{-1}\int_C dz\,\exp(izr)F(z) = \lambda_1 r + \left(\frac{2\pi}{v2!}\right)\lambda_2 r^2 + \left(\frac{2\pi}{v4!}\right)\lambda_1 r^4,$$

or, if account is taken of Eq. (2.171):

$$C(r) = -\lambda_1 - \frac{\pi\lambda_2}{v}r - \left(\frac{\pi}{12v}\right)\lambda_1 r^3, \quad r<\sigma. \qquad (2.185)$$

On the basis of Eqs. (2.185) and (2.181) we obtain a system of equations linear in λ_1 and λ_2:

$$\lambda_1 = A = 1 - \frac{4\pi}{v}\int_0^\sigma ds\, s^2 C(s) = 1 - \frac{4\pi}{3v}\sigma^3\lambda_1 + \frac{\pi^2}{v^2}\sigma^4\lambda_2 + \frac{1}{18v^2}\pi^2\sigma^6\lambda_1,$$
$$\lambda_2 = -2F(0) - \frac{2\pi i}{3v}F'''(0) = \int_0^\sigma ds\left\{2sC(s) + \frac{2\pi i}{3v}(is)^3 sC(s)\right\}$$
$$= -\sigma^2\left\{1 + \frac{\pi\sigma^3}{6v} + \frac{\pi^2\sigma^6}{144v^2}\right\}\lambda_1 - \left\{\frac{2\pi\sigma^3}{3v} + \frac{\pi^2\sigma^6}{9v^2}\right\}\lambda_2,$$

whence

$$\lambda_1 = \frac{(1-2\eta)}{(1-\eta)^4},$$
$$\lambda_2 = -\frac{-\sigma^2(1+\frac{1}{2}\eta)}{(1-\eta)^4}, \qquad (2.186)$$

where

$$\eta = \frac{\pi\sigma^3}{6v}. \qquad (2.187)$$

Thus,

$$c(x) = (1-\eta)^{-4}\{-(1+2\eta)^2 + 6\eta(1+\tfrac{1}{2}\eta)^2 x - \tfrac{1}{2}\eta(1+2\eta)^2 x^2\},$$
$$x = r/\sigma. \tag{2.188}$$

From this expression we can determine $F(z)$ and later also $G(z)$, the Laplace transform of the pair distribution function.

We determine the equation of state in the Percus–Yevick approximation. Starting from Eq. (2.161), we have

$$\frac{d}{dr}e^{-\beta\Phi(r)} = -\beta\frac{d\Phi}{dr}e^{-\beta\Phi(r)} = \delta(r-\sigma),$$

where $\delta(x)$ is the Dirac delta function. Taking this relation into account, we have from Eq. (2.156),

$$\frac{pv}{\theta} = 1 - \frac{2\pi}{3\theta}\sigma^3 C(\sigma) \tag{2.189}$$

and, using Eq. (2.188),

$$\frac{pv}{\theta} = \frac{1+2\eta+3\eta^2}{(1-\eta)^2}. \tag{2.190}$$

The equation of state can be obtained from the expression for the compressibility,

$$-\frac{\theta}{v^2}\left(\frac{\partial v}{\partial p}\right)_\theta = 1 + \frac{4\pi}{v}\int_0^\infty [\mu(r) - 1]r^2 dr \tag{2.191}$$

in the form

$$\frac{pv}{\theta} = \frac{1+\eta+\eta^2}{(1-\eta)^3}. \tag{2.192}$$

The inequality of Eq. (2.190) to Eq. (2.191) attests to the perfunctory character of the introduced Percus–Yevick approximation.

13. Solution of the Bogolyubov equation chain for a crystal

A consistent crystal-lattice theory was developed in the harmonic approximation in Ref. 50. In many cases, however, this approximation is inadequate at present. Account must be taken of the anharmonic terms in the potential energy of the interaction between the crystal atoms. A direct solution of this problem is extremely difficult. This explains why many approximate methods are used to allow for the influence of the anharmonic terms on various crystal properties.

One of the first among them is traditional perturbation theory,[27] with the harmonic approximation as the starting point and the anharmonic terms regarded as small perturbations. The small parameter in the expansion of the potential energy is in this case the ratio of the average vibration energy to the binding energy of the atoms in the crystal. Crystal investigations in a wide range of temperatures and pressures,[51] however, have shown this parameter to reach values of order unity at sufficiently high temperatures (on the order of half the melting temperature and higher). A new method of allowing for the anharmonicities, valid at all temperatures, was therefore needed. This turned out to be the self-consistent-field method.[33] The correlation theory developed in it permits a good description of an anharmonic crystal.[39,52] A self-consistent dynamic theory based on the Green's function method was also developed for anharmonic crystals.[53]

The contribution of the correlations of the atom motion to the thermodynamic potential of a crystal is small, but is significant in properties described in terms of derivatives of potentials.[52] It will be shown below that when correlations are taken into account the natural first choice, for fastest convergence of the successive-approximation series, is the self-consistent-field approximation. This approximation is valid for a crystal because the regions in which the particles move near the crystal sites do not overlap and are smaller than the nearest-neighbor distances.[33,34] This means that the atom single-particle distribution functions are strongly localized near the lattice sites. Moreover, the distribution function, by definition, is greatly decreased by the strong correlations when any two particles come close together. Thus, when considering a chain of Bogolyubov equations, the main contribution to the integral terms is made by integration near the lattice sites, where the correlations other than structural (i.e., those determining the form of the lattice) are small. A Bogolyubov-equation chain can therefore have two types of small-parameter series solutions, based either on a transformation of the potential of the integrand[34,55] or on a transformation of an $(s+1)$-particle distribution function.[56] Note that the second approach includes the first as a particular case. The system of Bogolyubov equations is uncoupled by direct use of the approximation

$$F_{s+1}(q_1,\ldots,q_{s+1}) = F_s(q_1,\ldots,q_s)F_1(q_{s+1}) \tag{2.193}$$

under the integral sign.

As stated earlier, particles in a crystal are strongly localized almost all the way to the melting temperature. It will therefore be convenient to introduce a system of cells, each containing one localized particle (the validity of this approach was proved in Ref. 34). This excludes the possibility of simultaneous presence of two, three, etc., particles in one cell. At small deviations from the equilibrium position, the particle motions are described by a certain self-consistent potential. At large deviations, however, when the correlations are large, the self-consistent-field approximation becomes less accurate, but the contribution of these terms to the integrals of the Bogolyubov-chain equations is small, since the distribution functions decrease rapidly as the cell boundary is approached. It is therefore natural to rewrite Eq. (1.144) in the form

$$\frac{\partial \rho_s}{\partial q_1^\alpha} + \frac{1}{\theta}\frac{\partial \overline{U}_s}{\partial q_1^\alpha}\rho_s + \frac{1}{\theta} \times \sum_{i \neq 1,s} \int_{v_i} \frac{\partial \Phi(|q_1 - q_i|)}{\partial q_1^\alpha} [\rho_{s+1}(q_1,\dots,q_i) - \rho_s(q_1,\dots,q_s)\rho_1(q_i)]\, dq_i = 0, \tag{2.194}$$

where

$$\overline{U}_s = U_s + \sum_{\substack{1 \leqslant i \leqslant s \\ j \neq 1,s}} \int_{v_j} \Phi(|q_i - q_j|)\rho_1(q_j)\, dq_j = U_s + \sum_{\substack{1 \leqslant i \leqslant s \\ j \neq 1,s}} \Phi(|q_i - \overline{q}_j|). \tag{2.195}$$

The sum of the integral terms in Eq. (2.194) is proportional to a small parameter that is smaller the stronger the particle localization. For an arbitrary temperature (up to the melting point) the localization is determined mainly by the character of the interaction (and, in the quantum case, also by the particle mass), and is high for a short-range potential.

We introduce the functions

$$\varphi_{ij} = \Phi(|q_i - q_j|) + \Phi(|q_i{}^0 - \overline{q}_j|) + \Phi(|\overline{q}_i - q_j{}^0|) - \Phi(|q_i - \overline{q}_j|) - \Phi(|\overline{q}_i - q_j|) - \Phi(|q_i^0 - q_j^0|), \tag{2.196}$$

where q_i^0 is the equilibrium position of the ith atom.

Equation (2.194) now takes the form

$$\frac{\partial \rho_s}{\partial q_1^\alpha} + \frac{1}{\theta}\frac{\partial \overline{U}_s}{\partial q_1^\alpha}\rho_s + \frac{1}{\theta} \sum_{i \neq 1,s} \int_{v_i} \frac{\partial \varphi_{1i}}{\partial q_1^\alpha} [\rho_{s+1}(q_1,\dots,q_i) - \rho_s(q_1,\dots,q_s)\rho_1(q_i)]\, dq_i = 0, \tag{2.197}$$

since addition of single-particle functions to the two-particle potential under the differentiation sign does not alter Eq. (2.194).

In addition, Eq. (2.197) remains unchanged if Eq. (2.195) for U_s is replaced by

$$\overline{U}_s = \sum_{1 \leqslant i < j \leqslant s} [\Phi(|q_i - q_j|) - \Phi(|q_i^0 - q_j^0|)] + \sum_{\substack{1 \leqslant j \leqslant s \\ j \neq 1,s}} [\Phi(|q_i - \overline{q}_j|) - \Phi(|q_i^0 - \overline{q}_j|)] \,.$$

If the two-particle potential near the crystal sites is approximated by an expression such as Eq. (2.194), where ρ_1 on the right-hand side is taken to be a Dirac δ function, the Bogolyubov equation chain can be represented as[34,55]

$$\frac{\partial \rho_s}{\partial q_1^\alpha} + \frac{1}{\theta}\frac{\partial \overline{U}_s}{\partial q_1^\alpha}\rho_s + \frac{1}{\theta} \sum_{i \neq 1,s} \int_{v_i} \frac{\partial \varphi_{1i}}{\partial q_1^\alpha} \rho_{s+1}(q_1,\dots,q_s)\, dq_s = 0, \tag{2.198}$$

where

$$\overline{U}_s = \sum_{1 \leqslant i < j \leqslant s} [\Phi(|q_i - q_j|) - \Phi(|q_i^0 - q_j^0|)]$$

$$+ \sum_{\substack{1 \leqslant i \leqslant s \\ j \neq 1,s}} [\Phi(|q_i - q_j^0|) - \Phi(|q_i^0 - q_j^0|)]. \tag{2.199}$$

The mean squared displacement calculated here from the obtained solutions will agree with the analogous value calculated by the free-volume theory.[55]

We see from Eq. (2.197) that in this case the principal part of the mean squared displacement is determined by averaging over the single-particle distribution function obtained in the self-consistent-field approximation:

$$\theta \ln C\rho_1(q_1) + \sum_{i \neq 1} \int_{v_i} \Phi(|q_1 - q_i|)\rho_1(q_i)dq_i = 0. \tag{2.200}$$

The mean squared displacement calculated with ρ_1 from Eq. (2.200) does not exceed the analogous quantity calculated by the free-volume theory. As the melting temperature is approached, the discrepancy for strongly anharmonic crystals can become appreciable. In the investigation of crystal properties it is necessary thus to use the solution of either Eq. (2.197) or Eq. (2.198), depending on whether it is easier to find the solution of the self-consistent-field equation and then obtain a rapidly converging series from Eq. (2.197), or to use the series of Eq. (2.198) with a large number of terms. Only the series in Eq. (2.198) can have a reasonable number of terms in the case of strong anharmonicity, although the use of Eq. (2.197) is effective in many cases.[34,55]

We seek the solution of the equation chain (2.197) in the form of the series

$$\rho_s = \rho_s{}^0 + \rho_s{}^1 + \rho_s{}^2 + \cdots, \tag{2.201}$$

where ρ_s^i is proportional to the ith power of a small parameter to which the integral terms in the system (2.96) are proportional. Substituting Eq. (2.201) in Eq. (2.197) and equating terms of like order, we get

$$\frac{\partial \rho_s^0}{\partial q_1^\alpha} + \frac{1}{\theta}\frac{\partial \overline{U}_s}{\partial q_1^\alpha}\rho_s^0 = 0,$$

$$\frac{\partial \rho_s^1}{\partial q_1^\alpha} + \frac{1}{\theta}\frac{\partial \overline{U}_s}{\partial q_1^\alpha}\rho_s^1 + \sum_{i \neq 1,s} \int_{v_i} \frac{\partial \varphi_{1i}}{\partial q_1^\alpha}\rho_{s+1}^0(q_1,\dots,q_i)dq_i = 0. \tag{2.202}$$

The boundary conditions are

$$\rho_s^0(q_1,\dots,q_s) - \prod_{1 \leqslant i \leqslant s} \rho_i^0(q_i) \to 0, \quad |q_i - q_j| \to \infty,$$

$$\rho_s^1(q_1,\dots,q_s) - \sum_{1 \leqslant i \leqslant s} \rho_1^1(q_i) \prod_{\substack{1 \leqslant i \leqslant s \\ j \neq i}} \rho_1^0(q_j) \to 0,$$

$$\vdots \tag{2.203}$$

The normalization conditions are

$$\int_v \rho_1^0(q)dq = 1, \quad \int_v \rho_1^1(q)dq = 0, \dots . \tag{2.204}$$

Putting

$$\rho_s{}^0 = C_s(q_1,\dots,q_s)\exp[-\overline{U}_s/\theta],$$

it follows from Eqs. (2.202) that

$$\frac{\partial C_s^0}{\partial q_1^\alpha} = 0,$$

i.e., C_s^0 does not depend on q_i. Since, however, C_s^0 is symmetric in $q_1,\dots,q_s$, it is a constant and the normalization conditions (2.204) yield

$$C_s{}^0 = (C_1{}^0)^s,$$

$$(C_1^0)^{-1} = \int_v \exp\left[-\frac{1}{\theta}\sum_{i\neq 1}[\Phi(|q_1 - \bar{q}_i|) - \Phi(|q_1^0 - \bar{q}_i|)]\right]dq_1. \tag{2.205}$$

Thus

$$\rho_s{}^0 = (C_1{}^0)^s \exp[-\overline{U}_s/\theta]. \tag{2.206}$$

We rewrite now the second equation of Eq. (2.202) in the form

$$\frac{\partial \rho_s^1}{\partial q_1^\alpha} + \frac{1}{\theta}\frac{\partial \overline{U}_s}{\partial q_1^\alpha}\rho_s^1 + \frac{1}{\theta}\sum_{i\neq 1,s}\int_{v_i}\frac{\partial \varphi_{1i}}{\partial q_1^\alpha}(C_1^0)^{s+1}\exp[-\overline{U}_{s+1}/\theta]dq_i = 0.$$

Note that

$$\frac{C_1^0}{\theta}\frac{\partial \varphi_{1i}}{\partial q_1^\alpha}\exp[-\overline{U}_{s+1}/\theta]$$

$$= \frac{1}{\theta}\frac{\partial \varphi_{1i}}{\partial q_1^\alpha}\exp[-\overline{U}_s/\theta]\exp\left[-\sum_{1\leqslant k\leqslant s}\varphi_{ki}/\theta\right]\rho_1^0(q_i)$$

$$= -\exp[-\overline{U}_s/\theta]\frac{\partial}{\partial q_1^\alpha}\exp\left[-\sum_{1\leqslant k\leqslant s}\varphi_{ki}/\theta\right]\rho_1^0(q_i)$$

$$= -\exp[-\overline{U}_s/\theta]\rho_1^0(q_i)\frac{\partial}{\partial q_1^\alpha}\prod_{1\leqslant k\leqslant s}(1+f_{ki}),$$

where we put for brevity

$$f_{ki} = \exp[-\varphi_{ki}/\theta] - 1.$$

We have taken it into account here that

$$\exp[-\overline{U}_{s+1}/\theta] = \exp[-\overline{U}_s/\theta]\exp\left[-\sum_{1\leqslant k\leqslant s}\varphi_{ki}/\theta\right]$$

$$\times \exp\left\{-\sum_{e\neq i}[\Phi(|q_l - q_i|) - \Phi(|q_l^0 - \bar{q}_i|)]/\theta\right\}.$$

For a short-range potential $\Phi(r)$ between the particles in the crystal, the functions f_{ij} decrease rapidly with increase of $|q_i - q_j|$, so that the integrals considered are absolutely convergent, and we can write

$$\frac{\partial}{\partial q_1^\alpha}\int_v\left[1+\sum_{1\leqslant k\leqslant s}(\overline{\exp[-\varphi_{ki}/\theta]}-1)\right]\rho_1^0(q_i)dq_i=0,$$

where

$$\overline{\exp[-\varphi_{ki}/\theta]}=\int_v\exp[-\varphi_{ki}/\theta]\rho_1^0(q_k)dq_k. \qquad (2.207)$$

Consequently,

$$\frac{\partial\rho_s^1}{\partial q_1^\alpha}+\frac{\rho_s^1}{\theta}\frac{\partial\overline{U}_s}{\partial q_1^\alpha}$$
$$=\exp[-\overline{U}_s/\theta](C_1^0)^s\sum_{i\neq 1,s}\frac{\partial}{\partial q_1^\alpha}\int_{v_i}\left\{\prod_{1\leqslant k\leqslant s}(1+f_{ki})-1\right.$$
$$\left.-\sum_{1\leqslant k\leqslant s}(\overline{\exp[-\varphi_{ki}/\theta]}-1)\right\}\rho_1^0(q_i)dq_i.$$

Putting now again

$$\rho_s^1=(C_1^0)^s\exp[-\overline{U}_s/\theta]C_s^1(q_1,\ldots,q_s), \qquad (2.208)$$

we obtain

$$\frac{\partial C_s^1}{\partial q_1^\alpha}=\frac{\partial}{\partial q_1^\alpha}\sum_{i\neq 1,s}\int_{v_i}\left\{\prod_{1\leqslant k\leqslant s}(1+f_{ki})-1\right.$$
$$\left.-\sum_{1\leqslant k\leqslant s}(\overline{\exp[-\varphi_{ki}/\theta]}-1)\right\}\rho_1^0(q_i)dq_i.$$

Since C_s^0 should be symmetric with respect to the arguments $q_1,\ldots,q_s$, we get

$$C_s^1=\sum_{1\neq 1,s}\int_{v_i}\left\{\prod_{1\leqslant k\leqslant s}(1+f_{ki})-1\right.$$
$$\left.-\sum_{1\leqslant k\leqslant s}(\overline{\exp[-\varphi_{ki}/\theta]}-1)\right\}\rho_1^0(q_i)dq_i+k_s,$$

$k_s=\text{const}$.

But in view of the rapid decrease of f_{ki} as $|q_i-q_j|\to\infty$ we have in this case

$$\int_{v_i}\left\{\prod_{1\leqslant k\leqslant s}(1+f_{ki})-1-\sum_{1\leqslant k\leqslant s}(\overline{\exp[-\varphi_{ki}/\theta]}-1)\right\}\rho_1^0(q_i)dq_i\to 0.$$

Therefore, taking Eqs. (2.203), (2.204), and (2.208) into account we find that $k_s=0$. Thus,

$$\rho_s^1=(C_1^0)^s\exp[-\overline{U}_s/\theta]\sum_{i\neq 1,s}\int_{v_i}\left\{\prod_{1\leqslant k\leqslant s}(1+f_{ki})\right.$$
$$\left.-1-\sum_{1\leqslant k\leqslant s}(\overline{\exp[-\varphi_{ki}/\theta]}-1)\right\}\rho_1^0(q_i)dq_i. \qquad (2.209)$$

The terms of next lower order can be similarly obtained. In this case, in view of the resultant expressions, we have for ρ_s,

$$\rho_s = (C_1^0)^s \exp[-\bar{U}_s/\theta]\Big\{1 + \int_v K_1(q_1,\dots,q_s,q_{s+1})\rho_1^0(q_{s+1})dq_{s+1}$$

$$+ \tfrac{1}{2}\int_v\int_v K_2(q_1,\dots,q_s,q_{s+1},q_{s+2})\rho_1^0(q_{s+1})\rho_1^0(q_{s+2})dq_{s+1}\,dq_{s+2} + \cdots\Big\}$$

$$(s = 1,2,\dots), \tag{2.210}$$

where

$$K_1(q_1,\dots,q_s,q_{s+1})$$
$$= \sum_{i\neq 1,s}\Big\{\exp\Big[-\sum_{1\leqslant i\leqslant s}\varphi_{ki}/\theta\Big] - 1 - \sum_{1\leqslant k\leqslant s}\overline{(\exp[-\varphi_{ki}/\theta]} - 1)\Big\},$$

$$K_2(q_1,\dots,q_s,q_{s+1},q_{s+2})$$
$$= \sum_{i\neq j\neq 1,s}\Big\{\exp\Big[-\sum_{1\leqslant k\leqslant s}(\varphi_{ki}+\varphi_{kj})/\theta - \varphi_{ij}/\theta\Big] - 1$$
$$- \sum_{1\leqslant k\leqslant s}(\exp[-(\varphi_{ki}+\varphi_{kj})/\theta] - 1)\exp[-\varphi_{ij}/\theta]$$
$$-\Big(\exp\Big[-\sum_{1\leqslant k\leqslant s}\varphi_{ki}/\theta\Big] - 1 - \sum_{1\leqslant k\leqslant s}(\exp[-\varphi_{ki}/\theta] - 1)\Big)$$
$$\times \sum_{1\leqslant k\leqslant s}(\exp[-\varphi_{ki}/\theta] - 1) + (\overline{\exp[-\varphi_{ij}/\theta]})$$
$$-\Big(\exp\Big[-\sum_{1\leqslant k\leqslant s}\varphi_{ki}/\theta\Big] - 1 - \sum_{1\leqslant k\leqslant s}(\exp[-\varphi_{kj}/\theta] - 1)\Big)$$
$$\times\Big(\sum_{1\leqslant k\leqslant s}\overline{(\exp[-\varphi_{ki}/\theta]} - 1) + \exp[-\varphi_{ij}/\theta]\Big)\Big\}. \tag{2.211}$$

In Eq. (2.211) we use

$$q_i = a_i + q_{s+1},$$
$$q_j = a_i + q_{s+2},$$

where a_i is the radius vector of the ith lattice site.

We have thus obtained a solution, accurate to second order, of the chain of Bogolyubov equations. The principal approximation was the self-consistent-field equation. This solution is easily generalized also to include the case when φ_{ij} in Eq. (2.196) is determined on the basis of ρ_1 not only in the zeroth but also in the first, second, etc., approximation, i.e., the solution becomes essentially nonlinear.

The expressions obtained for the distribution function permit development of a thermodynamic theory of crystals. Mention must be made here of the remarks made in Sec. 4 concerning partial summation of the series due to the harmonic part of the interaction potential.

We proceed now to solve the quantum chain of equations.[57] We write it in the form

$$[\overline{H}_s(1,\ldots,s), R_s(1,\ldots,s)] + \sum_{i\neq 1,s} \operatorname*{Tr}_i \left[\sum_{1\leqslant i\leqslant s} \Phi(i,j), R_{s+1}(1,\ldots,s,i) - R_s(1,\ldots,s)R_1(i) \right] = 0, \tag{2.212}$$

where

$$\overline{H}_s(1,\ldots,s) = \sum_{1\leqslant j\leqslant s} T(p_i) + \sum_{1\leqslant j<r\leqslant s} [\Phi(|q_j - q_r|) - \Phi(|q_j^0 - q_r^0|)] + \sum_{1\leqslant j\leqslant s} \sum_{i\neq 1,s} \left\{ \operatorname*{Tr}_i [\Phi(|q_j - q_i|) - \Phi(|q_j^0 - q_i|)] R_1(i) \right\}. \tag{2.213}$$

Just as in the classical case, we seek the solution in the form

$$R_s = R_s^0 + R_s^1 + \cdots, \tag{2.214}$$

where R_s^i is proportional to the ith power of a small parameter.

Equation (2.212) is equivalent to

$$[\overline{H}_s(1,\ldots,s), R_s(1,\ldots,s)] + \sum_{i\neq 1,s} \operatorname*{Tr}_i \left[\sum_{1\leqslant j\leqslant s} \varphi(j,i), R_{s+1}(1,\ldots,s,i) \right] = 0, \tag{2.215}$$

where

$$\varphi(j,i) = \Phi(|q_j - q_i|) - \Phi(|q_i^0 - q_j^0|) - \operatorname*{Tr}_i [\Phi(|q_j - q_i|) - \Phi(|q_j^0 - q_i|)] R_1(i) - \operatorname*{Tr}_j [\Phi(|q_j - q_i|) - \Phi(|q_j - q_i^0|)] R_1(j). \tag{2.216}$$

We obtain the solution of the chain (2.215), accurate to first order inclusive (this approach yields also the higher approximations). For R_s^0 and R_s^1 we have the system of equations

$$[\overline{H}_s(1,\ldots,s), R_s^0(1,\ldots,s)] = 0, \tag{2.217}$$

$$[\overline{H}_s(1,\ldots,s), R_s^1(1,\ldots,s)] + \sum_{i\neq 1,s} \operatorname*{Tr}_i \left[\sum_{1\leqslant j\leqslant s} \varphi(j,i), R_{s+1}^0(1,\ldots,s,i) \right] = 0. \tag{2.218}$$

From Eq. (2.217) we have[56]

$$R_s^0 = \exp[-\overline{H}_s/\theta]/Q_s, \tag{2.219}$$

where

$$Q_s = \operatorname{Tr} \exp[-\overline{H}_s/\theta](N-s)!/N!. \tag{2.220}$$

Since the condition that the correlations weaken as $|q_i - q_j| \to \infty$ $(1\leqslant i<j\leqslant s)$ calls for

$$R_s^0 \to \prod_i R_1^0(i), \tag{2.221}$$

we have

$$Q_s = (Q_1)^s. \tag{2.222}$$

Here

$$O_1 = \operatorname*{Tr}_1 \exp[-\overline{H}_1/\theta]. \tag{2.223}$$

The trace in Eq. (2.223) is taken over one unit cell. Since

$$\sum_{1 \leqslant j \leqslant s} \varphi(j,i) = \overline{H}_{s+1}(1,...,s,i) - \overline{H}_s(1,...,s) - \overline{H}_1(1), \tag{2.224}$$

Eq. (2.218) is rewritten in the form

$$\left[\overline{H}_s(1,...,s), R_s^1 - \sum_{i \neq i,s} \operatorname*{Tr}_i R_{s+1}^0(1,...,s,i)\right] = 0, \tag{2.225}$$

because $|H_{s+1}, R_{s+1}^0| = 0$,

$$\operatorname*{Tr}_i [H_1(i), R_{s+1}^0(1,...,s,i)] = 0.$$

In view of the correlation-weakening condition, the solution (2.225) can be represented in the form

$$R_s^1 = \sum_{i \neq 1,s} \operatorname*{Tr}_i \Bigg\{ R_{s+1}^0(1,...,s,i) - R_s^0(1,...,s) \times \sum_{1 \leqslant j \leqslant s} \left[\operatorname*{Tr}_j R_2^0(j,i) - R_1^0(i)\right] + R_1^0(i) \Bigg\}. \tag{2.226}$$

From Eqs. (2.219), (2.222), and (2.226) we have the final expression for the s-particle distribution function, accurate to first order:

$$R_s = \frac{1}{Q_1^s}\Bigg\{ \exp[-\overline{H}_s/\theta] + \frac{1}{Q_1} \sum_{i \neq 1,s} \operatorname*{Tr}_i \Bigg[\exp[-\overline{H}_{s+1}/\theta] - \exp[-\overline{H}_s/\theta] \Bigg(\sum_{1 \leqslant j \leqslant s} \Big[\operatorname*{Tr}_j \exp[-\overline{H}_2(j,i)/\theta]/Q_1 - \exp[-\overline{H}_1(j)/\theta] \Big] + \exp[-\overline{H}_1(i)/\theta] \Bigg) \Bigg] \Bigg\}. \tag{2.227}$$

The expression obtained for the s-particle density matrix depends on the one-particle matrix. The most laborious here is, undoubtedly, the solution of the self-consistent-field equation (and, in the next approximations, also of more complicated nonlinear equations). This problem was solved in the self-consistent-field approximation in Ref. 58. The approach proposed makes it possible to take consistently into account in the quantum region the correlations for a crystal in the first and succeeding approximations.

Chapter 3
Cluster form of the equation of state of a real gas

We consider in this chapter, for a real gas, the equation of state in the form of a series in powers of its density. We precede the treatment of the traditional virial equation with a discussion of the equation of state obtained on the basis of the series expansion of $q(\theta,v) = Q_N{}^{A/N}$, proposed in Sec. 7, which we shall show to be rapidly convergent. On the whole, this form of the equation of state of a real gas, for the pressure (free energy, etc.) and for $Q(\theta,v)$ in powers of the density, will be called the cluster form.

14. Hard-sphere system

The interaction in a system of N particles is characterized in general by a potential function $U(q_1,\ldots,q_N)$. This function is frequently chosen in the form of a sum of two-particle interactions:

$$U = \sum_{i<j} \Phi(|q_i - q_j|)\,, \tag{3.1}$$

which is justified in many cases. The pair-interaction potential is usually of the form shown in Fig. 2: attraction forces predominate at large distances and repulsion at short ones. At high temperatures, the principal role is played by the repulsive part of the potential. It is therefore important in statistical physics to determine the equation of state for a system of hard spheres with an interaction potential:

$$\Phi(r) = \begin{cases} 0 & (r>\sigma) \\ \infty & (r<\sigma) \end{cases}, \tag{3.2}$$

where σ is the sphere diameter.

This system is frequently used as basic, and the attractive part of the potential is taken into account in the correction terms.

In view of the importance of this system in theoretical investigations, we consider in turn its one- , two- , and three-dimensional variants.

14.1 Hard-rod system (one-dimensional "spheres")

Consider an aggregate of N hard rods of diameter σ, located on a line of length L. The problem of statistically describing such a system is exactly solvable and leads to the following equation of state:

$$\frac{pv}{\theta}=\frac{1}{1-\sigma/v}, \tag{3.3}$$

where $v=L/N$.

Before we describe this solution let us see what the two cluster-expansion variants give for such a system.

The first virial coefficients for it are quite simple to calculate:

$$B_2=-\tfrac{1}{2}\int f(|q|)dq=\sigma,$$

$$B_3=-\tfrac{1}{3}\int\int f(|q-q'|)f(|q'|)f(|q|)dq\,dq'=\sigma^2. \tag{3.4}$$

Therefore, accurate to the third virial coefficient, the virial equation of state (2.10) of the system considered takes the form

$$\frac{pv}{\theta}=1+\frac{\sigma}{v}+\frac{\sigma^2}{v^2}+\cdots. \tag{3.5}$$

On the other hand, the expression (2.18) obtained for the free energy on the basis of the $q(\theta,v)$ expansion in powers of the density yields the equation of state

$$\frac{pv}{\theta}=\frac{1+\Sigma_{i=2}^{\infty}(i-2)b_i/v^{i-1}}{1-\Sigma_{i=2}^{\infty}b_i/v^{i-1}}, \tag{3.6}$$

in which we have for the considered system

$$b_2=B_2=\sigma,$$

$$b_3=(B_3-B_2^2)/2=0,$$

which leads, if only the second and third virial coefficients B_2 and B_3 are known, to the exact equation of state (3.3) of the hard-rod system:

$$\frac{pv}{\theta}=1/(1-\sigma/v). \tag{3.7}$$

This demonstrates convincingly the efficacy of the new cluster expansion.

We obtain now an exact solution for the hard-rod system. In view of the short-range character of the potential (3.2), each particle interacts only with two nearest neighbors:

$$U_N=\sum_{i=1}^{N}\Phi(|q_{i+1}-q_i|).$$

Since the positions of the particles on the line remain unchanged, they can be numbered in sequence:

$$0\leqslant q_1 < q_2 < \cdots < q_N \leqslant L\,.$$

The configuration integral of such a system is therefore equal to $N!Q_N$. Following Ref. 37, we represent it in the form

$$N!\,Q_N(L) = \int\cdots\int_{0\leqslant q_1<q_2\cdots<q_N\leqslant L} \exp[-\beta\,[\Phi(|q_2-q_1|) + \Phi(|q_3-q_2|) + \cdots]]dq_1\cdots dq_N\,. \tag{3.8}$$

We place between 0 and q_1 an additional particle at a fixed point q_0. The configuration integral of the system is then

$$P_N(q_0,N) = \int\cdots\int_{q_0<q_1<\cdots<q_N\leqslant L} \exp[-\beta\,[\Phi(|q_1-q_0|) + \Phi(|q_2-q_1|) + \cdots]]dq_1\cdots dq_N\,.$$

We repeat this operation, placing between 0 and q_0 a particle at a fixed point q'. Then

$$P_{N+1}(q',L) = \int\cdots\int_{q'<q<\cdots\leqslant L} \exp[-\beta\,[\Phi(|q'-q_0|) + \cdots]]dq_0\cdots dq_N$$

and hence

$$P_{N+1}(q',L) = \int_{q'<q_0} \exp[-\beta\Phi(|q_0-q'|)]dq_0 \times \int\cdots\int_{q_0<q_1<\cdots\leqslant L} \exp[-\beta\,[\Phi(|q_1-q_0|) + \cdots]]dq_1\cdots dq_N$$

$$= \int\cdots\int_{q'<q_0\leqslant L} P_N(q_0,L)\exp[-\beta\Phi(|q_0-q'|)]dq_0\,. \tag{3.9}$$

We construct the grand partition function of the system, using $P_N(q_0,L)$:

$$\exp[-\beta G] = \sum_N \exp[-\beta\mu N] \times \frac{1}{N!}\left(\frac{m\theta}{2\pi\hbar^2}\right)^{N/2} Q_N$$

or

$$\exp[-\beta G] = \sum_N \exp(\beta\mu_1 N)P_N(q_0,L)\,, \tag{3.10}$$

where

$$\mu_1 = \mu + \theta \ln \sqrt{\frac{m\theta}{2\pi\hbar^2}} \,. \tag{3.11}$$

But $G = -pL$, therefore Eq. (3.10) takes the form

$$\exp[\beta p(L - q_0)] = \sum_N \exp[\beta\mu_1 N] P_N(q_0, N) \,.$$

Multiplying Eq. (3.9) by $\exp[(N+1)\mu_1\beta]$ and summing over N, we get

$$\exp[p(L - q')\beta] = \exp[\mu_1\beta] \int_{q'}^{L} \exp[p(L - q_0)\beta] \exp[\beta\Phi(|q_0 - q'|)] dq_0$$

or

$$\exp[-\beta\mu_1] = \int_{q'}^{L} \exp[-p(q_0 - q')\beta] \exp[-\beta\Phi(|q_0 - q'|)] dq_0 \,.$$

In the statistical limit ($L \to \infty$) we have

$$\exp[-\beta\mu_1] = \int_0^{\infty} \exp[-p\beta x] \exp[-\beta\Phi(x)] dx \,.$$

For the interaction potential (3.2),

$$\exp[-\beta\mu_1] = \int_0^{\infty} \exp[-p\beta x] dx = -\frac{1}{p\beta} \exp[-p\beta x] \Big|_{\sigma}^{\infty} = \frac{\exp[-p\beta\sigma]}{p\beta} \,,$$

whence

$$\mu_1 = -\frac{1}{\beta} \ln\left\{\frac{\exp[-p\beta\sigma]}{p\beta}\right\} = p\sigma + \theta \ln\frac{p}{\theta} \,.$$

Recognizing that μ_1 is connected with the chemical potential μ by relation (3.11), we have

$$\mu = \theta \ln\frac{p}{\theta} + p\sigma - \ln \sqrt{\frac{m\theta}{2\pi\hbar^2}} \,. \tag{3.12}$$

But $v = -\partial\mu/\partial p$, therefore we obtain from Eq. (3.12) the exact equation of state of the system of hard rods:

$$\frac{pv}{\theta} = \frac{1}{1 - \sigma/v} \,. \tag{3.13}$$

It follows from the expression for the stability coefficient

$$\left(\frac{\partial p}{\partial v}\right)_\theta = -\frac{\theta}{(v - \sigma)^2} < 0 \tag{3.14}$$

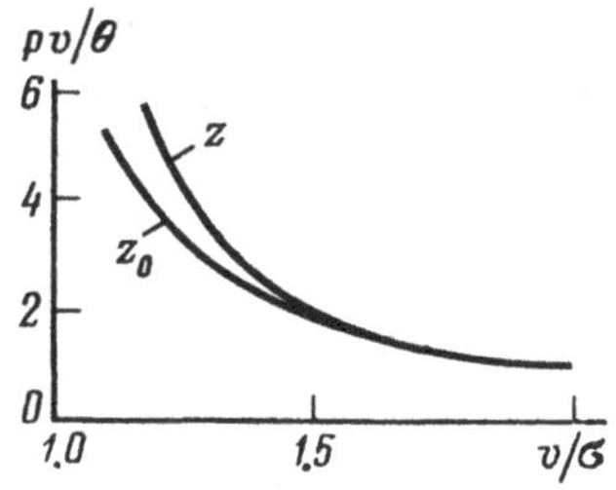

Figure 10.

that a one-dimensional system of hard rods is always stable (if $\theta \neq 0$), so that no phase transition of the vapor-condensation or crystallization type is possible for it when the stability coefficient vanishes.

As first shown by Van Hove, this impossibility of a phase transition applies to all one-dimensional systems with arbitrary large but finite action radius of the intermolecular forces.

Figure 10 shows the values of the quantity $z = pv/\theta$ obtained using Eq. (3.7) and of the analogous quantity z_0 obtained from the usual virial expansion (3.4), with eight virial coefficients taken into account. We see how much slower the usual virial expansion converges than the equation of state obtained from Eq. (3.6).

14.2 System of hard disks (two-dimensional system)

We determine first the virial coefficients of a system of hard disks. Direct calculations yield

$$B_2 = -\frac{1}{2}\int f(|q|)dq = \frac{1}{2}\int_0^{2\pi} d\varphi \int_0^{\sigma} dr\cdot r = \frac{\pi}{2}\sigma^2 \tag{3.15}$$

(σ is the disk diameter).

An exact value is known also for N_3 (Ref. 59):

$$B_3/B_2^2 = 0.782. \tag{3.16}$$

The other coefficients (to the sixth, inclusive) were determined numerically (Ref. 60):

$$\begin{aligned} B_4/B_2^3 &= 0.5324 \pm 0.0003, \\ B_5/B_2^4 &= 0.3338 \pm 0.0005, \\ B_6/B_2^5 &= 0.1992 \pm 0.0008. \end{aligned} \tag{3.17}$$

In the two-dimensional case, the area per close-packed disk is

$$v_0 = \frac{\sqrt{3}}{2}\sigma^3. \tag{3.18}$$

We shall find it convenient to express the virial coefficients (3.15)–(3.17) in units of this area (raised to the appropriate power):

$$\begin{aligned} B_2 &= 1.8138v_0\,, \\ B_3 &= 2.5727v_0^{\,2}\,, \\ B_4 &= 3.1769v_0^{\,3}\,, \\ B_5 &= 3.6128v_0^{\,4}\,, \\ B_6 &= 3.9105v_0^{\,5}\,. \end{aligned} \tag{3.19}$$

Substituting these quantities in Eq. (2.17), we obtain for h_i the numerical values

$$\begin{aligned} b_2 &= 1.8138v_0\,, \\ b_3 &= -0.3586v_0^{\,2}\,, \\ b_4 &= -0.2797v_0^{\,3}\,, \\ b_5 &= -0.1799v_0^{\,4}\,, \\ b_6 &= -0.0915v_0^{\,5}\,. \end{aligned} \tag{3.20}$$

Comparing Eq. (3.19) with Eq. (3.20) we see that the series (2.19) for b converges better than the virial series: B_i increases with increase of i, whereas b_i decreases, and furthermore substantially.

To improve the convergence of the virial expansion, Ree and Hoover[61] attempted to use a Padé approximant.[62] This method was used earlier in the solution of the Ising problem.[63,64]

In the general case the Padé approximant takes for series in powers of $1/v$ the form

$$P(n,m) = \sum_{i=1}^{n} \alpha_i/v^{i-1} \Big/ \sum_{i=1}^{m} \gamma_i/v^{i-1}\,. \tag{3.21}$$

The expression obtained for Pv/θ, with account taken of the six virial coefficients of the system considered, is usually represented in the form

$$\frac{pv}{\theta} = 1 + \frac{B_2/v[1 - 0.196\,703B_2/v + 0.017\,329(B_2/v)^2]}{1 - 0.561\,493B_2/v + 0.081\,313(B_2/v)^2}\,. \tag{3.22}$$

This relation describes well the results of a computer calculation by the molecular-dynamics system for a hard-disk gas, but does not lead to any singularities whatever on its phase diagram. This approach can therefore not describe the phase transition, obtained by the molecular-dynamics method (MDM), in a hard-disk system.[61] The Padé approximant estimates the asymptotic behavior of an infinite series from its first few coefficients. An estimate of the higher virial coefficients has shown that for hard disks all the virial coefficients are positive.[17] The virial coefficient, however, converges extremely slowly.

We proceed now to determine the equation of state from our Eq. (3.6), using the expressions (3.20) obtained for b. The results for $z = Pv/\theta$ are shown in Fig. 11, together with the analogous plots for the virial expansion (z_0) and of

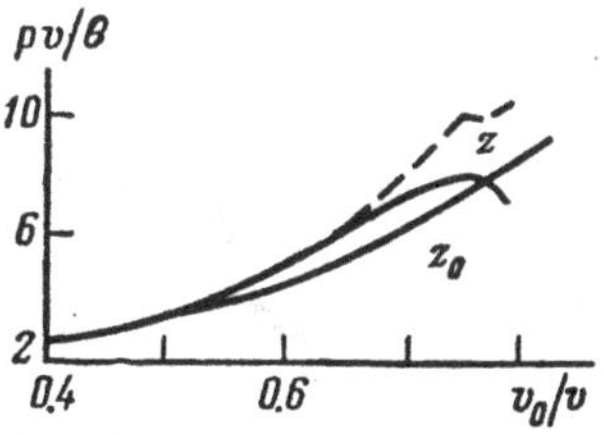

Figure 11.

the results of the MDM method (dashed line). We see that the MDM data are better described than those for z_0, and that z has a singularity that attests to a region of instability of the homogeneous particle distribution; this reveals the phase transition in the hard-disk system.

14.3 Hard-sphere system (three-dimensional system)

Just as in the preceding cases, we begin our analysis with a determination of the virial coefficients of a hard-sphere system. Analytic expressions are known for B_2, B_3, and B_4 (Refs. 65, 66, and 67). Thus

$$B_2 = -\frac{1}{2}\int f(|q|)dq = -\frac{1}{2}\int_0^{2\pi} d\varphi \int_0^{\pi} d\vartheta \int_0^{\infty} dr \cdot r^2 f(r) = \frac{2\pi}{3}\sigma^3, \tag{3.23}$$

and also

$$B_3/B_2{}^2 = 5/8,$$

$$B_4/B_2^3 = \frac{1283}{8960} + \frac{3}{2}\cdot\frac{73\sqrt{2} + 81\cdot 17(\arctan\sqrt{2} + \pi/4)}{32\cdot 35\pi} = 0.286\,95 \tag{3.24}$$

(σ is the sphere diameter).

In addition, numerical values are known for the following virial coefficients[24]:

$$\begin{aligned} B_5/B_2{}^4 &= 0.1104 \pm 0.000\,06, \\ B_6/B_2{}^5 &= 0.0386 \pm 0.0004, \\ B_7/B_2{}^6 &= 0.0138 \pm 0.0004, \\ B_8/B_2{}^7 &\approx 0.005. \end{aligned} \tag{3.25}$$

The virial equation obtained on the basis of these coefficients for the virial state Pv/θ (designated z_0) as a function of v is shown in Fig. 12 together with the plot obtained from the MDM data (dashed line).

The Padé approximant constructed with the first six virial coefficients of the hard-sphere system is given by[61]

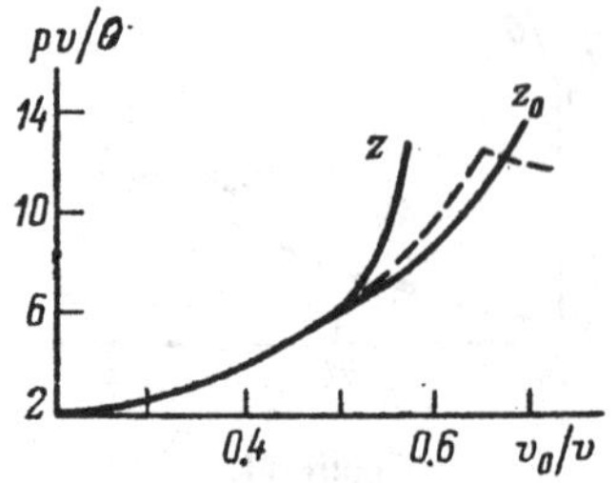

Figure 12.

$$\frac{pv}{\theta} = 1 + \frac{B_2/v[1 + 0.063\,507B_2/v + 0.017\,329(B_2/v)^2]}{1 - 0.561\,493B_2/v + 0.081\,313(B_2/v)^2}. \tag{3.26}$$

Just as in the two-dimensional case, it describes well the equation of state for a homogeneous phase, but has no singularities due to the presence of a phase transition in the hard-sphere system. The Padé approximant can therefore not be used to describe the phase transition that takes place in the system according the MDM data.

We obtain now from Eq. (2.17) the coefficients v_i expressed in terms of the volume

$$v_0 = \frac{\sqrt{2}}{2}\sigma^3, \tag{3.27}$$

per closed-packed sphere:

$$\begin{aligned} b_2 &= 2.9619v_0, \\ b_3 &= -1.6449v_0^2, \\ b_4 &= -1.3040v_0^3, \\ b_5 &= -0.1785v_0^4, \\ b_6 &= 0.7198v_0^5, \\ b_7 &= 0.8411v_0^6, \\ b_8 &= 0.4339v_0^7. \end{aligned} \tag{3.28}$$

Using these coefficients in Eq. (5.6) we find the equation of state of a system of hard spheres; it is plotted in Fig. 12 (curve z). At a certain volume per sphere, the curve has a singularity attesting to the presence of a phase transition in the system. Thus, in contrast to the virial equation of state, the one constructed by using the new cluster $q(\theta,v)$ expansion reveals the phase transition that takes place in this system in accord with the computer experiment.

Phase transitions in hard-disk and hard-sphere systems are described in Refs. 201 and 202 by an elaboration of the proposed new method of calculating the free energy.

Calculation of the virial coefficients is a complicated task even for the hard-sphere system. The higher-order virial coefficients are estimated by introducing the simpler system of rigid parallel cubes, according to which the molecules are cubes whose edges remain parallel after the interaction, i.e., the rotational degrees of freedom are disregarded. This system was first introduced by Geilikman[68] and later by Zwanzig.[69] The multidimensional integration is reduced here to evaluation of single integrals.

The coefficients B_2, B_3, B_4, and B_5 were introduced in Ref. 69, and B_6 and B_7 in Ref. 70:

$$\begin{gathered}B_2 = 4\sigma^3,\\ B_3/B_2^2 = 9/16 = 0.5625,\\ B_4/B_2^3 = 17/96 = 0.1771,\\ B_5/B_2^4 = 455/36\,864 = 0.0123,\\ B_6/B_2^5 = -2039/110\,592 = -0.0184,\\ B_7/B_2^6 = -169\,149\,119/15\,925\,248\,000 = -0.0106,\end{gathered} \tag{3.29}$$

where σ is the cube edge length.

Note that B_2 is four times larger than the molecule's own volume both in the hard sphere and in the parallel rigid cube system.

In the one- and two-dimensional cases,[69,70] all the virial coefficients up to B_7 inclusive are positive. The foregoing approach was generalized in Ref. 71 to include the four- and five-dimensional cases.

We have considered for the molecule interaction a model that takes into account only their mutual impenetrability: $\Phi = \infty$ at $r < \sigma$. This approximation of the repulsive part of the potential is undoubtedly quite crude. A more realistic potential is

$$\begin{gathered}\Phi(r) = \varepsilon(\sigma/r)^n = \gamma/r^n,\\ n > 3,\end{gathered} \tag{3.30}$$

where $\gamma = \varepsilon\sigma^n$ is a positive constant.

The second virial coefficient for such an interaction can be calculated directly:

$$\begin{aligned}B_2 &= \frac{1}{2}\int f(|q|)dq = \frac{1}{2}\int_0^{2\pi} d\varphi \int_0^{\pi} d\vartheta \int_0^{\infty} dr\, r^2\left\{\exp\left[-\frac{\varepsilon}{\theta}\left(\frac{r}{\sigma}\right)^{-n}\right]-1\right\}\\ &= \frac{2\pi}{3}\int_0^{\infty} dx\left[\exp\left(-\frac{\varepsilon}{\theta}x^{-n/3}\sigma^n\right)-1\right]\\ &= \frac{2\pi}{3}\left(\frac{\gamma}{\theta}\right)^{3/n}\int_0^{\infty} dx\, e^{-x}x^{-3/n}\\ &= \frac{2\pi}{3}\left(\frac{\gamma}{\theta}\right)^{3/n}\Gamma(1-3/n),\end{aligned} \tag{3.31}$$

where Γ is the gamma function.[72] For $3/n < 1$ we have

$$\Gamma(1-3/n) = \exp\left[\frac{3C}{n} + \frac{9S_2}{2n^2} + \cdots\right],$$

where

$$C = \lim_{p\to\infty}\left[-\ln p + 1 + \frac{1}{2} + \frac{1}{3} + \cdots + \frac{1}{p}\right] = 0.577\,2157$$

is the Euler constant and

$$S_p = 1 + \frac{1}{2^p} + \frac{1}{3^p} + \cdots = \zeta(p)\,.$$

$\Gamma(1-3/n)$ is of the order of unity for $n \geqslant 10$,[185] so that at such a value of n it can be assumed that Eq. (3.31) determines the second virial coefficient a system of hard spheres whose diameter $(\gamma/\theta)^{1/n}$ decreases with rise of temperature.

In the general case, B_i can be represented in the form of the hard-sphere value of B_i, multiplied by a factor

$$\nu_i(n)(1/\theta^*)^{3(i-1)/n}, \quad \theta^* = \theta/\varepsilon,$$

where ν_i is independent of temperature.[73]

A potential in the form (3.30) describes well gases at high temperatures.

The repulsive part of the potential is frequently chosen also in the form

$$\Phi(r) = \varepsilon \exp(-r/\sigma)\,, \tag{3.32}$$

where ε and σ are positive constants.

In this case we obtain in lieu of an analytic expression, even for the second virial coefficient, the asymptotic series[74–77]

$$B_2(\theta)/B_{20} = [\ln(\alpha/\theta^*)]^3 + c_1 \ln(\alpha/\theta^*) + c_2 + 6(\theta^*)^3 \exp(-1/\theta^*) + \cdots, \tag{3.33}$$

where

$$B_{20} = \frac{2\pi}{3}\sigma_3, \quad \theta^* = \theta/\varepsilon,$$
$$\alpha = 1.781\,072\cdots,$$
$$c_1 = 4.934\,804\cdots, \tag{3.34}$$
$$c_2 = 2.404\,111\cdots.$$

At small θ^* the series (3.33) converges well, but the model with the potential (3.32) describes inadequately the real system and the attraction forces must be taken into account. It can be shown that the series for $B_3(\theta)$ and $B_4(\theta)$ are of the form[78]

$$B_3(\theta)/B_{20}{}^2 = 0.625[\ln(\alpha/\theta^*)]^6 + \cdots,$$
$$B_4(\theta)/B_{20}{}^3 = 0.287[\ln(\alpha/\theta^*)]^9 + \cdots. \tag{3.35}$$

We see thus that if only the first term is retained in the series for $B_i(\theta)$ the system behaves as one of hard sphere having an effective diameter $\sigma \ln(\alpha/\theta^*)$.

Other systems, in which the repulsive part of the potential is even more complicated, are considered occasionally. The most useful, however, are potentials of exponential form and an inverse-power potential. We proceed now to consider systems having, besides the repulsive part, also a potential of attraction forces between the particles.

15. Particle system with Lennard-Jones interaction potential

One of the potentials most widely used in investigations of statistical systems is the Lennard-Jones potential

$$\Phi(r) = 4\varepsilon[(\sigma/r)^{12} - (\sigma/r)^6]\,, \tag{3.36}$$

where ε and σ are constants. It was introduced in more general form by Mie[79]:

$$\Phi(r) = \frac{Cmn}{n-m}\left[\frac{1}{n}\left(\frac{r_m}{r}\right)^n - \frac{1}{m}\left(\frac{r_m}{r}\right)^m\right], \tag{3.37}$$

where $n > m$ and r_m is determined from the condition

$$\frac{d\Phi}{dr} = 0\,.$$

It is easy to recast the potential (3.37) in the form (3.36):

$$\Phi(r) = \varepsilon C_1\left[\left(\frac{\sigma}{r}\right)^n - \left(\frac{\sigma}{r}\right)^m\right], \tag{3.38}$$

where

$$C_1 = [n/(n-m)](n/m)^{m/(n-m)}\,, \tag{3.39}$$

and σ is determined from the condition

$$\Phi(\sigma) = 0.$$

Putting here $n = 12$ and $m = 6$, we get Eq. (3.36).

Equation (3.38) was used in Refs. 80 and 81 to determine the static energy of a crystal lattice; minimization of this energy yielded the lattice constant. The bcc, fcc, and pc structures were considered and the fcc was found to be the most stable (the hcp structure was not considered). Note that initially (prior to quantum mechanics), the form of the interatomic interaction potential was introduced empirically. The choice $m = 6$ was rigorously corroborated only later. The reason for choosing $n = 12$ is that it leads to better agreement with experiment for most substances, although in individual cases another value of n may be more suitable.

We proceed now to calculate the second virial coefficient with the potential (3.36):

$$B_2 = \frac{1}{2}\int f(|q|)dq$$
$$= 2\pi\int_0^\infty dr\, r^2\left(\exp\left\{-\frac{4}{\theta^*}\left[\left(\frac{\sigma}{r}\right)^{12} - \left(\frac{\sigma}{r}\right)^6\right]\right\} - 1\right)$$
$$= B_{20}\sum_{i=1}^{\infty}\overline{B}_{2i}\theta^{*\,-(2i-1)/4}, \tag{3.40}$$

where

$$\overline{B}_{2i} = -\frac{2^{i+1/2}}{4i!}\Gamma\left(\frac{2i-1}{4}\right),$$
$$B_{20} = \tfrac{2}{3}\pi\sigma^3, \quad \theta^* = \theta/\varepsilon. \tag{3.41}$$

This result was obtained by expanding

$$\exp\left[\frac{4}{\theta^*}\left(\frac{\sigma}{r}\right)^6\right]$$

in an infinite series and integrating it analytically. Substituting in Eq. (3.41) the tabulated values for the gamma function, we obtain for the first few terms of the expansion the expression

$$B_2 = 2B_{20}\{1.226 - 1.812/\theta^{*1/2} - 0.612/\theta^* - 0.302/\theta^{*3/2} - 0.154/\theta^{*2}$$
$$- 0.077/\theta^{*5/2} - \cdots\}\sqrt{2/\theta^{*\,1/4}}. \tag{3.42}$$

An expression for the second virial coefficient can be obtained also with the more general potential (3.37):

$$B_2 = B_{20}\left(\frac{r_m}{\sigma}\right)^3\left(\frac{m}{n}y\right)^{3/n-m}\left[\Gamma\left(\frac{n-3}{n}\right) - \frac{3}{n}\sum_{i=1}^{\infty}\Gamma\left(\frac{im-3}{n}\right)\frac{y^i}{i!}\right], \tag{3.43}$$

where

$$y = \frac{C}{(1-m/n)\theta}\left[\frac{\theta}{C(n/m-1)}\right]^{m/n}.$$

These expressions permit a complete investigation of the behavior of the second virial coefficient as a function of temperature. The convergence of the series in Eq. (3.43) depends substantially on the ratio θ/C. Figure 13 shows $B_2^* = B_2/B_{20}$ as a function of the temperature T for the 6-12 potential. B^* is negative at low temperatures, vanishes at $\theta = \theta_B^* = 3.42$ (the reduced Boyle temperature), and has a maximum at $\theta^* = 25$. This behavior can be understood by considering the form of the potential function. At low temperatures the average kinetic energy of the molecules is of the same order as the depth of the potential well, and they stay most of the time in the region of the minimum potential energy. This decreases the pressure, meaning also the second virial coefficient. At high temperatures, the molecules "do not notice" the attractive part of the potential, and the principal role is played by its repulsive part. This enhanced role of the mutual repulsion increases the pres-

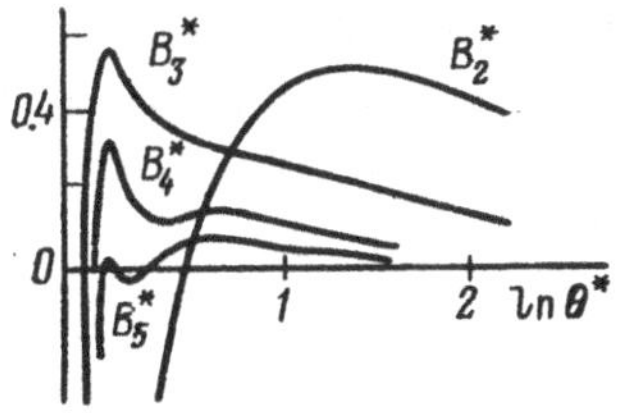

Figure 13.

sure, meaning also the second virial coefficient. At very high temperatures, the effective volume of the molecules increases, and B_2^* begins to decrease after reaching a maximum.

We have noted above that eight virial coefficients are known for the hard-sphere system. It is very difficult to calculate the higher-order virial coefficients for a potential of the Lennard-Jones type. Furthermore, although we can write down an analytic expression for the third virial coefficient, it is too unwieldy to be useful. We list therefore in Table I their tabulated values $B_3^* = B_3/B_{20}^2$, $B_4^* = B_4/B_{20}^3$, $B_5^* = B_5/B_{20}^4$ (Ref. 82), and also of $B_2^* = B_2/B_{20}$. Plots of these functions are shown in Fig. 13. The coefficient B_3^* has, just as B_2^*, one maximum and tends to $-\infty$ for small θ^*. The coefficient B_3^* has two maxima and one minimum. The same is true also of B_5^*. But these coefficients, too, become negative for small θ^* and increase in absolute value. The virial expansion is therefore not effective at low temperatures.

We obtain now the coefficients b_i from Eqs. (2.17). The coefficients b_2 and B_2 are equal, while the values of $b_3^* = b_3/B_{20}^2$, $b_4^* = b_4/B_{20}^3$, and $b_5^* = b_5/B_{20}^4$ are listed in Table II. The coefficient b_3^* is much smaller than B_3^* in the region of large θ_3^* (see Fig. 14), while b_4^* and b_5^* vary similarly with B_4^* and B_5^*, respectively (see Figs. 15 and 16). The excess of b_2^* over b_4^* and b_5^* is particularly appreciable in the region where b_2^* reaches its maximum. Thus, at high temperatures (compared with ε/k) the equation of state (3.6) is determined by a relation that contains the series for b and its derivative with respect to v, a series that converges well in a wide range of densities $1/v$.

At low temperatures, the virial coefficients can yield a series that converges well enough only at low densities, when the increase of the coefficients B_i is offset by the increase of v_i.

As a specific example of the use of the equation of state (3.6) and of its comparison with the virial equation of state (2.10), we consider argon at 25 °C and at various pressures. This is cited in Ref. 24 as an example of the poor convergence of the usual virial expansion at high pressures. Table III lists the contributions made to pv/θ by the first few terms of the series $z_0 = pv/\theta$ by the virial expansion (2.10) and by series z from our Eq. (3.6). It can be seen from Table III that the z_0 and z series converge equally well at low density, but z converges much better than z_0 with increase of pressure. A similar situation is observed also for other gases. Thus, in a number of cases z is a convergent series at high density, in contrast to z_0.

Table I. Tabulated values of $B_3^* = B_3/B_{20}^2$, $B_4^* = B_4/B_{20}^3$, $B_5^* = B_5/B_{20}^4$, and $B_2^* = B_2/B_{20}$.

θ^*	B_2^*	B_3^*	B_4^*	B_5^*
0.625	−5.757 8	−8.235 5	−121.08	−213
0.750	−4.175 9	−1.791 5	−18.840	−185.90
0.800	−3.735 2	−0.849 1	−9.3880	−77.910
0.875	−3.199 0	0.078 7	−3.2430	−22.460
0.950	−2.774 9	0.205 2	−9.4400	−6.6080
1.000	−2.538 1	0.429 7	−0.2769	−2.8600
1.050	−2.330 2	0.510 8	0.0691	−1.1650
1.100	−2.146 4	0.557 7	0.2379	−0.4162
1.125	−2.062 2	0.572 2	0.2838	−0.2231
1.150	−1.982 6	0.582 3	0.3126	−0.1011
1.200	−1.835 9	0.592 4	0.3359	0.0150
1.225	−1.768 2	0.593 8	0.3365	0.0360
1.250	−1.703 8	0.593 3	0.3324	0.0434
1.3	−1.524 1	0.588 2	0.3157	0.0361
1.4	−1.375 8	0.568 3	0.2695	−0.0022
1.5	−1.200 9	0.543 40	0.2250	−0.0303
1.6	−1.051 9	0.518 0	0.1894	−0.0413
1.7	−0.923 6	0.494 2	0.1630	−0.0403
1.8	−0.812 0	0.472 7	0.1442	−0.0326
1.9	−0.714 10	0.453 7	0.1313	−0.0217
2.0	−0.627 6	0.437 1	0.1218	−0.0100
2.1	−0.550 6	0.422 6	0.1175	0.0015
2.2	−0.481 7	0.400 9	0.1145	0.0120
2.3	−0.419 7	0.398 9	0.1131	0.0214
2.4	−0.363 6	0.389 3	0.1128	0.0295
2.5	−0.312 6	0.381 0	0.1131	0.0365
3.0	−0.115 2	0.352 3	0.1198	0.0578
4.0	0.115 4	0.326 6	0.1311	0.0668
5.0	0.243 3	0.315 1	0.1341	0.0620
7.5	0.396 4	0.298 8	0.1268	0.049
10	0.490 9	0.286 1	0.1156	0.039
15	0.510 6	0.264 2	0.0960	0.0275
20	0.525 4	0.246 4	0.0832	0.0209

We have considered particle systems with a Lennard-Jones interaction potential. This potential is most frequently used in theoretical studies of inert gases (He, Ne, Ar, Kr, Xe). The quantum corrections can be neglected at sufficiently high temperature, but they must be taken into account at low ones. The most sensitive to allowance for quantum correction is He.

Besides the Lennard-Jones potential, a widely used two-particle interaction potential is that of Buckingham:

$$\Phi(r) = \begin{cases} \dfrac{\varepsilon}{1-(6/\alpha)}\left\{\dfrac{6}{\alpha}\exp\left(\alpha\left[1-\dfrac{r}{r_m}\right]\right)-\left(\dfrac{r_m}{r}\right)^6\right\} & (r \geqslant r_{\max}) \\ \infty & (r < r_{\max}) \end{cases}, \quad (3.44)$$

where $r_{\max}$ is the value of r at which $\Phi(r)$ has a fictitious maximum. In contrast to the Lennard-Jones potential, the repulsive part of the potential (3.44)

Table II. Values of $b_3^* = b_3/B_{20}^2$, $b_4^* = b_4/B_{20}^3$, and $b_5^* = b_5/B_{20}^4$.

θ^*	b_3^*	b_4^*	b_5^*
0.625	20.694	−95.883	−888.96
0.750	−9.614 0	−22.157	−93.581
0.800	−7.400 4	−13.401	−42.328
0.875	−5.077 5	−6.411 3	−13.236
0.950	−3.747 4	−6.423 1	−12.464
1.000	−3.006 1	−2.272	−2.0094
1.050	−2.459 5	−1.490 6	−8.0526
1.100	−2.024 7	−0.070 2	−0.2148
1.125	−1.840 2	−0.777 1	−0.0468
1.150	−1.674 2	−0.617 4	0.0674
1.200	−1.389 1	−0.375 6	0.1913
1.225	−1.266 94	−0.284 2	0.2201
1.250	−1.154 8	−0.208 1	0.2351
1.3	−0.867 34	−0.036 5	0.2429
1.4	−0.662 3	0.046 7	0.2023
1.5	−0.449 4	0.112 6	0.1548
1.6	−0.294 3	0.141 6	0.1148
1.7	−0.179 4	0.151 2	0.0847
1.8	−0.093 3	0.150 8	0.0628
1.9	−0.028 1	0.145 1	0.0471
2.0	0.021 6	0.136 6	0.0357
2.1	0.059 7	0.127 7	0.0278
2.2	0.084 4	0.116 1	0.0223
2.3	0.111 4	0.100 1	0.0176
2.4	0.128 6	0.100 4	0.0142
2.5	0.141 6	0.092 2	0.0117
3.0	0.169 5	0.059 97	0.0047
4.0	0.156 7	0.025 1	−0.0006
5.0	0.128 0	0.008 8	−0.0030
7.5	0.070 8	−0.006 6	−0.0050
10	0.036 8	−0.011 1	−0.0049
15	0.001 7	−0.013 0	−0.0040
20	0.014 8	−0.012 8	−0.0031

is exponential. The potential (3.44) complicates substantially the calculation of the virial coefficients. The constants ε and r_m denote, respectively, the depth of the potential well and the point where the minimum of the potential is reached, while α is a measure of the repulsion slope: the larger α for given ε and r_m, the steeper the repulsive part. Thus, the Buckingham potential contains three parameters. This potential, just as that of Lennard-Jones, has an undesirable property: with change of α (of n in the case of the Lennard-Jones potential), the attractive part of the potential also changes, so that the repulsive and attractive parts of the potential cannot be varied independently.

The virial coefficients known for the Buckingham potential are the second B_2 (Ref. 22) and the third B_3 (Ref. 24). The expression for B_2 can be expressed in the form

$$B_2 = b_m B_2^*(\alpha,\theta^*)\,, \tag{3.45}$$

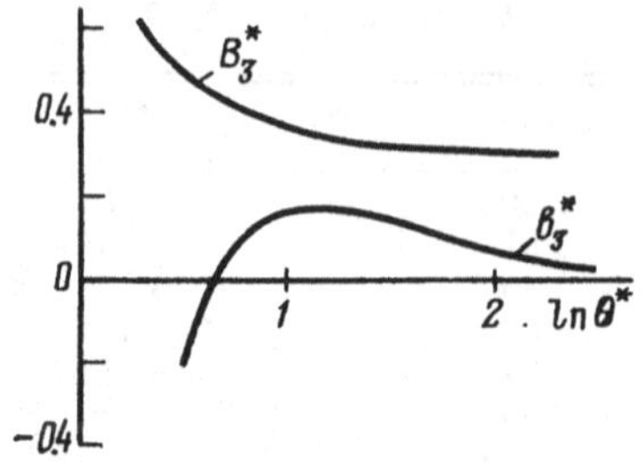

Figure 14.

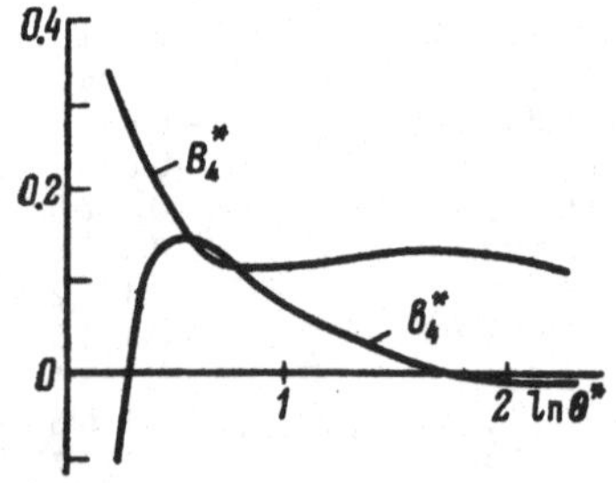

Figure 15.

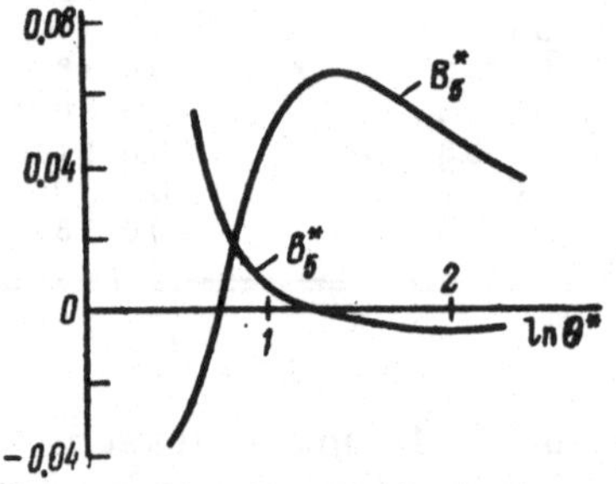

Figure 16.

where $b_m = \frac{2}{3}\pi r_m^3$, $\theta^* = \theta/\varepsilon$. The tabulated values of $B^*(\alpha,\theta^*)$ are given in Ref. 22 (for α equal to 12.0, 12.5, 13.0, 13.5, 14.0, 14.5, and 15.0 and for θ^* from 0.40 to 400.0). The same reference contains expressions for the parameters ε, r_m, and α for various gases (H_2, He, Ne, Ar, Kr, Xe, CH_4, N_2, CO), determined using the viscosity coefficients, the second virial coefficient, and the properties of the crystalline state. The numerical values of $B_2^*(\alpha,\theta^*)$ for α in the range from 16 to 300 are given in Ref. 83. Reference 24 contains a table of B_3 for $\alpha = 12$ and θ^* from 0.8333 to 20 000, and also a discussion of the published sources of the values of B_3 for α from 12 to 300.

Table III. Contributions made to pv/θ by the first few terms of the series $z_0 = pv/\theta$ by the virial expansion (2.10) and by series z from Eq. (3.6).

P_{atm}	pv/θ	
	z_0	z
1	$1 - 0.000\,64 + \cdots$ $+ 0.000\,00 + \cdots$	$\dfrac{1 - 0.000\,00 + \cdots}{1 + 0.000\,64 + 0.000\,00 + \cdots}$
10	$1 - 0.006\,48$ $+ 0.000\,20 + \cdots$	$\dfrac{1 + 0.000\,08 + \cdots}{1 + 0.006\,48 - 0.000\,08 + \cdots}$
100	$1 - 0.067\,54$ $+ 0.021\,27\ldots$	$\dfrac{1 + 0.008\,35\ldots}{1 + 0.067\,54 - 0.008\,354 + \cdots}$
1000	$1 - 0.384\,04$ $+ 0.637\,88 + \cdots$	$\dfrac{1 + 0.270\,20 + \cdots}{1 + 0.384\,04 - 0.270\,20 + \cdots}$

Besides the Buckingham potential, use is made also of the Buckingham–Korner and Buckingham–Korner–Konovalov potentials,[24] which contain exponential and power-law functions. Also used are the 12-6-4 potential, which contains a polynomial of these powers of r_m/r, a Morse potential (in the form of a combination of two exponential), the Rydberg potential, which is a product of a linear function and an exponential function of r, and others. We shall not dwell here in detail on the calculation of the virial coefficients for these potentials. We note only that the second virial coefficient does not depend strongly on the form of the interaction potential (if the latter is realistic).

Advances in computer science have permitted the use of potentials of complicated multiparameter power-law plus exponential form.

The two-particle potentials considered are spherically symmetric. If the potential is not spherically symmetric, the calculation of the virial coefficients for systems with such an intermolecular interaction is significantly more complicated. An interaction involving such a potential, however, is typical of a large class of molecular systems. Therefore, virial coefficients were investigated in many cases also for such interactions between particles.

Virial coefficients play an important role. Their knowledge permits all the thermodynamic properties of a gas to be determined (if the series converge rapidly enough). Particularly significant is the role of the second virial coefficient, which is used to determine the interaction-potential parameters of various substances.

16. Determination of intermolecular-interaction potential parameters from the second virial coefficient

We have considered above the calculation of virial coefficients on the basis of known interaction potentials. As a rule, these potentials depend on several parameters which are not known beforehand; therefore, the virial coefficients also depend on these parameters. The virial coefficients can be determined from experimental equations of state; this allows us to choose the parameters that provide a best fit (in one gauge or another) of the theoretical and experimental virial coefficients. The most reliably determined is the second virial coefficient, which is customarily used to find the interaction-potential parameters.

We consider first the procedure of determining the virial coefficients from experimental data.

The virial equation of state, written in the general case as

$$p = \frac{\theta}{v} + \theta \sum_{i=2}^{\infty} B_i / v^i , \tag{3.46}$$

is chosen for comparison with experiment in a form that takes into account the finite number m of terms

$$p = \frac{\theta}{v} + \theta\left(\frac{B_2}{v^2} + \frac{B_3}{v^3} + \cdots + \frac{B_m}{v^m}\right). \tag{3.47}$$

The experimental data are used next to determine the coefficients B_i ($i = \overline{2,m}$). Of course, in this approach these coefficients differ from those obtained as $m \to \infty$. To exclude a dependence of B_i on m it is necessary to use the experimental data for the lowest possible densities, and to determine the virial coefficients with the aid of the limiting states[24]:

$$\begin{aligned} B_2 &= \lim_{v\to\infty} [pv/\theta - 1]v , \\ B_3 &= \lim_{v\to\infty} [pv/\theta - 1 - B_2/v]v^2 , \\ B_4 &= \lim_{v\to\infty} [pv/\theta - 1 - B_2/v - B_3/v^2]v^3 , \end{aligned} \tag{3.48}$$

or in a somewhat different form:

$$\begin{aligned} \theta B_2 &= \lim_{v\to\infty}\left[\frac{\partial(pv)}{\partial\rho}\right], \\ \theta B_3 &= \frac{1}{2}\lim_{v\to\infty}\left[\frac{\partial^2(pv)}{\partial^2\rho^2}\right], \\ \theta B_4 &= \frac{1}{6}\lim_{v\to\infty}\left[\frac{\partial^3(pv)}{\partial\rho^3}\right], \end{aligned} \tag{3.49}$$

where $\rho = 1/v$.

Only B_2 can be obtained from the experimental data with good accuracy. The virial coefficients B_i $(i\geqslant 3)$ can be determined with much lower accuracy.

Complete data on the equations of state of inert gases in virial form are contained in Ref. 84.

The most effective is thus the determination of the intermolecular-potential constants using the most accurately determined second virial coefficient.

The parameters ε and σ, for example, are frequently determined by the following procedure.[22] The quantity

$$t=\left[\frac{B_2(\theta_2)}{B_2(\theta_1)}\right]_{\rm exp} \tag{3.50}$$

or

$$t=\left[\frac{B_2^*(\theta_2/\varepsilon)}{B_2^*(\theta_1/\varepsilon)}\right] \tag{3.51}$$

is introduced and used to determine ε. The zeroth approximation of ε can be the value determined from the boiling, melting, or critical temperature.[22] We next determine σ from the relation

$$1.2615\sigma^3=\left[\frac{B_2(\theta)}{B_2^*(\theta)}\right]=B_{20}\,, \tag{3.52}$$

where B_{20} is measured in cm^3/mol and σ in angstroms.

The Lennard-Jones potential parameters calculated by this method for the inert gases N_2, O_2, CH_4, and CO_2 are given in Ref. 22.

One more feature of the determination of the interaction-potential parameters using the second virial coefficient must be emphasized. Beside the fact that B_2 is determined from the experimental data with less than absolute accuracy, the very empirical form of the potential (even a logarithmic one) does not guarantee an exact description of the available experimental data. The calculated interaction-potential parameters must not be treated as absolute.

A good survey of the data on the intermolecular interaction potential of inert gases is given in Ref. 85.

17. Theory of the liquid state. Critical point

Attempts to describe the liquid state, and also of the liquid–vapor phase transition and of the critical point, were undertaken long ago. They are due in many respects to the fact that the virial expansion with even only the third virial coefficient yields a typical van der Waals loop that is capable, supplementing the Maxwell rule, of describing at least qualitatively the phase transition and the critical point. Moreover, when the succeeding virial coefficients[199] are taken into account, the critical parameters converge to certain stable values.[82] It is seen from the foregoing, however, that there is no assurance that the critical parameters obtained in this way are correct. First, the region of the critical point is strongly singular,[86] therefore, the critical parameters cannot be obtained by solving the set of equations[17]

$$p = p(\theta,v), \quad \left(\frac{\partial p}{\partial v}\right)_{\theta_c} = 0, \quad \left(\frac{\partial^2 p}{\partial v^2}\right)_{\theta_c} = 0, \tag{3.53}$$

and second, the virial expansion

$$\frac{pv}{\theta} = 1 + \sum_{i=2}^{\infty} B_i / v^{i-1}$$

is not valid in this region, since it holds true, strictly speaking, at low density $1/v$. The virial coefficients (see Fig. 13) are nonmonotonic near the critical point (except B_2) and have large negative values at low θ^*. Thus, lowering the temperature increases strongly the absolute values of the virial coefficients, thus decreasing greatly the convergence radius of the virial series.

Analysis of the equation of state (3.6) shows that when b_2 and b_3 are taken into account the phase diagram does not have a critical point as a solution of Eqs. (3.57). This indicates that the van der Waals loop obtained on the basis of the usual virial expansion (3.58) is not well founded. Analytic calculations with allowance for the succeeding virial coefficients become more cumbersome, and a numerical analysis becomes much more effective. It can be concluded in the general case that since the virial expansion is a polynomial in $x = v_0/v$, it cannot describe the region of the singular point—these can be deduced only from the presence of unphysical regions (or of other singularities) in the equation of state (for regions where first-order phase transitions take place in the free energy of the system, and there are no derivatives of second and higher order).

If the free energy of the system has continuous derivatives of only first order, it can be represented as an expansion in Chebyshev polynomials of the first kind,[87] but in this case we must have information on the entire function, or else be able to reconstruct the entire function from its known parts with some degree of accuracy.

In the case of dense gases the free energy can have discontinuities not of the second but of higher derivatives. In this case a virial power-law expansion cannot be used.

Thus, an interpolation phase diagram of a statistical system in which a first-order phase transition takes place can be effectively constructed on the basis of an expansion in Chebyshev polynomials of the first kind.

Chebyshev polynomials are orthonormal on the segment $[-1, 1]$. In our case, however, $0 \leqslant x \leqslant 1$. We use, therefore, the following theorem (Ref. 87, p. 40): if the polynomials $\{P_n(x)\}$ are orthonormal with weight $h(x)$ on a segment $[a,b]$, the polynomials

$$Q_n(x) = \sqrt{p} P_n(px + q), \quad n = 1,2,...$$

are orthonormal at $p > 0$ with weight $h(px + q)$ on the segment $[A,B]$, which goes over into the segment $[a,b]$ as a result of the linear transformation $px + q$. We obtain hence

$$Q_0 = \sqrt{\frac{2}{\pi}},$$

$$Q_1 = \frac{2}{\sqrt{\pi}}(2x-1), \quad Q_2 = \frac{2}{\sqrt{\pi}}(8x^2-8x+1),$$

$$Q_3 = \frac{2}{\sqrt{\pi}}(32x^3-48x^2+18x-1), \tag{3.54}$$

$$Q_4 = \frac{2}{\sqrt{\pi}}(128x^4-256x^3+160x^2-32x-1),$$

etc. The sought-for quantities must therefore be expanded in terms of the polynomials (3.54).

The formulation of the theory of fluids is a complicated problem which is frequently solved by using empirical and semiempirical equations. The best known is the Tait equation,[24] whose differential form is

$$-\frac{dV}{dp} = \frac{K}{L+p}, \tag{3.55}$$

where K is a constant and L is a function of the temperature. Integrating Eq. (3.55) and denoting by p_0 some standard pressure value corresponding to the volume V_0, we obtain

$$\ln\left(\frac{p+L(\theta)}{p_0+L(\theta)}\right) = n\left(\frac{V_0-V}{V_0}\right). \tag{3.56}$$

This equation describes well the experimental data for liquids in a wide range of temperatures and pressures (up to 1000 atm). To be able to use Eq. (3.56) also at higher pressures (this equation is incorrect for high pressure, for as $p \to \infty$ the left-hand side tends to ∞ and the right-hand side cannot exceed n) we modify it. We propose to use the equation[24]

$$\ln\left(\frac{p+A(S)}{A(S)}\right) = n\ln\left(\frac{V(0,S)}{V(p,S)}\right), \tag{3.57}$$

where the independent variable is chosen to be the entropy S. For water, for example, this equation leads to good agreement with experiment up to 25 000 atm.[88]

We transform Eq. (3.57) into

$$[p+A(S)]\,V^n = F(S), \tag{3.58}$$

where $F(S) = A(S)[V(0,S)]^n$. Then, integrating the relation $(\partial U/\partial V)_S = -p$, we obtain for the internal energy

$$U(p,S) - U(0,S) = \left(\frac{pV}{n-1}\right) - \left(\frac{n}{n-1}\right)A(S)[V(0,S) - V]\,. \qquad (3.59)$$

In the ideal-gas approximation $A(S) = 0$ and $pV = RT$ (per mole), hence $n = C_p/C_V$. In the general case this is not so.

Equation (3.58) is sometimes written in an approximate and not sufficiently convenient form:

$$pV^3 = F(S)\,, \qquad (3.60)$$

i.e., it is assumed in Eq. (3.58) that $n = 3$, and the term containing $A(S)$ is neglected.

As noted above, at low temperatures the absolute values of the virial coefficients increase strongly, and when the temperature is raised they decrease and at sufficiently high temperatures the principal role is assumed by the repulsive part of the distribution. This is confirmed also by an investigation of the real distribution function: Its behavior at high temperatures for systems with a Lennard-Jones potential, for example, is similar to the behavior of the radial function for a system of hard spheres with an appropriately chosen diameter. This fact was indeed used to develop one of the variants of the theory of fluids.[89–91] To describe the liquid, one defines the initial system as a system of hard spheres with a diameter that depends on density and temperature. This is how the repulsion is approximated. The thermodynamic quantities are next expanded in a perturbation-theory series in which the attractive part of the potential is treated as a perturbation. This approach results in a good description of the experimental data. Earlier theory of liquids, in which the zeroth approximation was either the cell or the hole model, described the experimental data poorly, since the degree of ordering in them was too high. Thus, one of the criteria of the applicability of the equation of state to a description of a homogeneous state is the requirement that it have an asymptotic behavior at low densities. The cell model yields for the virial coefficient a zero value, while the hole model yields values that agree satisfactorily with experiment. The reason for the latter is that in the hole model the liquid structure is regarded as containing many "holes," about 0.5% in a normal liquid and about 50% near the critical point. On the other hand, even in this model it is impossible to describe the liquid state satisfactorily, for here, too, the degree of ordering is too high.[92] Reference 24 contains, besides a detailed discussion of these models, a comparison of the calculated critical points, the boiling point, and the evaporation entropy at the boiling point with experiment.

At the same time, the hard-sphere model is a good zeroth approximation for the description of the liquid state, since it is the limiting state of a liquid at high densities. A great advantage of this model is that it has been thoroughly investigated. Its approach is the basis of a most complete study of a liquid with a Lennard-Jones interaction potential. Good agreement with experiment is obtained up to the triple point.

18. Thermodynamic functions in the form of a cluster expansion

In Chap. 2 we obtained two forms of the cluster expansion (in power of the density) for the free energy, viz., the virial expansion (2.11):

$$F = N\theta\,[\ln(\lambda^3/v) - 1] + N\theta \sum_{i=2}^{\infty} \frac{1}{i-1}\cdot\frac{B_i}{v^{i-1}} \tag{3.61}$$

and the $q(\theta,v)$ expansion (2.18):

$$F = N\theta(\ln\lambda^3 - 1) - N\theta \ln\left(v - \sum_{i=2}^{\infty} \frac{b_i}{v^{i-2}}\right). \tag{3.62}$$

We now obtain expressions for the other thermodynamic functions in cluster form. To obtain these expressions in virial form, we start from Eq. (3.61).

The virial equation of state is given by Eq. (2.10):

$$\frac{pv}{\theta} = 1 + \sum_{i=2}^{\infty} \frac{B_i}{v^{i-1}}. \tag{3.63}$$

The entropy, the internal energy, the heat capacity at constant volume, and the isothermal compression modulus are, respectively, equal to

$$S = -k\left(\frac{\partial F}{\partial\theta}\right)_V = Nk\ln\frac{(2\pi m\theta)^{3/2} v e^{5/2}}{(2\pi\hbar)^3} - Nk\sum_{i=2}^{\infty}\frac{1}{i-1}\,\frac{B_i + \theta B_i'}{v^{i-1}} \times \left(B_i' = \frac{dB_i}{d\theta}\right), \tag{3.64}$$

$$U = -\theta^2\frac{\partial}{\partial\theta}\left(\frac{F}{\theta}\right) = \frac{3}{2}N\theta - \theta^2\sum_{i=2}^{\infty}\frac{1}{i-1}\,\frac{B_i'}{v^{i-1}}, \tag{3.65}$$

$$C_V = -k\theta\left(\frac{\partial^2 F}{\partial\theta^2}\right)_V = \frac{3}{2}Nk - \sum_{i=2}^{\infty}\frac{k}{i-2}\,\frac{\theta^2 B_i'' + 2\theta B_i'}{v^{i-1}}, \tag{3.66}$$

$$\varepsilon_\theta = -V\left(\frac{\partial p}{\partial V}\right)_\theta = -\theta v\frac{\partial}{\partial v}\left\{\frac{\theta}{v} + \sum_{i=2}^{\infty}\frac{B_i}{v^i}\right\} = \frac{\theta}{v} + \sum_{i=2}^{\infty}\frac{iB_i}{v^i}. \tag{3.67}$$

The free energy (the Helmholtz energy) is the thermodynamic potential in the variables θ and V, and in the variables θ and p the thermodynamic potential is the Gibbs energy

$$G = F + pV.$$

To determine it, we invert Eq. (2.10) and represent it as a series in powers of the pressure. We get

$$\frac{pv}{\theta} = 1 + \sum_{i=2}^{\infty} B_i p^i, \tag{3.68}$$

where

$$\overline{B}_2 = B_2/\theta\,,$$
$$\overline{B}_3 = (B_3 - B_2{}^2)/\theta^2\,,$$
$$\overline{B}_4 = (B_4 - 3B_2B_3 + 2B_2{}^3)/\theta^3\,,$$
$$\overline{B}_5 = (B_5 - 4B_4B_2 - 2B_3{}^2 + 10B_3B_2{}^2 - 5B_2{}^4)/\theta^4\,, \tag{3.69}$$

etc.

Since

$$V = \left(\frac{\partial G}{\partial p}\right)_\theta\,,$$

...ith allowance for Eq. (3.68) yields an expression ...y function of θ. We obtain this function from the ...$p \to 0$. We ultimately have

$$\ldots = G_0 + N\sum_{i=2}^{\infty}\frac{\overline{\overline{B}}_i p^{i-1}}{i-1}\,. \tag{3.70}$$

...t pressure is therefore

$$\ldots\left(\frac{\ldots G}{\ldots\theta^2}\right)_p = \frac{5}{2}Nk - Nk\sum_{i=2}^{\infty}\frac{\theta\overline{\overline{B}}_i'' p^{i-1}}{i-1}\,. \tag{3.71}$$

... the thermodynamic function, starting from the ...y (3.62). The internal energy is in this case

$$\ldots N\theta - \frac{\theta^2\Sigma_{i=2}^{\infty} b_i'/v^{i-2}}{v - \Sigma_{i=2}^{\infty} b_i/v^{i-2}} \tag{3.72}$$

...tate is of the form

$$\ldots\frac{\ldots}{\theta} = \frac{1 - \Sigma_{i=2}^{\infty}(i-2)b_i/v^{i-1}}{1 - \Sigma_{i=2}^{\infty} b_i/v^{i-1}}\,. \tag{3.73}$$

It is easy to obtain from this the isothermal compression modulus

$$\varepsilon_\theta = -V\left(\frac{\partial p}{\partial V}\right)_\theta = -v\frac{\partial}{\partial v}\left(\frac{1 - \partial b/\partial v}{v - b}\right)$$
$$= v\frac{(v-b)(\partial^2 b/\partial v^2) + [1 - (\partial b/\partial v)]^2}{(v-b)^2}\,, \tag{3.74}$$

where b is defined by the relation

$$b = \sum_{i=2}^{\infty} b_i/v^{i-2}\,. \tag{3.75}$$

The heat capacity at constant volume is

$$C_V = \frac{3}{2}Nk - k\theta\frac{(v-b)[\theta(\partial^2 b/\partial\theta^2) + 2(\partial b/\partial\theta)] + \theta(\partial b/\partial\theta)^2}{(v-b)^2}\,. \tag{3.76}$$

Whereas in the case of the virial expansion the series for the pressure can always be inverted, in this expansion it is impossible, in general, to obtain an inversion in analogous form, since it leads to worse convergence of the resultant series.

To determine the heat capacity at constant pressure, we therefore use the equation

$$C_p = C_V + \theta\left(\frac{\partial p}{\partial \theta}\right)_V\left(\frac{\partial V}{\partial \theta}\right)_p \tag{3.77}$$

and, in view of the identity

$$\left(\frac{\partial V}{\partial \theta}\right)_p\left(\frac{\partial \theta}{\partial p}\right)_V\left(\frac{\partial p}{\partial V}\right)_\theta = -1,$$

we obtain the expression

$$C_p = C_V - \theta\left[\left(\frac{\partial p}{\partial \theta}\right)_V\right]^2 \Big/ \left(\frac{\partial p}{\partial V}\right)_\theta, \tag{3.78}$$

where $(\partial p/\partial V)_\theta$ is determined from Eq. (3.74), and

$$\left(\frac{\partial p}{\partial \theta}\right)_V = \frac{(\partial b/\partial\theta)(1 - \partial b/\partial v) - (\partial^2 b/\partial\theta\partial v)}{(v-b)^2}. \tag{3.79}$$

We have thus found expressions for the basic thermodynamic functions by using the virial expansion and an expansion in powers of the density of the function $q(\theta,v)$ [or using the $q(\theta,v)$ expansion].

The coefficients $\{B_i\}$ and $\{b_i\}$ obtained in the present chapter for a number of systems lead to suitably approximated thermodynamic functions of these systems.

Chapter 4
Integral equations in the theory of the liquid state

The development of a theory of the liquid state is at present one of the fundamental problems of statistical mechanics. Notwithstanding the abundant experimental material on the thermodynamics of various liquids, development of a theory of the liquid state encounters great difficulties. In contrast to gases that are not very dense and to ordered structures, it is difficult to find a parameter small enough for perturbation-theory expansion of the thermodynamic functions and for obtaining in zeroth order a fundamental approximation suitable for a qualitative description of the most important properties of the liquid.

Advances in the theory of the liquid state are made possible by the Bogolyubov distribution function method. In view of the difficulty of finding a small parameter, the chain of Bogolyubov equations is closed by approximation of an s-particle distribution function (usually $s = 3$–4) with a lower-order function deduced from physical considerations. Another approach is also used, based on introducing the concept of "direct" and "indirect" correlation. Both cases yield for the sought functions integral equations that describe the liquid state to some degree of accuracy. The present chapter is devoted to the integral equations in the theory of the liquid state.

19. Initial stage of investigation. Superposition approximation

The first attempt at an approximate description of a fluid date back to Newton, Euler, Bernoulli, Saint-Venant, and others, who regarded a fluid as a continuous medium. Laplace succeeded in obtaining expressions for the latent heat of evaporation and of the surface tension, in terms of integrals of molecular-interaction functions,[93] i.e., he attempted to relate the macroscopic and microscopic properties of a liquid.

The modern approach to the liquid state is associated with van der Waals' 1873 work.[94] Its phenomenological character notwithstanding, the van der Waals approach was highly successful.

The van der Waals equation

$$p = \frac{N\theta}{V - Nb} - \frac{aN^2}{V^2} \tag{4.1}$$

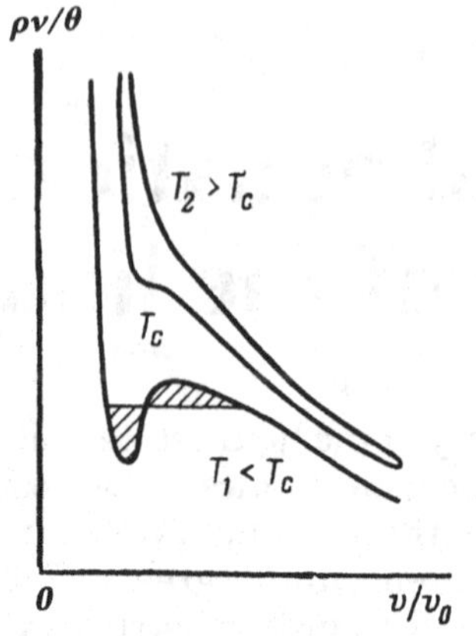

Figure 17.

introduces into the equation of state of an ideal liquid two corrections to account for the molecules' proper volume and for the attraction force between them. The two phenomenological constants a and b are determined by comparing Eq. (4.1) with experiment.

The van der Waals equation can be obtained from the Gibbs distribution only under certain additional assumptions.[95]

Consider the form of the isotherms of a van der Waals gas (see Fig. 17). At sufficiently low temperatures, the isotherm plotted on the V–p diagram has wavelike shape rather than the horizontal corresponding to the coexistence of two phases. A region with $-(\partial p/\partial V)<0$ exists, i.e., the system is unstable. This shortcoming of the van der Waals equation was eliminated by Maxwell, who proposed to replace the wavelike section of the isotherm by a horizontal straight line drawn on the basis of the equality of the chemical potentials of the liquid and gas in equilibrium:

$$\mu_l - \mu_\Gamma = \int_l^\Gamma d\mu = \int_l^\Gamma V\,dp = \int_{V_l}^{V_\Gamma} V\left(\frac{\partial p}{\partial V}\right)_\theta dV = 0\,, \tag{4.2}$$

i.e., equality of the shaded areas in Fig. 17.

The van der Waals equation (4.1) is actually an interpolation that reflects the main features of the behavior of the pressure at high and low densities. At intermediate densities, however, it leads to an unphysical region that is eliminated with the aid of Maxwell's rule. With rise of temperature, the line drawn in accordance with Maxwell's rule is shortened and vanishes at some temperature. In this case the crest and trough of the wavelike part of the isotherm merge into one point, indicating that the gas and liquid are no longer different. Thus, the van der Waals equation predicts the existence of a critical point with coordinates:

$$V_k = 3Nb, \quad p_k = \frac{a}{27b^2}, \quad \theta_k = \frac{8a}{27b}\,. \tag{4.3}$$

At $\theta > \theta_k$ the isotherms agree well with the experimental data at not too high densities.

Using the critical parameters (4.3), Eq. (4.1) can be written in the form

$$\pi = \frac{8\tau}{3\omega - 1} - \frac{3}{\omega^2}, \tag{4.4}$$

where

$$\pi = p/p_k, \quad \omega = V/V_k, \quad \tau = \theta/\theta_k . \tag{4.5}$$

The reduced van der Waals equation (4.4) shows that in terms of the reduced variables (4.5) all the substances are described by one and the same equation of state. This result is called the law of corresponding states. It can be written in a more general form[96–98]: the connection between the reduced pressure π, volume ω, and temperature τ for a substance of a given class is described by one universal function:

$$\varphi(\pi,\omega,\tau,\alpha_1,\alpha_2,...,\alpha_n) = 0 , \tag{4.6}$$

where α_i are nondimensional parameters.

For a number of substances only one parameter suffices, and can therefore be written in the form

$$\varphi_1(\pi,\omega,\tau,\alpha_1) = 0 . \tag{4.7}$$

Further studies of the equations of the corresponding states led to the development of the theory of dynamic scaling, which permits an effective analysis and interpretation of the experimental material and predicts the properties of new substances.

Following the publication of van der Waals' papers, many empirical equations of state were proposed for the description of the liquid state, with various degrees of correspondence to the experimental data. In the microscopic theory of the liquid state, prior to Kirkwood's work, most progress was made in the papers of Frenkel.[99] He proposed a mechanism of diffusion in a liquid, considered the statistical thermodynamics of the solid–liquid phase transition, investigated the temperature dependence of the surface tension of a liquid, and dealt with a large number of other problems. Frenkel formulated in his papers the basic features of the character of thermal motion.

Kirkwood was the first to make extensive use of the method of integral equations in the investigation of the liquid state, including the superposition approximation. Since his approach[100] is somewhat different from the one used to derive the integral equation (2.153), we shall repeat it here briefly.

Let the interaction of N particles have a potential:

$$U_N = \sum_{i<j} \Phi(|q_i - q_j|) . \tag{4.8}$$

We introduce a parameter ξ that varies in the range $0 \leqslant \xi \leqslant 1$, and define on the basis of Eq. (4.8) a new function:

$$U_N(\xi) = \xi \sum_{i=2}^{N} \Phi(|q_1 - q_i|) + \sum_{j=2}^{N} \sum_{k=j+1}^{N} \Phi(|q_j - q_k|) . \tag{4.9}$$

For $\xi = 1$ we have $U_N(1) = U_N$, corresponding to a system having a potential energy (4.8). At $\xi = 0$ the first particle does not interact with any other.

For a system with a function U_N, all the N-particle distribution functions will depend on the parameter ξ.

Differentiating the Gibbs distribution with respect to ξ and integrating the resultant equation with respect to $q_3,\dots,q_N$ we get

$$\theta \frac{\partial}{\partial \xi} \ln F_2(q_1,q_2,\xi) = -\Phi(|q_1 - q_2|)\, F_2(q_1,q_2,\xi)$$

$$-\frac{1}{v}\int \Phi(|q_1 - q_3|)\left[\frac{F_3(q_1,q_2,q_3,\xi)}{F_2(q_1,q_2,\xi)} - 1\right] F_2(q_1,q_2,\xi)dq_3\,. \tag{4.10}$$

We use for $F_3(q_1,q_2,q_3)$ the superposition approximation

$$F_3(q_1,q_2,q_3) = F_2(q_1,q_2)\, F_2(q_1,q_3)\, F_2(q_2,q_3) \tag{4.11}$$

and obtain in this approximation from Eq. (4.10), after integration, the Kirkwood integral equation

$$\theta \ln F_2(q_1,q_2,\xi) = -\xi\Phi(|q_1 - q_2|)$$

$$-\frac{1}{v}\int_0^{\xi}\int \Phi(|q_1 - q_3|)\, F_2(q_1,q_3,\xi)[\, F_2(q_2,q_3) - 1\,]dq_3\, d\xi\,. \tag{4.12}$$

Let us compare Eq. (4.12) with Eq. (2.153). To this end, we differentiate both halves of Eq. (4.12) with respect to q_1^{α}:

$$\theta \frac{\partial}{\partial q_1^{\alpha}} \ln F_2(q_1,q_2,\xi) = -\xi \frac{\partial \Phi(|q_1 - q_2|)}{\partial q_1^{\alpha}}$$

$$-\frac{1}{v}\int_0^{\xi} \frac{\partial}{\partial q_1^{\alpha}}[\Phi(|q_1 - q_3|)\, F_2(q_1,q_2,\xi)]\, F_2(q_2,q_3)dq_3\, d\xi\,. \tag{4.13}$$

But

$$\frac{\partial}{\partial q_1^{\alpha}}[\Phi(|q_1 - q_3|)\, F_2(q_1,q_3,\xi)]$$

$$= F_2(q_1,q_3,\xi)\frac{\partial \Phi(|q_1 - q_3|)}{\partial q_1^{\alpha}} + \Phi(|q_1 - q_3|)\frac{\partial F(q_1,q_3,\xi)}{\partial q_1^{\alpha}}\,. \tag{4.14}$$

Substituting Eq. (4.14) in Eq. (4.13) we get

$$\theta \frac{\partial}{\partial q_1^{\alpha}} \ln F_2(q_1,q_2,\xi) = -\xi \frac{\partial \Phi(|q_1 - q_2|)}{\partial q_1^{\alpha}}$$

$$-\frac{1}{v}\int_0^{\xi}\int \frac{\partial \Phi(|q_1 - q_3|)}{\partial q_1^{\alpha}}\, F_2(q_1,q_3,\xi)\, F_2(q_2,q_3)dq_3\, d\xi$$

$$-\frac{1}{v}\int_0^{\xi}\int \Phi(|q_1 - q_3|)\frac{\partial F_2(q_1,q_3,\xi)}{\partial q_1^{\alpha}}\, F_2(q_2,q_3)dq_3\, d\xi\,. \tag{4.15}$$

We express Eq. (2.148) in terms of the interaction parameter and carry out small transformations. Equation (2.153) differs from Eq. (2.148) only by a number of identity transformations:

$$\theta \frac{\partial F_2(q_1,q_2,\xi)}{\partial q_1^\alpha} = -\xi \frac{\partial \Phi(|q_1-q_2|)}{\partial q_1^\alpha}$$

$$-\frac{\xi}{v}\int \frac{\partial \Phi(|q_1-q_3|)}{\partial q_1^\alpha} F_2(q_1,q_3,\xi)\, F_2(q_2,q_3,\xi) dq_3 . \qquad (4.16)$$

Comparing Eq. (4.15) with Eq. (4.16) we see that they are not equal. If $F_2(q_1,q_2)$ does not depend on ξ, Eq. (4.15) differs from Eq. (4.16) by an additional term:

$$-\frac{1}{v}\xi \int \Phi(|q_1-q_3|) \frac{\partial F_2(q_1,q_3)}{\partial q_1^\alpha} F_2(q_2,q_3) dq_3 .$$

At the same time, as will be shown below, the solution of Eq. (4.12) does not differ greatly from the solution of Eq. (2.153).

The solution of the integral equations is highly complicated. We investigate first the solutions at low densities, and then seek a solution for the general case. Using the obtained two-particle distribution function, we can determine the thermal equation of state

$$\frac{pv}{\theta} = 1 - \frac{2\pi}{3\theta v}\int_0^\infty \frac{d\Phi(r)}{dr}\mu(r) r^3\, dr \qquad (4.17)$$

and an expression for the compressibility

$$-\frac{\theta}{v^2}\left(\frac{\partial v}{\partial p}\right)_\theta = 1 + \frac{4\pi}{v}\int_0^\infty [\mu(r)-1] r^2\, dr . \qquad (4.18)$$

Integrating Eq. (2.158), we get

$$4\pi \int_0^\infty [\mu(r)-1] r^2\, dr = 4\pi \int_0^\infty C(r) r^2\, dr$$

$$+ \left[\frac{4\pi}{v}\int_0^\infty C(r) r^2\, dr\right]\left\{4\pi \int_0^\infty [\mu(r)-1] r^2\, dr\right\} .$$

Hence

$$4\pi \int_0^\infty [\mu(r)-1] r^2\, dr = \frac{4\pi \int_0^\infty C(r) r^2\, dr}{1-(4\pi/v)\int_0^\infty C(r) r^2\, dr} .$$

Substituting this expression in Eq. (4.17), we get the equation

$$\frac{v^2}{\theta}\left(\frac{\partial p}{\partial v}\right)_\theta = 1 - \frac{4\pi}{v}\int_0^\infty C(r) r^2\, dr ,$$

which can be represented, using the Fourier transform of the function $C(r)$,

$$\tilde{C}_k = 4\pi \int_0^\infty \exp(-ikr) C(r) r^2\, dr$$

in the form

$$-\frac{v^2}{\theta}\left(\frac{\partial p}{\partial v}\right)_\theta = 1 - \frac{\tilde{C}_0}{v}.$$

This relation is used if the direct correlation function $C(r)$ is simpler than the two-particle distribution function.

We derive now relation (4.18). We determine first the fluctuation of the number of particles in the system. We assume to this end that the system considered, with volume V, is a subsystem of a large system with volume Ω and with M particles. We introduce a function $f_V(q)$ equal to unity when q is inside V and to zero when q is outside V.[2] The total number of molecules in the volume V will be

$$N_V = \sum_{1\leq i\leq M} f_V(q_i),$$

and we obtain the average number of particles from the equation

$$\bar{N}_V = \frac{M}{\Omega}\int_\Omega f_V(q)\,F_1(q)dq = \frac{M}{\Omega}\int_V F_1(q)dq.$$

To determine the mean squared deviation we write

$$\overline{(N_V - \bar{N}_V)^2} = \int_\Omega\cdots\int_\Omega D_M\left\{\sum_{1\leq i\leq M}[f_V(q_i) - \alpha]\right\}^2 dq_1\cdots dq_M,$$

where

$$\alpha = \frac{1}{\Omega}\int_V F_1(q)dq.$$

Therefore,

$$\overline{(N_V - \bar{N}_V)^2} = \int_\Omega\cdots\int_\Omega D_M \sum_{\substack{1\leq i\leq M\\ 1\leq j\leq M}} [f_V(q_i) - \alpha]\,[f_V(q_i) - \alpha]dq_1\cdots dq_M.$$

Hence

$$\overline{(N_V - \bar{N}_V)^2} = 2\int_\Omega\cdots\int_\Omega D_M \sum_{1\leq i<j\leq M}[f_V(q_i) - \alpha]\,[f_V(q_j) - \alpha]dq_1\cdots dq_M$$
$$+ \int_\Omega\cdots\int_\Omega D_M \sum_{1\leq i\leq M}[f_V(q_i) - \alpha]^2 dq_1\cdots dq_M$$

or

$$\overline{(N_V - \bar{N}_V)^2} = \frac{M(M-1)}{\Omega^2}\int_\Omega\int_\Omega F_2(q_1,q_2)[f_V(q_1) - \alpha]\,[f_V(q_2) - \alpha]dq_1\,dq_2$$
$$+ \frac{M}{\Omega}\int_\Omega F_1(q)[f_V(q) - \alpha]^2 dq.$$

We simplify this expression:

$$\overline{(N_V - \overline{N}_V)^2} = \frac{M}{\Omega}\int_V F_1(q)dq\left\{1 - \frac{1}{\Omega}\int_V F_1(q)dq\right\}$$
$$+ \frac{M(M-1)}{\Omega^2}\int_V\int_V\{F_2(q_1,q_2) - F_1(q_1)F_1(p_2)\}dq_1\,dq_2$$
$$= \overline{N}_V\left(1 - \frac{\overline{N}_V}{M}\right) + \frac{M(M-1)}{\Omega^2}\int_V\int_V\{F_2(q_1,q_2)$$
$$- F_1(q_1)F_1(q_2)\}dq_1\,dq_2\,. \tag{4.19}$$

For a homogeneous phase (for gases and liquids), the binary function is, accurate to $O(1/N)$, radially symmetric:

$$F_2(q_1,q_2) = \mu(|q_1 - q_2|)$$

and $F = 1$. Let V be $\ll \Omega$ but still macroscopic (i.e., $V \sim N_A\sigma^3$). In this case $N_V/M \ll 1$, and the integral

$$\int_V\int_V\{F_2(q_1,q_2) - F_1(q_1)F_1(q_2)\}dq_1\,dq_2$$

is approximately equal to

$$V\int_V\{\mu(|q|) - 1\}dq = 4\pi V\int_0^\infty r^2\{\mu(r) - 1\}dr\,.$$

We put $\overline{N}_V = N$ and obtain on the basis of Eq. (4.19),

$$\overline{(N_V - N)^2} = N\left\{1 + \frac{4\pi}{v}\int_0^\infty r^2\{\mu(r) - 1\}dr\right\}, \tag{4.20}$$

i.e., the mean-square deviation is proportional to the number of particles in the system.

The method used to separate a small system of volume V from a large system is equivalent to the use of a grand canonical ensemble. Indeed, consider a system of fixed volume V, placed in a heat bath at a temperature θ and in contact with a particle reservoir with which it can exchange particles of one species. The chemical potential of these particles in a heat bath is equal to μ. The probability of the ith quantum state of a system with N_i particles and energy E_i in a grand canonical ensemble is given by

$$p_i = \frac{\exp[-(E_i - \mu N_i)/\theta]}{\Sigma_k \exp[-(E_k - \mu N_k)/\theta]}\,. \tag{4.21}$$

The average number of particles in the system is then

$$\overline{N}_V = \frac{\Sigma_i N_i \exp[-(E_i - \mu N_i)/\theta]}{\Sigma_i \exp[-(E_i - \mu N_i)/\theta]}\,.$$

Let us differentiate this expression with respect to μ:

$$\left(\frac{\partial \overline{N}_V}{\partial \mu}\right)_\theta = \frac{1}{\theta}\cdot\frac{\Sigma_i N_i^2 \exp[-(E_i-\mu N_i)/\theta]}{\Sigma_i \exp[-(E_i-\mu N_i)/\theta]} - \frac{1}{\theta}\left\{\frac{\Sigma_i N_i \exp[-(E_i-\mu N_i)/\theta]}{\Sigma_i \exp[-(E_i-\mu N_i)/\theta]}\right\}^2 = \frac{1}{\theta}[\overline{N_V^2} - (\overline{N}_V)^2] = \frac{1}{\theta}\overline{(N_V - \overline{N}_V)^2}\,.$$

On the other hand,

$$\left(\frac{\partial \overline{N}_V}{\partial \mu}\right)_{V,\theta} = \left(\frac{\partial \overline{N}_V}{\partial p}\right)_{V,\theta}\left(\frac{\partial p}{\partial \mu}\right)_{V,\theta} = \left(\frac{\partial \overline{N}_V}{\partial p}\right)_{V,\theta}\left(\frac{\partial \overline{N}_V}{\partial V}\right)_{\theta,\mu}\,.$$

If the extensive variable N_V is written in the form of a function of V, θ, and μ, it can take only the form

$$\overline{N}_V = V_f(\theta,\mu)\,.$$

From this follows

$$\left(\frac{\partial \overline{N}_V}{\partial \mu}\right)_{V,\theta} = \frac{\overline{N}_V}{V}\,,$$

meaning that

$$\left(\frac{\partial \overline{N}_V}{\partial \mu}\right)_{V,\theta} = \frac{\overline{N}_V}{V}\left(\frac{\partial \overline{N}_V}{\partial p}\right)_{V,\theta} = \overline{N}_V\left(\frac{\partial(\overline{N}_V/V)}{\partial p}\right)_{\theta,V}\,.$$

If $\overline{N}_V$ is regarded as a function of the intensive variables p and θ, then

$$\overline{N}_V = V_\varphi(\theta,p)$$

so that N_V/V depends only on θ and p. Therefore,

$$\left(\frac{\partial \overline{N}_V}{\partial \mu}\right)_{\theta,V} = -\frac{\overline{N}_V}{v^2}\left(\frac{\partial v}{\partial p}\right)_\theta\,,$$

meaning

$$\frac{1}{\theta}\overline{(N_V - \overline{N}_V)^2} = -\frac{\overline{N}_V}{v^2}\left(\frac{\partial v}{\partial p}\right)_\theta\,.$$

Putting here $N_N = N$ and substituting in Eq. (4.20), we get

$$-\frac{1}{v^2}\left(\frac{\partial v}{\partial p}\right)_\theta = 1 + \frac{4\pi}{v}\int_0^\infty r^2\{\mu(r)-1\}dr\,,$$

which is identical with Eq. (4.18).

Thus, solving the integral equations (2.153) and (4.12) in the superposition approximation we obtain a two-particle distribution function that enables us to determine the thermal equation of state (4.17) and the expression (4.18) for the compressibility. The latter, in turn, also makes it possible to determine the thermal equation. We have thus obtained two thermal equations of state which coincide if the exact expression is used for the binary function $F_2(q_1,q_2)$. Since, however, the integral equations yield $F_2(q_1,q_2)$ only approximately, the equation of state obtained from Eq. (4.17) (designated p_g) and that obtained from Eq. (4.18) (designated p_c) are in general not the same. Various methods were developed to reconcile p_g with p_C. The degree of proximity of the two is in a certain sense a measure of the approximation used in the definition of the free energy.

We turn now to the results of the solution of integral equations in the superposition approximation.

20. Equations of state in the superposition approximation

The superposition approximation yielded the first integral equations for the binary distribution function. While these are less accurate than the presently used integral equations, they have a unique property, viz., a mathematical singularity which Kirkwood identified with a liquid–solid phase transition. No other integral equations for a binary distribution function, used for the description of liquids, have singularities.

As already noted, the solution of the integral equations must begin with an analysis of the solution at low densities.

We shall seek the solution of Eqs. (2.153) and (4.12) in the form

$$\mu(r) = \exp[-\beta\Phi(r)]\left[1 + \sum_{i=1}^{\infty} \beta_i(r)/v^i\right] \tag{4.21'}$$

[the parameter ξ must be introduced for Eqs. (4.12) and (4.21)]. Substituting Eq. (4.21) in Eqs. (2.153) and (4.12) and then equating terms with equal powers of the density, we determine the functions β_i. The functions β_1 are the same in both equations [we must put $\xi = 1$ in the solution (4.12)], and $\mu(r)$ is equal in this case to

$$\mu(r) = \mu(|q|) = \exp[-\beta\Phi(|q|)]\left[1 + \frac{1}{v}\int f(|q-q'|)f(|q'|)dq'\right].$$

For a hard-sphere system, $\mu(r)$ is of the form (see Fig. 18)

$$\mu(\bar{r}) = \begin{cases} 0, & \bar{r}<1 \\ 1+8\eta\left(1-\frac{3}{4}\bar{r}+\frac{1}{16}\bar{r}^2\right), & 1<\bar{r}<2, \\ 1, & \bar{r}>2 \end{cases} \tag{4.22}$$

where $\bar{r} = r'/\sigma$ and σ is the sphere diameter. There exists thus a region $1<r<2$ in which $\mu(\bar{r}) > 1$, i.e., at these distances the probability of finding two particles is larger than the average probability. This points to the presence of an effective attraction between the particles, although there are no attracting interactions in the Hamiltonian. We have in this case a collective effect, which can be regarded as a screening effect. Indeed, if the second particle is

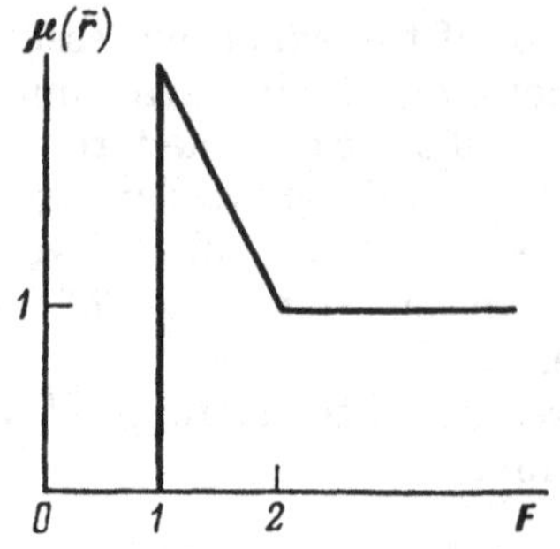

Figure 18.

located at a distance $1<\bar{r}<2$, no third particle can be placed between the central and the second, so that the number of collisions produced from opposite sides by the second particle is unequal—it is smaller from the side facing the central particle. This explains the effective attraction.

The succeeding functions $\beta_i (i>1)$ in the expansions (4.21) are unequal in Eqs. (2.153) and (4.11). The equation of region $1<r<2$ in which $\mu(\bar{r})>1$, i.e., at these distances the probability of finding two particles is larger than the average probability. This points to the presence of an effective attraction between the particles, although there are no attracting interactions in the Hamiltonian. We have in this case a collective effect, which can be regarded as a screening effect. Indeed, if the second particle is located at a distance $1<\bar{r}<2$, no third particle can be placed between the central and the second, so that the number of collisions produced from opposite sides by the second particle is unequal—it is smaller from the side facing the central particle. This explains the effective attraction.

The succeeding functions $\beta_i (i>1)$ in the expansions (4.21) are unequal in Eqs. (2.153) and (4.11). The equation of state, accurate to the third virial coefficient, obtained by using the distribution function from the Kirkwood equation from the expression for the pressure, takes the form[17]

$$\frac{pv}{\theta} = 1 + \frac{b}{v} + \frac{0.625b^2}{v^2} + \frac{0.1400b^3}{v^3}, \tag{4.23}$$

and from the equation for the compressibility

$$\frac{pv}{\theta} = 1 + \frac{b}{v} + \frac{0.625b^2}{v^2} + \frac{0.4418b^3}{v^3}, \tag{4.24}$$

where

$$b = \tfrac{2}{3}\pi\sigma^3.$$

On the other hand, the virial equation of state with exact virial coefficients is

$$\frac{pv}{\theta} = 1 + \frac{b}{v} + \frac{0.625b^2}{v^2} + \frac{0.2869b^3}{v^3}. \tag{4.25}$$

Similar equations can be obtained by solving Eq. (2.153) (Refs. 101 and 102):

$$\frac{p_g v}{\theta} = 1 + \frac{b}{v} + \frac{0.625b^2}{v^2} + \frac{0.2252b^3}{v^3}, \tag{4.26}$$

$$\frac{p_C v}{\theta} = 1 + \frac{b}{v} + \frac{0.625b^2}{v^2} + \frac{0.3424b^3}{v^3}. \tag{4.27}$$

This shows that at low densities Eq. (2.153) is more accurate than Eq. (4.12).

Mathematical methods that permit Eqs. (2.153) and (4.12) to be reduced to a form convenient for numerical solution are analyzed in Ref.103, and a survey of these methods is given in Ref. 104.

Equation (4.12) at $v = 1.24v_0$ and Eq. (2.153) at $v = 1.48v_0$ are no longer integrable[17] (here v_0 is the volume per closed-packed particle). The molecular-dynamics method yields for the maximum density of a homogeneous phase a value $v = 1.5v_0$ (Ref. 93). Whether the singularity of the solution of Eqs. (4.12) and (2.153) can be attributed to a phase transition is still under discussion.

The integral equation for the binary distribution leads, compared with other integral equations, to worse agreement between the calculated equations of state and those obtained by the molecular-dynamics method.

A method was developed of modifying Eq. (2.153) so as to preserve the singularity in the solution; this improved significantly the agreement between the theoretical and experimental data.[107]

Let us examine in greater detail, following Ref. 2, Eq. (2.153) for a hard-sphere system and discuss the questions connected with the linearization of this equation.

For the case of $\Phi(r)$ in the form

$$\Phi(r) = \begin{cases} \infty & (0 < r < \sigma) \\ 0 & (r > \sigma) \end{cases},$$

caution must be exercised when the function $E(t)$ is determined. We first introduce, therefore, the continuous function

$$\Phi(r) = \begin{cases} \Phi_0 & (0 < r < \sigma - \eta) \\ \Phi_0(\sigma - r)/\eta & (\sigma - \eta < r < \sigma). \\ 0 & (r > \sigma) \end{cases} \tag{4.28}$$

We go next to the limit in two steps. We first let $\Phi(0) \to 0$, so that

$$\Phi(r) = \begin{cases} \Phi_0 & (0 < r < \sigma) \\ 0 & (r > \sigma) \end{cases}. \tag{4.29}$$

We next go to the limit as $\Phi_0 \to \infty$.

For $\Phi(r)$ determined from Eq. (4.28) we have

$$E(t) = 0, \quad t > \sigma,$$

$$E(t) = E(0) = \int_{\sigma-\eta}^{\sigma} \frac{\Phi_0}{\eta} \mu(t) dt, \quad t < \sigma - \eta,$$

$$0 < E(t) < E(0), \quad \sigma - \eta < t < \sigma, \tag{4.30}$$

and from Eq. (4.17) we get

$$\frac{pv}{\theta}-1=\frac{2\pi}{3\theta v}\frac{\Phi_0}{\eta}\int_{\sigma-\eta}^{\sigma}t^3\mu(t)dt\,.$$

Therefore,

$$E(0)=\frac{3\theta v}{2\pi\sigma^3}\left(\frac{pv}{\theta}-1\right)(1+\delta_n),\quad \delta_n\to 0 \text{ as } \eta\to 0\,. \tag{4.31}$$

Starting from Eq. (2.153), we can write

$$\mu(r)=\nu(r)\exp\left\{-\frac{1}{\theta}\Phi(r)\right\},$$

$$\nu(r)=\exp\left\{-\frac{2\pi}{\theta v r}\int_{\substack{\rho>0\\ \rho>r-\sigma}}^{\sigma+r}\rho\left\{\int_{|r-\rho|}^{|r+\rho|}tE(t)dt\right\}\{\mu(\rho)-1\}d\rho\right\}. \tag{4.32}$$

We introduce the notation

$$E_0=\frac{3\theta v}{2\pi\sigma^3}\left(\frac{pv}{\theta}-1\right),$$

$$\nu_0=\exp\left\{-\frac{2\pi}{\theta v\sigma}\int_0^{2\sigma}\rho\left\{\int_{|\sigma-\rho|}^{\sigma}t\,dt\right\}[\mu(\rho)-1]d\rho\right\}, \tag{4.33}$$

where p and μ are taken for a system with an interaction potential (4.29). On the basis of Eq. (4.31) we have

$$E(0)\to E_0,\quad \eta\to 0\,.$$

In the interval $\sigma-\eta<r<\sigma$ we have as $\eta\to 0$, by virtue of Eqs. (4.30) and (4.31),

$$\nu(r)\to\nu_0\,,$$

therefore,

$$E_0=\lim_{\eta\to 0}\frac{\Phi_0}{\eta}\int_{\sigma-\eta}^{\sigma}\nu(r)\exp[-\Phi(r)/\theta]dr$$

$$=\nu_0\lim_{\eta\to 0}\frac{\Phi_0}{\eta}\int_{\sigma-\eta}^{\sigma}\exp\left(-\frac{\Phi(r)}{\theta}\right)dr=\nu_0\theta(1-\exp[-\Phi_0/\theta])\,.$$

$E(t)$ for Eq. (4.29) therefore takes the form

$$E(t)=\begin{cases}E_0 & (0<t<\sigma)\\ 0 & (t>\sigma)\end{cases},$$

where

$$E_0=\theta\nu_0[1-\exp(-\Phi_0/\theta)]\,. \tag{4.34}$$

In this case we can rewrite Eq. (2.155) in the form

$$\theta\ln\mu(r)=\Phi(r)+\frac{2\pi E_0}{rv}\int_{\substack{\rho>0\\ \rho>r-\sigma}}^{\sigma+r}\rho\left\{\int_{|\sigma-\rho|}^{\sigma}t\,dt\right\}[\mu(\rho)-1]d\rho\,. \tag{4.35}$$

Comparing this expression with the second equation in Eq. (4.32), we have

$$\nu_0=\mu(\sigma+0)\,.$$

We now make the transition $\Phi_0 \to 0$. Then

$$\mu(r) = 0, \quad 0 < r < \sigma;$$

we shall therefore consider Eq. (4.34) only for $r > \sigma$, and it takes in this case the form

$$-\theta \ln \mu(r) = \frac{2\pi E_0}{vr} \int_{r-\sigma}^{r+\sigma} \rho \left\{ \frac{\sigma^2 - (r-\rho)^2}{2} \right\} [\mu(\rho) - 1] d\rho . \tag{4.36}$$

According to Eqs. (4.33) and (4.34),

$$E_0 = \frac{3\theta v}{2\pi\sigma^3} \left(\frac{pv}{\theta} - 1 \right) = \theta\mu(\sigma + 0) . \tag{4.37}$$

We introduce a new function $\varphi(r/\sigma)$ defined by the relation

$$\mu(r) = 1 + \frac{\sigma}{r} \varphi(r/\sigma) .$$

From Eqs. (4.36) and (4.37) we have

$$x \ln\{1 + \varphi(x)/x\} = \lambda \int_{x-1}^{x+1} \{(x-\xi)^2 - 1\} \varphi(\xi) d\xi ,$$

$$1 < x < \infty , \tag{4.38}$$

where

$$\lambda = \frac{\pi\sigma^3}{v} [1 + \varphi(1+0)] ,$$

$$\varphi(x) = 0, \quad 0 < x < 1 .$$

Solving Eq. (4.27), we determine λ, meaning also the equation of state

$$\frac{pv}{\theta} = 1 + \frac{2}{3} \lambda . \tag{4.39}$$

Kirkwood's equation (4.12) can also be reduced to a form similar to Eq. (4.38). Kirkwood proposed also the approximation

$$x \ln\left\{1 + \frac{\varphi(x)}{x}\right\} = \varphi(x) , \tag{4.40}$$

which makes it possible to obtain equations from which to calculate $\varphi(x)$. It must be noted here that Eq. (4.40) makes the solution worse at low densities. For example, the fourth virial coefficient determined from the thermal equation of state is decreased by a factor 7.5 and becomes an order of magnitude smaller than the true value [we are considering in this case Eq. (4.38) linearized with the aid of relation (4.40)]. The situation in the case of the Kirkwood equation will be even worse.

On the other hand, the approximation (4.40) still does not indicate directly that the solution becomes much worse at high densities, inasmuch as at low densities Eq. (4.40) yields correct values of the second and third virial coefficients. This question calls for the special treatment given in Ref. 108 where, in addition, a hard-sphere system was investigated by using the correlation function g an integral equation determined by introducing the approximations[109,110]

$$F_2(q_1,q_2) = \exp[-\Phi(|q_1-q_2|)][1+g(|q_1-q_2|)],$$
$$F_3(q_1,q_2,q_3) = \exp\{-[\Phi(|q_1-q_2|)+\Phi(|q_1-q_3|) + \Phi(|q_2-q_3|)]/\theta\}[1+g(|q_1-q_2|)+g(|q_1-q_3|) + g(|q_2-q_3|)]. \qquad (4.41)$$

This resulted in the equation

$$g(r) = -\rho\int_1^{r+1}\frac{r'[1-(r-r')^2]}{r}g(r')dr' + \frac{1}{4}\rho\int_r^2(4-r'^2)g(r')dr' + \rho\lambda\left(\frac{r^3}{12}-r+\frac{4}{3}\right), \quad 1\leqslant r\leqslant 2;$$
$$g(r) = -\rho\int_{r-1}^{r+1}\frac{r'[1-(r-r')^2]}{r}g(r')dr', \quad r>2, \qquad (4.42)$$

where $\rho = \pi/v$ (it is assumed that $\sigma = 1$).

While this equation is less accurate at low densities than Eq. (2.153), it is much better than its linearized analog as well as the linearized Kirkwood equation.

We turn now to solutions of the equations in the superposition approximation for more realistic potentials.

There are many published results of measurements of the parameters of the equations of state and of the distribution functions in liquids and gases. Comparison of the results of the integral-equation solutions with the experimental data is made complicated, however, by the fact that the form of the potential employed by us is generally speaking inexact, so that it is difficult to ascertain whether the primary cause of the disparity between the theory and experiment is the inaccuracy of the theory or of the chosen potential. Moreover, good agreement between the theoretical and experimental data cannot ensure that the theory is correct, since this agreement can be due to mutual cancellation of errors in the potential and to the choice of the approximation used to obtain the integral equations. To check on the theory, the calculated distributions as well as the equations of state were compared with the molecular-dynamics-method (MDM) data. Whereas the distribution functions for hard-sphere systems can be calculated by the MDM method relatively simply, the difficulties increase strongly for more complicated potentials. Nonetheless, extensive data were obtained for a potential of the Lennard-Jones type. The superposition approximation leads for such systems to a qualitative agreement between the equations of state and the MDM data.

Integral equations in the superposition approximation permit a description of the critical point, and solution of Eq. (2.153) yields the following values of the reduced critical parameters[22]:

$$p_k^* = 0.199; \quad v_k^* = 2.585;$$
$$\theta_k^* = 1.433; \quad p_k v_k / \theta_k = 0.358. \tag{4.43}$$

Here $\theta^* = \theta/\epsilon$, $p_k^* = p\sigma^3/\epsilon$, and $v_k^* = v/\sigma^3$.

The results for the Kirkwood equation and for Eq. (2.153) are compared in Ref. 22. While the results of both equations are close enough, the unsatisfactory agreement between the theoretical and experimental data in the equations of state for high densities has stimulated a search for other integral equations capable of yielding more accurate equations of state. The most widely used and effective among them is the Percus–Yevick equation, the numerical solutions of which will be discussed in the next section.

To conclude this section, we consider one more aspect of the superposition approximation. At short distances between three particles, the behavior of the three-particle function $F_3(q_1,q_2,q_3)$ is determined by the factor

$$\exp\left\{ -\frac{1}{\theta}[\Phi(|q_1 - q_2|) + \Phi(|q_1 - q_3|) + \Phi(|q_2 - q_3|)] \right\}. \tag{4.44}$$

If the distance between the particles is large, we have for F_3:

$$F_3 \to F_1(q_2)\, F_1(q_2)\, F_1(q_3)\,. \tag{4.45}$$

For a homogeneous phase we have $F_1 \equiv 1$, therefore,

$$F_3 \to 1\,.$$

On the other hand, $F_1 \not\equiv 1$ for ordered structures, i.e., structures with higher density. Therefore, the superposition approximation leads in this case to the expression

$$F_3 \to [\, F_1(q_1)\, F_1(q_2)\, F_1(q_3)]^2\,, \tag{4.46}$$

which differs drastically from Eq. (4.45). It is this, apparently, which causes the solution of the equation to diverge in the superposition approximation at high densities. Although the singularity in the solution of the equation can indeed be regarded as the presence of a phase transition, a complete description of this transition presupposes also the possibility of describing both phases, and not only the homogeneous or ordered one. The foregoing shortcoming of the superposition approximation at high densities can be eliminated by introducing the following approximation:

$$F_2 = \exp\left[-\frac{1}{\theta}\,\Phi(|q_1 - q_2|) \right] p_2(q_1,q_2)\,,$$
$$F_3 = \exp\left\{ -\frac{1}{\theta}[\Phi(|q_1 - q_2|) + \Phi(|q_1 - q_3|) + \Phi(|q_2 - q_3|)] \right\}$$
$$\times [p_2(q_1,q_2) p_2(q_1,q_3) p_2(q_2,q_3)]^{1/2}\,. \tag{4.47}$$

It allows us to write the second equation of the Bogolyubov chain in the form

$$2\frac{\partial\sqrt{p_2(q_1,q_2)}}{\partial q_1^\alpha}+\frac{1}{\theta v}\int\frac{\partial\Phi(|q_1-q_3|)}{\partial q_1^\alpha}\exp\left\{-\frac{1}{\theta}[\Phi(|q_1-q_3|)\right.$$

$$\left.+\Phi(|q_2-q_3|)]\right\}\sqrt{p_2(q_1,q_3)p_2(q_2,q_3)},\quad dq_3=0\,. \tag{4.48}$$

Using the radial symmetry of the two-particle distribution function and introducing the symbol

$$z(|q|)=\sqrt{p_2(|q|)}\,, \tag{4.48'}$$

we can repeat the procedure that led in the superposition approximation to Eq. (2.153). In our case we get

$$z(r)=1-\frac{\pi}{\theta v r}\int_0^\infty\rho\left\{\int_{|r-\rho|}^{r+\rho}E(t)t\,dt\right\}\left\{\exp\left[-\frac{1}{\theta}\Phi(\rho)\right]z(\rho)-1\right\},$$

$$E(t)=\int_\infty^t\exp\left[-\frac{1}{\theta}\Phi(s)\right]z(s)\frac{d\Phi(s)}{ds}\,ds\,. \tag{4.49}$$

At low densities, Eq. (4.49) yields an equation of state that is no less accurate than that obtained with the approximation (4.41).

An additional investigation is needed to determine the behavior at high densities.

The superposition approximation considered in the present section has yielded thus for the binary distribution function an equation whose solution describes qualitatively the liquid state.

21. The Percus–Yevick equation

As already noted, the Percus–Yevick equation (2.160) is regarded at present as one of the most accurate equations in the theory of the liquid state. If it is recognized that it has an analytic solution for the hard-sphere system (see Sec. 13), the special role that this equation plays in the description of liquids becomes clear. Namely, by obtaining an exact solution for the basis, which is chosen to be the hard-sphere system, we can construct in this case the perturbation-theory series in the most effective manner. In the functional approach, the approximation (2.150) which leads to the Percus–Yevick equation (2.160) is the first term of the expansion of a functional into a functional Taylor series. This makes it possible to construct an approximate second-order integral equation (PY-2, Ref. 17).

Let us discuss the solution obtained in Sec. 13 for the Percus–Yevick equation for a hard-sphere system. We have noted there that the equations of state obtained from the expression for the pressure and from the expression for the compressibility differ. This points to a thermodynamic inconsistency of the Percus–Yevick equation. Let us estimate the degree of this inconsistency at low densities. To this end we expand Eqs. (2.100) and (2.102) in powers

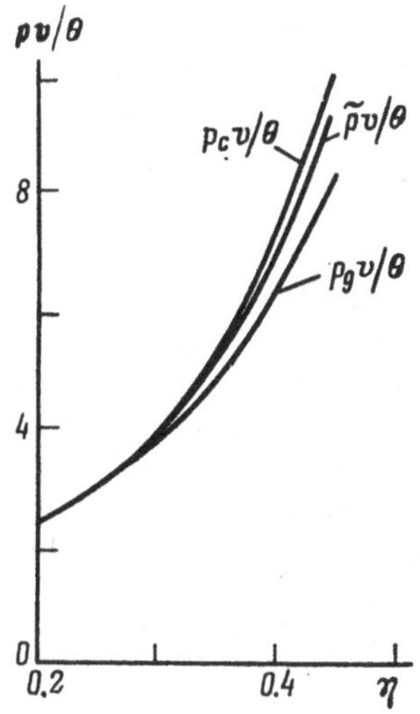

Figure 19.

of η and designate, as above, the pressure p obtained from the thermodynamic equation by p_g and that from the expression for the compressibility by p_C:

$$\frac{p_g v}{\theta} = 1 + 4\eta + 10\eta^2 + 16\eta^3 + 22\eta^4 + 28\eta^5 + \cdots, \tag{4.50}$$

$$\frac{p_C v}{\theta} = 1 + 4\eta + 10\eta^2 + 19\eta^3 + 31\eta^4 + 46\eta^5 + \cdots. \tag{4.51}$$

The virial equation of state, accurate to the sixth virial coefficient, takes the form (3.23)–(3.25):

$$\frac{pv}{\theta} = 1 + 4\eta + 10\eta^2 + 18.36\eta^3 + 28.2\eta^4 + 39.5\eta^5. \tag{4.52}$$

Therefore, the equation of state (4.50) underestimates the virial coefficients, while Eq. (4.51) overestimates them.

Attempts were made to reconcile these two equations of state. In particular, a semiempirical approach[111] yielded the equation of state

$$\frac{\tilde{p}v}{\theta} = \frac{v}{3\theta}[2p_C + p_g] = \frac{1 + \eta + \eta^2 - \eta^3}{(1-\eta)^3}, \tag{4.53}$$

which, when expanded in a series, takes the form

$$\frac{pv}{\theta} = 1 + 4\eta + 10\eta^2 + 18\eta^3 + 28\eta^4 + 40\eta^5 + \cdots,$$

i.e., it approximates Eq. (4.52) with good accuracy. Figure 19 shows a comparison of the results of the MDM for a system of hard spheres[93] and of the equations of state (4.50), (4.51), and (4.53). At low densities, the agreement between the equations of state $p_g v/\theta$ and $p_C v/\theta$ is good, but at large η the disparity becomes appreciable (the difference between $p_g v/\theta$ and $p_C v/\theta$ at $\eta = \sqrt{2}\pi/9$ is of the order of 20%). Equation (4.53) describes well the MDM data all the way to the point $\eta = \sqrt{2}\pi/9$ that characterizes (according to the MDM data) the start of a phase transition from a homogeneous to an ordered structure.

The Percus–Yevick equation for a hard-sphere system yields for $S(k)$ an analytic expression that can be determined from Eq. (2.155). This $S(k)$ is in fair quantitative agreement with the experimental values of this function for liquefied noble gases and for a number of liquid metals. Thus, at high densities the principal role in the formation of the short-range order is played by the repulsive part of the potential. The structure factor of a number of liquid metals was compared in Ref. 112 with calculations in the Percus–Yevick approximation for the hard-sphere model. The calculated value of the factor $S(k)$ accords with all the characteristic features of the experimental data, and describes the first peak particularly well. With increase of K, the difference between the experimental and calculated structure factors increases, since no account is taken of the attractive part of the potential.

Even though the Percus–Yevick equation describes well the homogeneous phase, and the agreement between the equations of state calculated on the basis of the thermal equation and of compressibility can be improved in the higher-order approximations (PY-2 and others), it does not account at all for the presence of the phase transition reliably established in the MDM calculations.

Many papers are devoted to the solution of the Percus–Yevick equation for a system of particles with a Lennard-Jones interaction potential. The calculation results are compared with the MDM data. At low densities the equation leads to exact values of the second and third virial coefficients. The fourth, determined from the solution of the Percus–Yevick equation differs from that obtained by group expansion of the partition function, but at $\theta < \epsilon$ the difference is insignificant. The disparity becomes larger for the fifth virial coefficient.[112] At high densities the two-particle distribution function $\mu(r)$ and the structure factor $S(k)$ determined from the solution of the Percus–Yevick equation with a Lennard-Jones potential lead to results that agree well both with the MDM calculations and with the experimental data for liquid argon.

If the overall qualitative structure of a liquid is considered, it can be seen that the calculated $\mu(r)$ for a Lennard-Jones potential agree with the MDM data in a wide range of temperatures and densities, and the agreement improves at high densities.

The Percus–Yevick equation describes a liquid–gas transition and the critical point. There is no analytic solution of this equation for a Lennard-Jones potential, so that an analysis of the system near the critical point is difficult. The available analytic solution for a system of hard spheres with sticking, which can be regarded as effective attraction, has not been analyzed in detail.[113] All that is known is that this system has near the critical point a classical behavior similar to that described by the elementary thermodynamic theory. One can expect the behavior to be the same for a Lennard-Jones potential. The phase transition is manifested by the fact that at certain values of the temperature and density there is no solution of the Percus–Yevick equation.

We present the critical parameters of a system with a Lennard-Jones potential as calculated from the equations for the pressure

$$\theta_k^* = \frac{\theta_k}{\epsilon} = 1.25 \pm 0.02; \quad \rho_k^* = \frac{\sigma^3}{v_k} = 0.29 \pm 0.03;$$

$$\left(\frac{pv}{\theta}\right)_k = 9.30 \pm 0.02; \quad p_k^* = \frac{p_k \sigma^3}{\epsilon} = 0.11 \tag{4.54}$$

and the compressibility

$$\theta_k^* = 1.32 \pm 0.02; \quad \rho_k^* = 0.28 \pm 0.03;$$

$$\left(\frac{pv}{\theta}\right)_k = 0.36 \pm 0.02; \quad \rho_k^* = 0.13 \text{ (Refs. 113 and 114)}. \tag{4.55}$$

The corresponding values obtained by the molecular-dynamics method are

$$\theta_k^* = 1.32 - 1.36; \quad \rho_k^* = 0.32 - 0.36;$$

$$(pv/\theta)_k = 0.30 - 0.36; \quad p_k^* = 0.13 - 0.17; \tag{4.56}$$

and the experimental data for argon[115]:

$$\theta_k^* = 1.26; \quad \rho_k^* = 0.316;$$

$$(pv/\theta)_k = 0.293; \quad p_k^* = 0.117. \tag{4.57}$$

Calculations in the region of the critical point by the MDM method are highly complicated, so that the range of possible values in Eq. (4.56) is large. The experimental data for argon differ from the MDM data, but this is apparently due to failure to take multiparticle interactions into account. The agreement between the critical parameters calculated from the solution of the Percus–Yevick equation with the MDM data is satisfactory. There is a discrepancy between the values determined from the equation for the pressure (4.54) and from the equation for the compressibility (4.55). The quantitative results of Eqs. (4.54) and (4.55) for the coexistence curve of the liquid and gas phases below the critical point are not in satisfactory agreement.

The behavior of the function $\mu(r)$ influences substantially the calculated pressure, and is furthermore most sensitive to the $\mu(r)$ behavior details near the first peak. The Percus–Yevick equation is not accurate enough to describe the equations of state of a liquid. The disparity of the data obtained from the thermal equation of state and from the expression for the pressure becomes very large. They also agree poorly with the MDM data.

The expression for the system energy

$$E = \frac{3}{2}\theta N + \frac{2\pi}{v} N \int_0^\infty \Phi(r)\mu(r)r^2\, dr$$

is less sensitive to the singularities of the behavior of the function $\mu(r)$. The agreement between the internal interaction energy values obtained for a

Lennard-Jones potential by the Percus–Yevick method and by the MDM is good: the difference does not exceed 3% even at low temperatures. The results for the entropy are also satisfactory.

Using the fact that the Percus–Yevick equation yields the caloric equation of state with good accuracy, the results for the pressure can be reconciled with the aid of the identity

$$\left(\frac{\partial E}{\partial V}\right)_\theta = \theta\left(\frac{\partial p}{\partial \theta}\right)_V - p\,. \tag{4.58}$$

At present this is the most effective method of determining the thermal equation of state. It leads to better agreement between the theoretical and experimental data.

The Percus–Yevick equation can be used also to determine the effective pair potential if the experimental data for $\Phi(r)$ are employed. The result is an expression for $\mu(r)$, which agrees with the general variation of the two-particle interaction potential. In this case, however, $\Phi(r)$ has also too strong a dependence on the density.

An important role in the theory of liquid metals is played by the perturbation-theory treatment of the Percus–Yevick equation.[113] The interaction potential is taken in the form

$$\Phi(r) = \Phi_0(r,v) + \Phi_1(r,v,\theta)\,,$$

where $\Phi_0(r, v)$ is the potential of a system of hard spheres having a density-dependent diameter, and $\Phi_0(r, v, \theta)$ is the attractive part of the potential. We regard Φ_1 as a correction to Φ_0. On the basis of the exact solution of the equation for Φ_0 it is possible to construct a perturbation theory for the determination of μ in any order in Φ_1. We choose the hard-sphere diameter $\sigma = \sigma(v)$ such that the perturbation-theory series converges at a maximum rate. This leads to a number of important results.

We consider now a somewhat different approach to the Percus–Yevick equation, involving consideration of the density fluctuations in the system. Let a volume V contain N particles located at points R_i ($i = 1,\ldots,N$). The particle-number density is then given by

$$\rho(r) = \sum_{i=1}^{N} \delta(|R_i - r_i|)\,, \tag{4.59}$$

where $\delta(x)$ is the Dirac delta function.

We obtain now the Fourier components of ρ. They are equal to

$$\rho_k = \sum_{i=1}^{N} \exp(ikR_i)\,. \tag{4.60}$$

For a radially symmetric two-particle distribution function we can write the equality

$$\rho\mu(r) = \frac{1}{N}\left\langle \sum_{i\neq j} \delta(|r - R_i + R_j|)\right\rangle, \tag{4.61}$$

where $\langle \cdots \rangle$ denotes averaging over a grand canonical ensemble. Consider the expression

$$\rho_k \rho_{-k} = \sum_{i=j}^{N} \exp(ikR_i) \sum_{j=1}^{N} \exp(-ikR_j), \tag{4.62}$$

which can also be written in the form

$$\rho_k \rho_{-k} - N = \sum_{i \neq j} \exp[ik(R_i - R_j)]. \tag{4.63}$$

It is known that the structure factor $S(k)$ is given by

$$S(k) = 1 + \rho \int [\mu(r) - 1] \exp(ikr) dr, \tag{4.64}$$

we have therefore on the basis of Eqs. (4.61) and (4.63),

$$S(k) = \frac{1}{N} \langle \rho_k \rho_{-k} \rangle. \tag{4.65}$$

From the second Bogolyubov-chain equation we find in the k representation

$$S(k) = 1 + \frac{1}{Nk^2} \sum_n \frac{\overline{\Phi}(n)}{\theta} \langle \rho_{k+n} \rho_k \rho_n \rangle_{kn}, \tag{4.66}$$ [116]

where

$$\overline{\Phi}(k) = \int dr \exp(ikr) \Phi(r), \tag{4.67}$$

i.e., it is assumed that $\Phi(r)$ has a Fourier transform. Here $\langle \rho_{k+n} \rho_k \rho_n \rangle$ is a three-particle correlation function.

To obtain a closed equation for $S(k)$ we must express $\langle \rho_{k+n} \rho_k \rho_n \rangle$ in terms of $S(k)$. The random-phase approximation is defined as an approximation in which only one term with $n = -k$ is left. Then $\langle \rho_{k+n} \rho_k \rho_n \rangle$ is connected with $S(k)$ directly. We have in this case

$$S(k) = \left[1 + \rho \frac{\overline{\Phi}(k)}{\theta} \right]^{-1}. \tag{4.68}$$

In the hyperchain approximation it is assumed that

$$h - C = \ln \mu + \Phi/\theta, \tag{4.69}$$

we have therefore at large r,

$$C(r) \sim -\Phi(r)/\theta \tag{4.70}$$

if $h^2 < |C|$. This condition is met for regions far from the critical point. From the Fourier transform of the equation

$$h(|r|) = C(|r|) + \rho \int C(|r - r'|) h(|r'|) dr'$$

we have

$$\overline{h}(k) = \overline{c}(k) + \overline{h}(k) \overline{c}(k),$$

or, with allowance for Eq. (4.64),

$$\overline{C}(k)=\frac{\bar{h}(k)}{1+\bar{h}(k)}=\frac{S(k)-1}{S(k)}.$$

Substituting here Eq. (4.68) we obtain

$$\overline{C}(k)=-\rho\overline{\Phi}(k)/\theta. \tag{4.71}$$

Taking the Fourier transform of this equation we have

$$C(r)=-\Phi(r)/\theta.$$

Thus, the random-phase approximation is equivalent to the hyperchain approximation. The foregoing analysis is valid if the interaction potential has a Fourier transform. For a liquid, the main part of the potential is taken to be a hard-sphere-system potential, for which no Fourier transform exist. We can attempt, however, to introduce an effective potential to replace $\Phi(k)$ in the relations above. We choose ρ_k to be the collective variable. In the initial Hamiltonian we have $3N$ coordinates $R_1,...,R_N$. The final system is a cube with side L and with periodic boundary conditions. The most convenient are the mass-center coordinates

$$R_\mu=\frac{1}{N}\sum_{i=1}^{N}R_i \tag{4.72}$$

and the $3N-3$ coordinates ρ_k.

We express the potential energy in the form

$$U=\tfrac{1}{2}\sum_k\overline{\Phi}(k)\langle[\rho_k\,\rho_k-N]\rangle. \tag{4.73}$$

The problem is to determine the $\overline{\Phi}_{\rm eff}(k)$ that yields the best approximation[116] and is determined from the condition that

$$\left\langle\left\{\overline{\Phi}_{\rm eff}(R_i-R_j)-\sum_k\overline{\Phi}_{\rm eff}(k)\exp[-ik(R_i-R_j)]\right\}^2\right\rangle \tag{4.74}$$

be a minimum. It was found[116] by successive approximations that

$$\overline{\Phi}_{\rm eff}(k)=\theta F\Big(\mu(r)\{\exp[\Phi(r)/\theta]-1\}\Big), \tag{4.75}$$

where $F(x)$ is the Fourier transform of x.

Substituting now $\overline{\Phi}_{\rm eff}$ for $\overline{\Phi}(k)$ in Eq. (4.71), we get

$$C(k)=-\rho F\Big(\mu(r)\{\exp[\Phi(r)/\theta]-1\}\Big),$$

or, taking the inverse Fourier transform,

$$C(r)=\mu(r)\{1-\exp[\Phi(r)/\theta]\}, \tag{4.76}$$

i.e., $C(r)$ coincides with the expression used to obtain the Percus–Yevick equation. The approach considered above was the first through which the Percus–Yevick approximation was introduced. It was superseded by a diagram approach that led to the approximation (4.76) for the direct correlation function

$C(r)$. The Percus–Yevick equation has made it possible to describe the liquid state qualitatively. In a number of cases, to eliminate incompatibility with the MDM method, it became necessary to use additional premises in the thermal equation of states. The discrepancy between the thermal equations obtained on the basis of the expressions for the pressure and the compressibility is unacceptably large. Elimination of these shortcomings calls for the development of a second-order approximation, to which we now proceed.

22. Second-order approximation for the Percus–Yevick equation

To obtain more accurate equations for the theory of the liquid state, use is made of the functional-differentiation method (Refs. 17, 93, and 117). The corresponding functionals are expanded in functional Taylor series. This provides more rigorous premises for the Percus–Yevick (PY) equation, the hyperchain (HC) approximation, and others. On the other hand, the choice of the functional itself is arbitrary. To justify such a choice it would be necessary to relate it to fundamental physical principles, say the minimum-free-energy principle.[17]

The second-order contribution to the direct correlation function for the Percus–Yevick equation can be represented in the form[17]

$$\begin{aligned} C_2(|q_1-q_2|) &= h(|q_1-q_2|) - \{\mu(|q_1-q_2|)\exp[\Phi(|q_1-q_2|)/\theta] - 1\} \\ &\quad + \frac{1}{2v^2}\int\int C(|q_1-q_2|)C(|q_1-q_3|)\mu(|q_3-q_4|) \\ &\quad \times [F_3(q_2,q_3,q_4) - \ln F_3(q_2,q_3,q_4) - h(q_2,q_3) + \ln \mu(|q_2-q_3|) \\ &\quad - \mu(|q_2-q_3|) + \ln \mu(|q_2-q_4|)]dq_3\, dq_4\,. \end{aligned} \tag{4.77}$$

We see that C_2 depends on a three-particle distribution function, so that the equations are no longer closed. Additional assumptions must therefore be made concerning the approximation of F_3. It is proposed to choose as the first approximation for F_3 the functional derivative of the expression[122]

$$\rho_2(q_2,q_3)\exp[\Phi(|q_1-q_2|)/\theta + \Phi(|q_1-q_3|)/\theta]\,, \tag{4.78}$$

regarded as a linear functional of $\rho_1(q_4)$. As a result, we obtain for F_3 the equation

$$\begin{aligned} &F_3(q_1,q_2,q_3)\exp[\Phi(|q_1-q_2|)/\theta + \Phi(|q_1-q_3|)/\theta] \\ &\quad = \mu(|q_2-q_3|) + \frac{1}{v}\int [F_3(q_2,q_3,q_4) - \mu(|q_2-q_3|)] \\ &\quad \times C(|q_1-q_4|)dq_4\,. \end{aligned} \tag{4.79}$$

We thus obtain all the equations necessary to determine the function $\mu(|q-q|)$. In contrast to the first-order approximation, the resultant equations are much more complicated. Only numerical results were obtained for them, and there are no calculations for the liquid state. At the same time, the appreciable complication of the equations can be justified only if they make it possible to describe the phase transition from an ordered to a homogeneous state, something not obtainable from the first approximation of the Percus–Yevick equation.

Consider equations of state based on a solution of the second-approximation Percus–Yevick equation (PY-2). As noted above, a substantial shortcoming of the Percus–Yevick equation in first-order approximation is that the equations of state determined from the expressions for the pressure and for the compressibility do not agree. The (PY-2) equation alleviates the situation to a considerable degree. Thus, the fifth virial coefficient calculated for a hard-sphere system from the expression for the pressure is

$$B_5 = 0.124\, B_2^{\,4}\,,$$

and from the expression for the compressibility

$$B_5 = 0.107\, B_2^{\,4}\,.$$

The virial coefficients for a Lennard-Jones potential were also calculated up to the fifth one, inclusive. The PY-2 equation results in virial coefficients that agree well with the exact values at high temperature, and the equations obtained from the expressions for the pressure and for the compressibility are in satisfactory agreement.

We examine now the critical constants for systems with Lennard-Jones potentials. They are given by the relations[118]

(a) from the expression for pressure:

$$\theta_k^* = 1.36; \quad B_2/v_k = 0.73; \quad \left(\frac{pv}{\theta}\right)_k = 0.31;$$

(b) from the expression for compressibility:

$$\theta_k^* = 1.33; \quad B_2/v_k = 0.77; \quad \left(\frac{pv}{\theta}\right)_k = 0.34\,.$$

Thus, the disparity of the various results is somewhat decreased in the second approximation. In general, however, the agreement between these results and those of the molecular-dynamics method can be regarded as satisfactory.

The second-order approximation for the Percus–Yevick equation can be obtained also by another method, using a diagram technique.[119]

The extent to which the second-order approximation determined by the functional-differentiation method is sensitive to the choice of a three-particle function[93] has not yet been investigated. Were this sensitivity to turn out to be strong, the second-order approximation could not be regarded as better than the first-order one.

Since Eq. (4.70) does not ensure symmetry of $F_3(q_1,q_2,q_3)$ with respect to permutation of the arguments, another equation was also proposed for this function[120,121]:

$$F_3(q_1,q_2,q_3) = \mu(|q_1 - q_2|)\mu(|q_1 - q_3|)\mu(|q_2 - q_3|)$$
$$\times\left[1 + \frac{1}{v}\int h(|q_1 - q_4|)h(|q_2 - q_4|)h(|q_3 - q_4|)dq_4\right]. \quad (4.80)$$

Since the integral equations in the second-order Percus–Yevick equations are complicated, they were not fully investigated, and a number of questions remain unanswered. It is practically impossible to determine in this approximation the interatomic potential from x-ray scattering data.

In both the first- and second-order approximation we assume that the interaction potential is of the form

$$U = \sum_{i<j} \Phi(|q_i - q_j|) = \sum_{i<j} \Phi_{ij}\,. \quad (4.81)$$

In the general case, however, it must be recognized that U contains many-particle potentials besides the two-potential one, and U is then given by

$$U = \sum_{i<j} \Phi_{ij} + \sum_{i<j<k} \Phi_{ijk} + \cdots\,. \quad (4.82)$$

The three-particle distribution function must be known to determine the influence of a three-particle potential. The importance of the contribution of the latter is usually demonstrated by the presence of a large maximum in the virial coefficient at low temperatures. The influence of the three-particle interactions should thus be manifested even in first order of the Percus–Yevick equation. Allowance for the three-particle interactions leads to replacement of the pair potential by an effective pair potential. At very low densities, many-particle interactions can be neglected. Their influence increases strongly with increase of density. X-ray scattering by liquid argon has shown that the depth of the two-particle potential decreases strongly with increase of density. It follows therefore that at liquid-state densities the three-particle potential must be positive. This agrees with conclusions drawn from other experiments. On the other hand, in view of the difficulty of describing the processes that take place in liquids, it is difficult to separate the effect of the three-particle interactions. Besides, MDM calculation of system properties with allowance for three-particle interactions are quite complicated.

We derive now the Percus–Yevick equation by the functional-differentiation method. We choose the potential energy in the form

$$U = \sum_{i<j} \Phi(|q_i - q_j|) + \sum_{1\leqslant i\leqslant N} \varphi(q_i)\,, \quad (4.83)$$

where $\varphi(q_i)$ is a single-particle potential that depends on the coordinate of the ith particle. The choice of a potential in two-particle form does not make the analysis less general, since the procedure can always be used also for more general forms of the potential, by analogy with procedure that follows.

In our case, the configuration integral Q is given by

$$Q[\varphi] = \int dq_1 \cdots dq_N \exp\left\{ -\beta \left[\sum_{i<j} \Phi(|q_i - q_j|) + \sum_{1 \leqslant i \leqslant N} \varphi(q_i) \right] \right\} . \quad (4.84)$$

We determine from this the configuration part of the system's free energy

$$F = -\theta \ln Q[\varphi] . \quad (4.85)$$

We calculate now the variation of F with respect to φ:

$$\begin{aligned} \delta F &= -\theta\delta \ln Q[\varphi] \\ &= Q^{-1}[\varphi] \int dq_1 \cdots dq_N \exp\left\{ -\beta \left[\sum_{i<j} \Phi(|q_i - q_j|) \right.\right. \\ &\quad \left.\left. + \sum_{1 \leqslant i \leqslant N} \varphi(q_i) \right] \right\} \left[\sum_{1 \leqslant j \leqslant N} \delta\varphi(q_i) \right] . \end{aligned} \quad (4.86)$$

Since the integrand is symmetric, all the terms of the sum are equal, and we have

$$\begin{aligned} \delta F &= Q^{-1}[\varphi] N \int dq_1 \cdots dq_N \exp[-\beta U] \delta\varphi(q_1) \\ &= \int dq_1\, \delta\varphi(q_1) \left\{ Q^{-1}[\varphi] N \int dq_2 \cdots dq_N \exp[-\beta U] \right. \\ &= \left. \int dq_1\, \delta\varphi(q_1) \rho_1(q_1, \varphi) \right\} , \end{aligned} \quad (4.87)$$

where $\rho_1(q_1, \varphi)$ is the one-particle distribution function in the presence of an external field φ.

From the definition of the functional derivative[2] we get

$$\frac{\delta F}{\delta\varphi(q_1)} = \rho_1(q_1, \varphi) , \quad (4.88)$$

and at $\varphi = 0$,

$$\left[\frac{\partial F}{\delta\varphi(q_1)} \right]_{\varphi=0} = \rho_1(q_1) . \quad (4.89)$$

The one-particle distribution function is thus determined by the trial functional derivative of the free energy with respect to the external field, and by putting $\varphi = 0$ in the final expression.

We obtain now the second functional derivative of F. To this end we calculate first the second variation of the free energy

$$\delta^2 F = \delta\rho_1(q_1,\varphi) = N\delta\,\frac{\int dq_2\cdots dq_N \exp(-\beta U)}{\int dq_1\cdots dq_N \exp(-\beta U)}$$

$$= -N\beta Q^{-1}[\varphi]\int dq_2\cdots dq_N \exp[-\beta U]\Bigl(\sum_{2<j<N}\delta\varphi_j\Bigr)$$

$$-N\beta Q^{-1}[\varphi]\delta\varphi(q_1)\int dq_2\cdots dq_N \exp[-\beta U]$$

$$+N\beta Q^{-2}[\varphi]\Bigl[\int dq_2\cdots dq_N \exp[-\beta U]\Bigr]$$

$$\times\Bigl\{\int dq_1\cdots dq_N \exp[-\beta U]\Bigl[\sum_{1<j<N}\delta\varphi(q_j)\Bigr]\Bigr\}\,.$$

Taking into account the symmetry property and the definitions of the one- and two-particle distribution functions, we get

$$\delta^2 F = -\beta\int dq_2\,\delta\varphi(q_2)N(N-1)Q^{-1}[\varphi]$$

$$\times\int dq_3\cdots dq_N \exp[-\beta U] - \beta\delta\varphi(q_1)\rho_1(q_1,\varphi)$$

$$+\beta n_1(q_1,\varphi)\int dq_2\,\delta\varphi(q_2)NQ^{-1}[\varphi]\int dq_1\,dq_3\cdots dq_N$$

$$\times\exp[-\beta U] = -\beta\int dq_2\,\delta\varphi(q_2)[\rho_2(q_1,q_2,\varphi)$$

$$+\rho_1(q_1,\varphi)\delta(q_1-q_2)-\rho_1(q_1,\varphi)\rho_1(q_2,\varphi)]\,.$$

This yields the second functional derivative of the free energy

$$\frac{\delta^2 F}{\delta\varphi(q_1)\delta\varphi(q_2)} = -\beta\,[\rho_2(q_1,q_2,\varphi)-\rho_1(q_1,\varphi)\rho_1(q_2,\varphi)+\rho_1(q_1,\varphi)\delta(q_1-q_2)]\,. \quad (4.90)$$

At $\varphi = 0$ we have

$$\frac{\delta^2 F}{\delta\varphi(q_1)\delta\varphi(q_2)} = \frac{\delta\rho_1(q_1,\varphi)}{\delta\varphi(q_2)} = -\beta\,[\rho_2(q_1,q_2)-\rho_1(q_1)\rho_1(q_2)+\rho_1(q_1)\delta(q_1-q_2)]\,. \quad (4.91)$$

The two-particle distribution function is therefore connected with the second functional derivative of the free energy. Relations for higher-order distribution functions can be derived in the same manner.

We define the first correlation function $C(q_1,q_2,\varphi)$ by the relation

$$-\beta\,\frac{\delta\varphi(q_2)}{\delta\rho_1(q_1,\varphi)} \equiv \frac{1}{\rho_1(q_1,\varphi)}\,\delta(q_1-q_2) - C(q_1,q_2,\varphi)\,. \quad (4.92)$$

By virtue of the definition of the inverse derivative we have

$$\int dq_3 \frac{\delta\rho_1(q_1,\varphi)}{\delta\varphi(q_3)} \frac{\delta\varphi(q_3)}{\delta\rho_1(q_2,\varphi)} = \delta(q_1 - q_2) \,. \tag{4.93}$$

We substitute here the corresponding expressions from Eqs. (4.91) and (4.92):

$$\begin{aligned}\int dq_3 [\rho_2(q_1,q_3,\varphi) &- \rho_1(q_1,\varphi)\rho_1(q_3,\varphi) \\ &+ \rho_1(q_1,\varphi)\delta(q_1 - q_3)]\,[\delta(q_3 - q_2)/\rho_1(q_2,\varphi) \\ &- C(q_3,q_2,\varphi)] = \delta(q_1 - q_2) \,.\end{aligned} \tag{4.94}$$

Simple transformations yield

$$\begin{aligned}\rho_2(q_1,q_2,\varphi)/\rho_1(q_1,\varphi) &- \rho_1(q_2,\varphi) \\ &= C(q_2,q_1,\varphi) + \int dq_3 [\rho_2(q_1,q_3,\varphi) \\ &\quad - \rho_1(q_1,\varphi)\rho_1(q_3,\varphi)]\,C(q_1,q_3,\varphi) \,.\end{aligned} \tag{4.95}$$

We put here $\varphi = 0$ and assume that the system is spatially homogeneous. We get ultimately

$$\begin{aligned}\mu(|q_1 - q_2|) &= C(|q_1 - q_2|) \\ &+ (1/v)\int dq_3\, \mu(|q_1 - q_3|)C(|q_2 - q_3|) \,.\end{aligned} \tag{4.96}$$

We have obtained the Ornstein–Zernike equation, which was used above to get the first-order Percus–Yevick equation.

We choose the generating functional in the form

$$F_{\text{gen}} = \rho_1(q_1,\varphi)\exp[\beta\varphi(q_1)] \,, \tag{4.97}$$

where

$$\varphi = \sum_{1 \leqslant i \leqslant N} \varphi(q_i) = \sum_{1 \leqslant i \leqslant N} \Phi(|q_0 - q_i|) \,, \tag{4.98}$$

i.e., the external potential is chosen to be the potential of the interaction between the additional particle at the point q_0 and all the remaining particles. We choose the independent functional to be the one-particle distribution function

$$F_{\text{ind}} = \rho_1(q_1,\varphi) \,. \tag{4.99}$$

We now expand the functional F_{gen} in powers of the functional F_{ind}:

$$\begin{aligned}F_{\text{gen}}(q_1,\varphi) = F_{\text{gen}}(q_1) &+ \int dq_2 [F_{\text{ind}}(q_2,\varphi) - F_{\text{ind}}(q_2)]\left(\frac{\delta F_{\text{gen}}(q_1,\varphi)}{\delta F_{\text{ind}}(q_2,\varphi)}\right)_{\varphi=0} \\ &+ \frac{1}{2!}\int dq_2\, dq_3 [F_{\text{ind}}(q_2,\varphi) - F_{\text{ind}}(q_2)]\,[F_{\text{ind}}(q_3,\varphi) \\ &- F_{\text{ind}}(q_3)]\left(\frac{\delta^2 F_{\text{gen}}(q_1,\varphi)}{\delta F_{\text{ind}}(q_2,\varphi)\delta F_{\text{ind}}(q_3,\varphi)}\right) + \cdots \,.\end{aligned} \tag{4.100}$$

In this case,

$$\rho_1(q_1,0) = \rho_1(q_1)\,,$$

$$\rho_2(q_1,\varphi) - \frac{N \int dq_2\cdots dq_N \exp[-\beta U]}{\int dq_1\cdots dq_N \exp[-\beta U]}$$

$$= \frac{N(N+1)Q_{N+1}^{-1} \int dq_2\cdots dq_N[-\beta\,\Sigma_{0\leqslant i<j\leqslant N}\,\Phi(|q_i - q_j|)]}{(N+1)Q_{N+1}^{-1} \int dq_1\cdots dq_N \exp[-\beta\,\Sigma_{0\leqslant i<j\leqslant N}\,\Phi(|q_i - q_j|)]}$$

$$= \frac{\rho_2(q_1,q_0)}{\rho_1(q_0)}\,, \tag{4.101}$$

where Q_{N+1} is the configuration integral of a system of $N+1$ particles.

We now rewrite Eq. (4.100) in the form

$$\rho_1(q_1,\varphi)\exp[\beta\varphi(q_1)] = \rho_1(q_1) + \int dq_2[\rho_1(q_2,\varphi) - \rho_1(q_2)]\frac{\delta\{\rho_1(q_1,\varphi)\exp[\beta\varphi(q_1)]\}}{\delta\rho_1(q_2,\varphi)} + \frac{1}{2!}\int dq_2\,dq_3[\rho_1(q_2,\varphi) - \rho_1(q_2)][\rho_1(q_3,\varphi) - \rho_1(q_3)] \times\frac{\delta^2\{\rho_1(q_1,\varphi)\exp[\beta\varphi(q_1)]\}}{\delta\rho_1(q_2,\varphi)\delta\rho_1(q_3,\varphi)} + \cdots\,. \tag{4.102}$$

Next,

$$\frac{\delta\{\rho_1(q_1,\varphi)\exp[\beta\varphi(q_1)]\}}{\delta\rho_1(q_2,\varphi)} = \int dq_3\,\frac{\delta\{\rho_1(q_1,\varphi)\exp[\beta\varphi(q_1)]\}}{\delta[-\beta\varphi(q_3)]}\,\frac{\delta[-\beta\varphi(q_3)]}{\delta\rho_1(q_2,\varphi)}\,. \tag{4.103}$$

Consider now the expression

$$\frac{\delta\rho_1(q_1,\varphi)\exp[\beta\varphi(q_1)]}{\delta[-\beta\varphi(q_3)]} = \exp[\beta\varphi(q_1)]\,\frac{\delta\rho_1(q_1,\varphi)}{\delta[-\beta\varphi(q_3)]} + \rho_1(q_1,\varphi)\,\frac{\delta\exp[\beta\varphi(q_1)]}{\delta[-\beta\varphi(q_3)]}\,. \tag{4.104}$$

Taking Eq. (4.90) into account, we get

$$\frac{\delta\rho_1(q_1,\varphi)\exp[\beta\varphi(q_1)]}{\delta[-\beta\varphi(q_3)]} = \exp[\beta\varphi(q_1)][\rho_2(q_1,q_3,\varphi) - \rho_1(q_1,\varphi)\rho_1(q_3,\varphi) + \rho_1(q_1,\varphi)\delta(q_1 - q_3) - \rho_1(q_1,\varphi)\delta(q_1 - q_3)]\,. \tag{4.105}$$

Substituting this expression in Eq. (4.103) we obtain, with allowance for Eq. (4.92),

$$\int dq_3\left\{\frac{\delta\rho_1(q_1,\varphi)\exp[\beta\varphi(q_1)]}{\delta[-\beta\varphi(q_3)]}\,\frac{\delta[-\beta\varphi(q_3)]}{\delta\rho_1(q_2,\varphi)}\right\}_{\varphi=0}$$

$$=\int dq_3\Big(\exp[\beta\varphi(q_1)][\rho_2(q_1,q_3,\varphi)-\rho_1(q_1,\varphi)$$

$$\times\rho_1(q_3,\varphi)]\{[\rho_1(q_2,\varphi)]^{-1}\delta(q_1-q_3)-C(q_2,q_3,\varphi)\}\Big)_{\varphi=0}$$

$$=\int dq_3[\rho_2(q_1,q_3)-\rho_1(q_1)\rho_1(q_3)][\delta(q_2-q_3)/\rho_1(q_2)-C(q_2,q_3)]\,. \tag{4.106}$$

Taking the Ornstein–Zernike equation (4.96) into account, we obtain

$$\int dq_3\left\{\frac{\delta\rho_1(q_1,\varphi)\exp[\beta\varphi(q_1)]}{\delta[-\beta\varphi(q_3)]}\,\frac{\delta[-\beta\varphi(q_3)]}{\delta\rho_1(q_2,\varphi)}\right\}_{\varphi=0}=\rho_1(q_1)C(q_2,q_1)\,. \tag{4.106'}$$

Substituting Eq. (4.106) in Eq. (4.102), retaining only the linear term, and allowing for Eq. (4.101), we get for a homogeneous system

$$\mu(|q_0-q_1|)\exp[\beta\varphi(q_3)]=1+\frac{1}{v}\int dq_2[\mu(|q_0-q_2|)-1]C(|q_1-q_2|)$$

$$=\mu(|q_0-q_1|)-C(|q_0-q_1|)\,.$$

Assuming $\varphi(q_i)=\Phi(|q_0-q_i|)$, we have

$$C(r)=\{1-\exp[\beta\Phi(r)]\}\mu(r)\,. \tag{4.107}$$

Substituting the $C(r)$ obtained in this manner in the Ornstein–Zernike equation, we obtain the Percus–Yevick equation in first order:

$$\mu(r)\exp[\beta\Phi(r)]$$

$$=1-\frac{1}{v}\int dq'[\mu(|q-q'|)-1]\mu(|q'|)\{\exp[\beta\Phi(|q'|)]-1\}\,, \tag{4.108}$$

where $r=|q|$.

Transformation of the quadratic term in Eq. (4.102) adds another term to the direct correlation function defined by Eq. (4.77).

This concludes the analysis of the Percus–Yevick equation, which is regarded at present as one of the best approximations.

23. Integral equations for one- and two-particle distribution functions

The integral equations considered in the preceding sections for the two-particle distribution functions described a homogeneous phase. On the other hand, for an ordered structure it is more effective to use the self-consistent-field equation.[33] A general description of the state of a statistical system therefore requires a system of equations which depends on the one- and two-particle distribution functions.

We determine first the integral equation for $\rho_1(q)$. To this end we substitute in the first Bogolyubov-chain equation

$$\theta \frac{\partial \rho_1(q_1)}{\partial q_1^\alpha} + \int \frac{\partial \Phi(|q_1 - q_2|)}{\partial q_1^\alpha} \rho_2(q_1,q_2) dq_2 = 0$$

an approximate expression for the two-particle distribution function, in the form[38]

$$\rho_2(q_1,q_2) = B\rho_1(q_1)\rho_1(q_2)\exp[-\beta\Phi(|q_1 - q_2|)]\,, \tag{4.109}$$

where

$$\beta^{-1} = \frac{1}{N^2}\int \rho_1(q_1)\rho_1(q_2)\exp[-\beta\Phi(|q_1 - q_2|)]dq_1\, dq_2\,.$$

We obtain then a closed equation for the one-particle distribution function

$$\theta \frac{\partial \rho_1(q_1)}{\partial q_1^\alpha} + B\int \frac{\partial \Phi(|q_1 - q_3|)}{\partial q_1^\alpha} \exp[-\beta\Phi(|q_1 - q_2|)]\rho_1(q_1)\rho_1(q_2) dq_2 = 0$$

or

$$\theta \frac{\partial \ln \rho_1(q)}{\partial q_1^\alpha} + B \frac{\partial}{\partial q_1^\alpha} \int \theta\{1 - \exp[-\beta\Phi(|q_1 - q'|)]\}\rho_1(q')dq' = 0\,.$$

Integrating this equation, we obtain a nonlinear integral equation for the one-particle function[123]

$$\theta \ln \lambda\rho_1(q) + B\int \kappa(|q - q'|)\rho_1(q')dq' = 0\,, \tag{4.110}$$

where

$$\kappa(|q - q'|) = \theta(1 - \exp)[-\beta\Phi(|q - q'|)] \tag{4.111}$$

and

$$\lambda = \text{const}\,.$$

Equation (4.110) with $B = 1$ and with the kernel (4.111),

$$\theta \ln \lambda\rho_1(q) + \int \kappa(|q - q'|)\rho_1(q')dq' = 0 \tag{4.112}$$

was obtained in Ref. 124 by introducing the approximating function $\rho_2(q_1, q_2)$ obtained on the basis of its asymptotic behavior at small and large distances. A similar equation was derived even earlier[125,126] from intuitive considerations. Integral equations for a one-particle distribution function has been investigated in many papers.[127–130] The first Bogolyubov-chain equation is represented in Refs. 128 and 130 in the form

$$\nabla_{q_1} \ln \rho_1(q_1) = -\beta\nabla_1 \int \rho_1(q_1)\psi(q_1,q_2)dq_2\,,$$

where the function $\psi(q_1, q_2)$ is expressed in terms of $\rho_1(q_2)$ under a number of simplifying assumptions. It must be noted that a similar approach was proposed earlier,[131] with $\psi(q_1, q_2)$ determined from experimental data.

To describe collective effects it is necessary to include, besides the two-particle kernel (4.111), also contributions from many-particle kernels[125]:

$$\kappa_{\text{com}} = \kappa + \int \kappa_3 \rho_1(q_3)dq_3 + \cdots$$
$$+ \int \cdots \int \kappa_{N-1}\rho_1(q_3)\cdots\rho_1(q_{N-1})dq_3\cdots dq_{N-1}\,. \tag{4.113}$$

An explicit form of the expansion (4.113) was obtained in Ref. 132 by the generating-functional method.

We introduce now the function $\bar{p}_2(q_1, q_2)$ defined by the relation

$$\rho_2(q_1,q_2) = B\bar{p}_2(q_1,q_2)\rho_1(q_1)\rho_1(q_2)\,, \tag{4.114}$$

where the constant B is determined from the condition

$$B = \frac{N^2}{\int\int \bar{p}_2(q_1,q_2)\rho_1(q_1)\rho_1(q_2)dq_1\,dq_2}\,.$$

For the case when ρ_1 is not necessarily constant, the superposition approximation can be represented in the form

$$\rho_3(q_1,q_2,q_3) = \rho_1(q_1)\rho_1(q_2)\rho_1(q_3)\bar{p}_2(q_1,q_2)\bar{p}_2(q_1,q_3)\bar{p}_2(q_2,q_3)C$$
$$= \rho_2(q_1,q_2)\rho_1(q_3)\bar{p}_2(q_1,q_3)\bar{p}_2(q_2,q_3)(C/B)\,, \tag{4.115}$$

where

$$C = \frac{N^3}{\int\int\int \rho_1(q_1)\rho_1(q_2)\rho_1(q_3)\bar{p}_2(q_1,q_2)\bar{p}_2(q_1,q_3)\bar{p}_2(q_2,q_3)dq_1\,dq_2\,dq_3}\,. \tag{4.116}$$

Substituting Eqs. (4.115) and (4.116) in the first two equations of the Bogolyubov chain, we obtain a system of equations for the unary and binary distribution functions

$$\theta\frac{\partial\rho_1(q_1)}{\partial q_1^\alpha} + B\int \frac{\partial\Phi(|q_1 - q_2|)}{\partial q_1^\alpha}\bar{p}_2(q_1,q_2)\rho_1(q_2)dq_2 \times \rho_1(q_1) = 0\,, \tag{4.117}$$

$$\theta\frac{\partial\rho_2(q_1,q_2)}{\partial q_1^\alpha} + \frac{\partial\Phi(|q_1 - q_2|)}{\partial q_1^\alpha}\rho_2(q_1,q_2)$$
$$+ \frac{C}{B}\int \frac{\partial\Phi(|q_1 - q_3|)}{\partial q_1^\alpha}\rho_2(q_1,q_2)\bar{p}_2(q_1,q_3)\bar{p}_2(q_2,q_3)\rho_1(q_3)dq_3\,. \tag{4.118}$$

We assume that the function $\bar{p}_2(q_1, q_2)$ satisfies the relation

$$\bar{p}_2(q + q', q') = \bar{p}_2(|q|)\,. \tag{4.119}$$

The second term of Eq. (4.117) is then transformed into

$$B\int \frac{\partial\Phi(|q_1-q_2|)}{\partial q_1^\alpha}\,\bar{p}_2(q_1,q_2)\rho_1(q_2)dq_2\,\rho_1(q_1)$$
$$= B\rho_1(q_1)\int \frac{\partial\Phi(|q_1-q_2|)}{\partial q_1^\alpha}\,\bar{p}_2(q_1-q_2+q_2,q_2)\rho_1(q_2)dq_2$$
$$= B\rho_1(q_1)\frac{\partial}{\partial q_1^\alpha}\int\left\{\int_\infty^{|q_1-q_2|}\bar{p}_2(r)\frac{d\Phi(r)}{dr}\,dr\right\}\rho_1(q_2)dq_2\,. \tag{4.120}$$

We can similarly transform in Eq. (4.118) the third term, after first expressing it in the form

$$\frac{C}{B}\int \frac{\partial\Phi(|q_1-q_3|)}{\partial q_1^\alpha}\,\rho_2(q_1,q_2)\bar{p}_2(q_1,q_3)\bar{p}_2(q_2,q_3)\rho_1(q_3)dq_3$$
$$= \frac{C}{B}\rho_2(q_1,q_2)\int \frac{\partial\Phi(|q_1-q_3|)}{\partial q_1^\alpha}\,\bar{p}_2(q_1-q_3+q_3,q_3)\{\bar{p}_2(q_2,q_3)-1\}\rho_1(q_3)dq_3$$
$$+\frac{C}{B}\rho_2(q_1,q_2)\int \frac{\partial\Phi(|q_1-q_3|)}{\partial q_1^\alpha}\,\bar{p}_2(q_1-q_3+q_3,q_3)\rho_1(q_3)dq_3\,. \tag{4.121}$$

Since $B=C=1$ in the statistical limit, we obtain ultimately from Eqs. (4.117) and (4.118) the following system of equations for $\rho_1(q_1)$ and $\rho_2(q_1,q_2)$:

$$\theta\frac{\partial\rho_1(q_1)}{\partial q_1^\alpha}+\rho_1(q_1)\int\left\{\int_\infty^{|q_1-q_2|}\bar{p}_2(r)\frac{d\Phi(r)}{dr}\,dr\right\}\rho_1(q_2)dq_2=0\,, \tag{4.122}$$

$$\theta\frac{\partial\rho_2(q_1,q_2)}{\partial q_1^\alpha}+\frac{\partial}{\partial q_1^\alpha}\Bigg[\Phi(|q_1-q_2)+\int\int_\infty^{|q_1-q_3|}\left\{\bar{p}_2(r)\frac{\partial\Phi(r)}{dr}\,dr\right\}\rho_1(q_3)dq_3$$
$$+\int\int_\infty^{|q_2-q_3|}\left\{\bar{p}_2(r)\frac{d\Phi(r)}{dr}\,dr\right\}\rho_1(q_3)dq_3+A\Bigg]=0\,, \tag{4.123}$$

where

$$A=\int\int_\infty^{|q_1-q_3|}\left\{\bar{p}_2(r)\frac{d\Phi(r)}{dr}\,dr\right\}\{\bar{p}_2(|q_2-q_3|)-1\}\rho_1(q_3)dq_3\,. \tag{4.124}$$

This system can be simplified by using Eqs. (4.114) and (4.122), and by recognizing that

$$\left.\begin{array}{l}\bar{p}_1(|q|)\to 1\\ \Phi(|q|)\to 0\\ A(|q|)\to 0\end{array}\right\}\quad\text{as } |q|\to\infty\,.$$

Integration yields then

$$\rho_1(q_1)=\alpha\exp\left[-\frac{1}{\theta}\,\bar{u}(q_1)\right], \tag{4.125}$$

$$\bar{p}_2(q_1,q_2) = \exp\left[-\frac{1}{\theta} W(q_1,q_2)\right], \tag{4.126}$$

where

$$\bar{u}(q_1) = \int\left[\int_{\infty}^{|q_1-q_2|}\left\{\bar{p}_2(r)\frac{d\Phi(r)}{dr}dz\right\}\rho_1(q_2)\right]dq_2, \tag{4.127}$$

$$W(q_1,q_2) = \Phi(|q_1-q_2|) + A(q_1,q_2) = \Phi(|q_1-q_2|)$$
$$+\int\left\{\int_{\infty}^{|q_1-q_2|}\bar{p}_2(r)\frac{d\Phi(r)}{dr}dr\right\}\{\bar{p}_2(|q_2-q_3|)-1\}\rho_1(q_3)dq_3, \tag{4.128}$$

and α is a normalization constant.

The single integral equation for the radially symmetric distribution function is replaced in the superposition approximation in our case by a system of nonlinear integral equations for the unary and binary distribution functions.

Equations (4.125)–(4.128) were derived without discarding terms whose smallness would be due to weak interaction forces, to low density, etc. The only assumption made was that $\rho_3(q_1, q_2, q_3)$ can be represented in the form (4.115).

Note that the kernels of the first equations of the system (4.125), (4.126) and the kernel proposed in Ref. 131 have the same structure.

24. Two approaches to the derivation of the integral equations. Some conclusions

The integral equations considered in the present chapter were obtained by two approaches. The first is based on uncoupling the chain of Bogolyubov equation and obtaining a closed system of equations. This is how the superposition approximation was obtained. We consider briefly certain uncouplings used to obtain the system of integral equations. The chain of equations was uncoupled in Refs. 43 and 133 in the following approximation of a four-particle distribution function $F_4(q_1, q_2, q_3, q_4)$:

$$F_3(q_1,q_2,q_3) = F_2(q_1,q_2)F_2(q_2,q_3)F_2(q_1,q_3)\times T_3(q_1,q_2,q_3),$$
$$F_4(q_1,q_2,q_3,q_4) = \frac{F_3(q_1,q_2,q_3)F_3(q_2,q_3,q_4)F_3(q_1,q_3,q_4)F_3(q_1,q_2,q_4)}{F_2(q_1,q_2)F_2(q_1,q_3)F_2(q_1,q_4)F_2(q_2,q_3)F_2(q_2,q_4)F_2(q_3,q_4)}. \tag{4.129}$$

In this case the second and third equations of the Bogolyubov chain become closed:

$$\theta \frac{\partial \ln F_2(q_1,q_2)}{\partial q_1^\alpha} + \frac{\partial \Phi(|q_1 - q_2|)}{\partial q_1^\alpha}$$

$$+ \frac{1}{v} \int \frac{\partial \Phi(|q_1 - q_3|)}{\partial q_1^\alpha} F_2(q_1,q_3) F_2(q_2,q_3) T_3(q_1,q_2,q_3) dq_3 = 0 \,,$$

$$\theta \frac{\partial \ln T_3(q_1,q_2,q_3)}{\partial q_1^\alpha} + \frac{1}{v} \int \frac{\partial \Phi(|q_1 - q_4|)}{\partial q_1^\alpha} F_2(q_1,q_4)$$

$$\times [F_2(q_2,q_4) F_2(q_3,q_4) T_3(q_1,q_2,q_4) T_3(q_2,q_3,q_4) T_3(q_2,q_3,q_4)$$

$$- F_2(q_2,q_4) T_3(q_1,q_2,q_4) - F_2(q_3,q_4) T_3(q_1,q_3,q_4)] dq_4 \,. \tag{4.130}$$

The condition for weakening of the correlations for the functions $F_2(q_1, q_2)$ and $F_3(q_1, q_2, q_3)$ takes the form

$$F_2(q_1,q_2) \to 1, \quad T_3(q_1,q_2,q_3) \to 1 \,,$$

if one of the particles goes off to infinity. Introduction of $T_3(q_1, q_2, q_3)$ is in principle not obligatory, and it suffices to use an approximation in the form of the second equality in Eq. (4.129) (Ref. 134).

Differentiation of the second equation of the chain yielded in Refs. 135–139 a tensor equation in which $\partial F_3(q_1,q_2,q_3)/\partial q_1^\alpha$ was replaced by an expression from the third equation, i.e., in this case the equation contains $F_2(q_1, q_2)$, $F_3(q_1, q_2, q_3)$, and $F_4(q_1, q_2, q_3, q_4)$. The approximation proposed for the uncoupling was

$$F_4(q_1,q_2,q_3,q_4) = F_3(q_1,q_2,q_3) F_2(q_1,q_4) F_2(q_2,q_4) F_2(q_3,q_4) \,. \tag{4.131}$$

It was assumed furthermore that the Kirkwood approximation is valid for $F_3(q_1, q_2, q_3)$. This requirement is not consistent, for in this case the superposition approximation does not generally follow from Eq. (4.131) upon differentiation with respect to q_4.

An expansion in powers of the density can be obtained in the equation for the binary distribution function in the superposition approximation. This is done by substituting in Eq. (2.148) the expression

$$\Phi(|q_1 - q_3|) \frac{\partial \Phi(|q_1 - q_3|)}{\partial q_1^\alpha}$$

from the equation

$$\theta \frac{\partial F_2(q_1,q_3)}{\partial q_1^\alpha} + \frac{\partial \Phi(|q_1 - q_3|)}{\partial q_1^\alpha} F_2(q_1,q_3)$$

$$+ \frac{1}{v} \int \frac{\partial \Phi(|q_1 - q_4|)}{\partial q_1^\alpha} F_3(q_1,q_3,q_4) dq_4 = 0 \,, \tag{4.132}$$

and making in the resultant equation a similar substitution for the expression $F_2(q_1,q_4) \partial \Phi(|q_1 - q_4|)/\partial q_1^\alpha$. The series obtained by iteration of this procedure can be represented as

$$-\theta\frac{\partial \ln F_2(q_1,q_2)}{\partial q_1^\alpha}=\frac{\partial\Phi(|q_1-q_2|)}{\partial q_1^\alpha}$$
$$+\theta\sum_{n=1}^{\infty}(-1)^n\frac{1}{v^n}\int\cdots\int\left\{\frac{\partial F_2(q_1,q_{n+2})}{\partial q_1^\alpha}\right\}F_2(q_{n+1},p_{n+2})$$
$$\times\prod_{i=3}^{n+1}\{F_2(q_1,q_i)F_2(q_{i-1},q_i)\}dq_3\cdots dq_{n+2}\,. \qquad (4.133)$$

Methods exist for improving the superposition approximation for the purpose of obtaining the maximum possible number of virial coefficients that coincide with the exact values. The problem here is to have as a result an equation that preserves the important property of the superposition approximation, viz., that it reveals the presence of a phase transition and describes at the same time the equation of state of a homogeneous phase.

The second approach to obtaining closed integral equations is based on the concept of direct and total correlation. Such an approach was proposed by Ornstein and Zernike. As the definition of a direct correlation function one can use either Eq. (4.134) itself or an expression for the inverse functional derivative for a definite type of functional, from which follows the Ornstein–Zernike equation

$$h(|q|=C|q|)+\frac{1}{v}\int C(|q-q'|)h(|q'|)dq'\,. \qquad (4.134)$$

In this approach it is assumed that the total correlation $h(|q|)$ consists of the direct correlation $C(|q|)$ due to the direct interaction of two particles, and an indirect correlation due to the action of one particle on another via a third. To obtain a closed equation we must express $C(|q|)$ in terms of $h(|q|)$ and the interaction potential. We have already considered the most important equation obtained by this approach—the Percus–Yevick equation. Another well-known equation is also obtained similarly, viz., the equation in the hyperchain approximation. In this case the direct correlation function is of the form

$$C(|q_1-q_2|)=h(|q_1-q_2|)-\ln\mu(|q_1-q_2|)-\frac{\Phi(|q_1-q_2|)}{\theta}\,. \qquad (4.135)$$

Substituting this approximation in the Ornstein–Zernike equation, we obtain a closed equation for $\mu(|q_1-q_2|)$:

$$\ln\mu(|q_1-q_2|)+\frac{\Phi(|q_1-q_2|)}{\theta}=\frac{1}{v}\int\Bigg[\mu(|q_1-q_3|)$$
$$-1-\ln\mu(|q_1-q_3|)-\frac{\Phi(|q_1-q_3|)}{\theta}\Bigg][\mu(|q_2-q_3|)-1]dq_3\,. \qquad (4.136)$$

The equation in the hyperchain approximation can be obtained also by using the functional-differentiation technique. To this end we choose a generating functional

$$F_{gen} = \ln\{\rho_1(q_1,\varphi)\exp[\beta\Phi(|q_0 - q_1|)]\}\,, \tag{4.137}$$

and an independent functional

$$F_{ind} = \rho_1(q_1,\varphi)\,; \tag{4.138}$$

equations can then be derived in both the first and second approximations.

Analysis of the hyperchain-approximation (HCA) equation and of the Percus–Yevick (PY) equation shows that they can be obtained by summing a definite class of diagrams. The solutions of both equations lead to exact expressions for the second and third virial coefficients; the higher coefficients are not exact, the PY equation leading to more accurate results than the HCA, even though the latter takes more diagrams into account. The PY equation gives better results not only for low densities but also for high ones, and for temperatures above critical. Consequently, summation of a large number of diagrams does not necessarily improve the result, since the diagrams disregarded in the PY approximation enter in the general equation for $C(|q|)$ with different signs and are partly canceled out. When a large number of diagrams is included in the HCA approximation, the cancellation becomes incomplete and the results are worse. Therefore, any approximation obtained by summing part of the diagrams cannot be regarded as strictly valid.

Thus, all the integral equations, no matter how obtained, can be closed only by introducing certain postulates. Neither the method of functional differentiation nor the diagram technique can corroborate these postulates. They permit, however, observation of certain characteristic features of the approximation made and permit their analysis.

The recent advances in the theory of the liquid state is due to the use of integral equations. Good quantitative agreement was obtained in a number of cases between the theoretical results and the molecular-dynamics-method data. Still, to obtain good equations of state at temperatures below critical it is necessary, beside resorting to solutions of the integral equations, also to introduce semiempirical rules. Improvement of the calculation results by using second-order integral equations (PY-2 and others) entails considerable mathematical difficulties even for simple liquids. It is therefore extremely important to obtain for the hard-sphere PY equation an analytic solution that describes well the homogeneous phase. This solution can be used successfully as a zeroth approximation for the investigation of real systems with the aid of perturbation theory. On the other hand, on the basis of a solution of the integral equations, a phase transition in a hard-sphere system can be revealed only by an equation in the superposition approximation, as attested by the absence of a solution at high density. A more accurate PY equation for the homogeneous phase provides no indication of a phase transition [to be sure, $\mu(r) < 0$ at high densities,[93] but this fact is usually disregarded]. Therefore, an important task of the theory of the liquid state is to obtain an equation that not only describes well the homogeneous phase, but also permits a description of the phase transition.

Chapter 5 Numerical methods in statistical physics

Advances in computer technology have made it possible to calculate the partition function for a sufficiently large number (50 to 1000) of particles. An aggregate of 1000 particles contained in a cube with an edge length on the order of ten molecules diameters can be regarded as a macroscopic system.

Computer determination of the numerical values of the system parameters made possible an analysis of the feasibility of various approximate approaches to the calculation of the partition function, for in contrast to a real experiment one can regard here the system potential as specified. The theoretical calculation for a system of hard spheres can, for example, be compared with the results of computer calculations.

Two methods can be used for the numerical calculations: the Monte Carlo method (MCM) and the molecular-dynamics method (MDM). Each has its own distinctive features. Among the advantages of the MCM is the possibility of calculating the parameters of quantum systems, while the MDM permits investigation of nonequilibrium processes. Let us examine these methods.

25. The Monte Carlo method

"Monte Carlo method" is a collective name used for methods in which the exact dynamic behavior of a system is replaced by a stochastic process. In the Monte Carlo method the system executes random walks in configuration space, with the initial state taken to be some regular arrangement of the particles. Each state is assigned a definite probability, and the system reaches equilibrium after a certain number of steps. The statistical mean values obtained in the MCM are averages over various configurations. The possibility of identifying averages over time and over an ensemble is determined in the MCM by the ergodic theorem. It is assumed that the system considered is subject to periodic boundary conditions. A particle departing from a cubic volume is replaced by a particle entering the volume from the opposite direction.

Assume a system of N particles with a regular initial order, say at the sites of a cubic lattice. A regular arrangement of the particles is chosen to ensure that the system be in a physically realizable state at the start of the calculation. The data on the locations of all the particles are fed to a computer. The first, second, etc., particles are next displaced in turn. After the displacement of the Nth particle, the procedure is repeated. The displacement of the ith particle is described by

$$q_i = (x_i, y_i, z_i) \rightarrow q_i + \Delta q_i$$
$$= (x_i + A\eta_{1i}, y_i + A\eta_{2i}, z_i + A\eta_{3i}), \qquad (5.1)$$

where A is the maximum displacement, η_{1i}, η_{2i}, η_{3i} is a set of random numbers, obtained with the computer, such that $-1 \leqslant \eta_{ij} \leqslant 1$. Random-number generation is a separate topic not touched upon here. A particle undergoing a displacement (5.1) has equal probability of being located in a cube $2A$ on a side and the existence of a periodic boundary condition allows it to remain inside the volume under consideration. Assume that at the initial state i the system energy was E_i and that after the transition described by Eq. (5.1) the system went over into a state j and the corresponding calculated energy became equal to E_j. The energy increment E_{ij} is thus

$$E_{ij} = E_j - E_i. \qquad (5.2)$$

If $E_{ij} < 0$, the particle remains in its previous state. In the case $E_{ij} \geqslant 0$, the transition can have only a probability $\exp[E_{ij}/\theta]$. To realize such a transition, the computer generates a random quantity $0 \leqslant \eta_{4i} \leqslant 1$, and the transition is implemented if $0 \leqslant \eta_{4i} \leqslant \exp[-E_{ij}/\theta]$, but not in the opposite case, i.e., $\exp[-E_{ij}/\theta] < \eta_{4i} < 1$. The particle remains in the preceding state, and the given configuration is considered again. The value of f is calculated in each step. If it is equal to f_i in the ith step, the mean value after k displacements is obtained from the equation

$$\langle f \rangle = \frac{1}{k} \sum_{i=1}^{k} f_i. \qquad (5.3)$$

Determination of the mean value by using Eq. (5.3) is equivalent to averaging over a Gibbs distribution, i.e., to ascribing a weight $\exp[-E_i/\theta]$ to the ith configuration. In fact, assume that we have a large ensemble of systems with N particles, and ν_i systems of the ensemble each have an energy E_i. If the particles are displaced in each of the ν_i systems, then p_{ij} is the probability that the system will go over from a state i to a state j, and in accordance with the conditions introduced above we have $p_{ij} = p_{ji}$. If $E_{ij} < 0$, then $\nu_i\, p_{ij}$ out of the ν_i systems will go over into the state j, and $\nu_i\, p_{ji} \exp[-E_{ji}/\theta]$ systems will go from j to i. The final number of systems going from i to j is equal to

$$p_{ij}\{\nu_i - \nu_j \exp[-E_{ji}/\theta]\}.$$

If

$$\nu_i - \nu_j \exp[-E_{ji}/\theta] > 0$$

or

$$\frac{\nu_i}{\nu_j} > \exp[-E_{ji}/\theta] = \frac{\exp[-E_i/\theta]}{\exp[-E_j/\theta]}$$

a large number of particles will go over on the average from i to j, and after a sufficient number of displacements the system reaches a stationary state in which

$$\nu_i \sim \exp[-E_i/\theta].$$

At small k, the quantity $< f >$ first fluctuates and then tends to an equilibrium value with increase of k. The optimal rate at which $< f >$ tends to an equilibrium value is reached by choosing the optimal value of A.

The Monte Carlo method is most convenient for isothermal processes, since the temperature is a fixed parameter in this method.

We examine now the determination of the thermodynamic function for an NVT ensemble. Since the systems considered are small, the development of the thermodynamic theory must be approached with certain caution, for not all the usual thermodynamic relations will be satisfied simultaneously. In ordinary thermodynamics, for example, the pressure p is an intensive parameter, but here it will depend not only on θ and V/N, but also on N.

For an NVT ensemble, the thermodynamic potential is the free energy $F_N(V,\theta)$ defined by the relations

$$F = -\theta \ln Z(N,V,\theta),$$

$$Z(N,V,\theta) = \frac{1}{\lambda^{3N}N!} Q(N,V,\theta), \quad \lambda = \left(\frac{h^2}{2\pi m\theta}\right)^{1/2}, \tag{5.4}$$

$$Q(N,V,\theta) = \int \cdots \int \exp[-\beta U_N] dq_1 \cdots dq_N. \tag{5.5}$$

Most frequently, the thermodynamic parameters are calculated in the Monte Carlo method not by using Eq. (5.4) directly, but by using the thermal and caloric equations of state. Thus, the system energy is

$$E = \tfrac{3}{2} N\theta + \langle U_N \rangle, \tag{5.6}$$

where $< U_N >$ is the average potential energy and can be calculated from Eq. (5.3) by putting in it $f = U_N$,

$$U_N = \sum_{i<j} \Phi(|q_i - q_j|) + \sum{}' \sum_{i<j} \Phi(|L\nu + q_i - q_j|), \tag{5.7}$$

$L = V^{1/3}$, $\nu = (\nu_1, \nu_2, \nu_3)$ is a vector whose coordinates are defined by a triad of integers. The prime on the sum over ν means exclusion of the vector (0, 0, 0) corresponding to the considered basis cell. The second sum in Eq. (5.7) is the result of the periodic boundary conditions.

Equation (5.7) can be used in calculation for only short-range potentials. The calculations are frequently simplified by introducing a "truncated" potential

$$\overline{\Phi}(r) = \Phi(r)\theta(r_k - r), \tag{5.8}$$

where $\theta(t)$ is the unit step function

$$\Phi(t) = \begin{cases} 0 & t<0 \\ 1 & t \geqslant 0 \end{cases}. \tag{5.9}$$

It is most frequently assumed that $r_k = L/2$, since this excludes effects due to the finite period of the system. In addition, there exist now for any (i, j) pair at least one ν for which the following relation holds:

$$|L\nu + q_i - q_j| < r_k.$$

We can then redefine $r_{ij} = q_i - q_j$ and set it equal to

$$r_{ij} = \min_{\nu} |L\nu + q_i - q_j|. \tag{5.10}$$

Equation (5.7) takes then the usual form

$$U_N = \sum_{i<j} \Phi(|r_{ij}|). \tag{5.11}$$

By choosing $r_k < L/2$ the periodic system can be represented in toroidal form. To this end we join mentally the opposite faces and introduce in a suitable manner the distance between the points. This representation is more convenient than consideration of an infinite periodic system.

To determine the heat capacity of the system at constant volume we begin with the expression

$$C_V = k \left(\frac{\partial E}{\partial \theta}\right)_{N,V} \tag{5.12}$$

and, using Eq. (5.6), obtain

$$C_V = \frac{3}{2} Nk + \frac{k}{\theta^2} [\langle U_N^2 \rangle - \langle U_N \rangle^2], \tag{5.13}$$

where $\langle U_N^2 \rangle$ and $\langle U_N \rangle$ are determined from Eq. (5.3).

We represent the expression for the pressure

$$p = -\left(\frac{\partial F}{\partial V}\right)_{N,\theta}$$

in the form

$$\frac{pv}{\theta} = 1 - \frac{\langle U_N' \rangle}{3N\theta}, \tag{5.14}$$

where

$$\langle U_N' \rangle = \sum_{i<j} |r_{ij}| \Phi'(|r_{ij}|) + \sum_{\nu}{}' \sum_{i<j} |L\nu + r_{ij}| \Phi'(|L\nu + r_{ij}|). \tag{5.15}$$

If the "truncated" potential (5.8) is used, Eq. (5.15) acquires a δ function whose contribution is neglected, with approximate account taken of the discarded term. If the interaction potential has discontinuities, expression (5.15) cannot be used directly. Therefore, say to determine the pressure of a hard-sphere system, relation (5.39) is used rather than Eq. (5.14). That is to say, it is necessary in this case to know the two-particle distribution function at the discontinuity of the potential. In our case, the two-particle distribution function is given by[139]

$$F(r_1, r_2) = \rho_1^{-1}(r_1)\rho_1^{-1}(r_2)\left\langle \sum_{\nu} \sum_{i<j} \sum \delta(q_i - r_1)\delta(L\nu + q_j - r_2) + \sum_{\nu} \sum_{i} \delta(r_i - r_1)\delta(L\nu + q_i - r_2) \right\rangle, \tag{5.16}$$

where $\rho_1(r)$ is the one-particle generic distribution function

$$\rho_1(r) = \left\langle \sum_i \delta(r_i - r) \right\rangle. \tag{5.17}$$

Assume that r_1 is located in cell $\nu = 0$ and r_2 in cell ν_1 or in any other cell (except $\nu = 0$). The second sum in Eq. (5.16) is due to the periodicity of the system and corresponds to the case when one and the same ith particle is located in both positions, has therefore no physical meaning, and is disregarded in the calculations.

Besides the described use for the calculation of an *NVT* ensemble, the Monte Carlo method can be used also for the calculation of other types of ensemble.[17,139]

A serious problem in the Monte Carlo method is violation of the ergodicity condition when states cannot be reached. We shall not dwell on this here. Aspects of the solution of this problem are discussed in a number of papers (see, for example, Refs. 17 and 139).

As has been already noted, an advantage of the Monte Carlo method is that it can be used to describe properties of quantum systems. Numerical calculations of the properties of the ground state of He^4 are given in Ref. 140. It is assumed there that the molecules are Bose particles with zero spin and that the potential energy is given by Eq. (5.7); moreover, the interaction has a Lennard-Jones potential with parameters ε and σ determined from data on the behavior of the virial coefficients at high temperatures. The Hamiltonian of the system in question is of the form

$$H_N = -\frac{\hbar^2}{2m} \sum_{i=1}^{N} \nabla_i^2 + U_N. \tag{5.18}$$

The trial wave function is chosen in the form

$$\psi = \prod_{1 \leqslant i < j \leqslant N} f(|q_i - q_j|), \tag{5.19}$$

where $f(r)$ is chosen in the form of the two-parameter function

$$f(r) = \exp[-(a_1/r)^{a_2}], \tag{5.20}$$

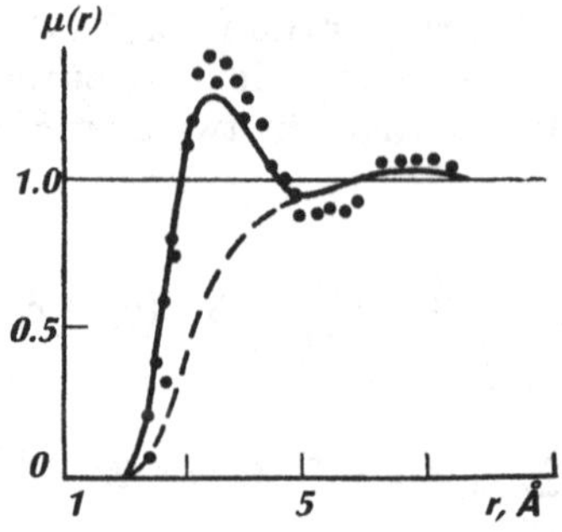

Figure 20.

while a_1 and a_2 are certain parameters. If a_1 and a_2 are fixed, the mean value of H_N is given by

$$E_N(a_1, a_2) = \langle H_N \rangle = \frac{\int \cdots \int \psi^* H_N \psi \, dq_1 \cdots dq_N}{\int \cdots \int \psi^* \psi \, dq_1 \cdots dq_N}. \tag{5.21}$$

In our case it can be represented in the form

$$E_N = (a_1, a_2) = \frac{\int \cdots \int F(q_1, \ldots, q_N) dq_1 \cdots dq_N}{\int \cdots \int \psi^2 \, dq_1 \cdots dq_N}, \tag{5.22}$$

where

$$F(q_1, \ldots, q_N) = \sum_{1 \leqslant i < j \leqslant N} \left\{ -\frac{\hbar^2}{2m} \nabla_1^2 \ln f(|q_i - q_j|) \right\} + U_N. \tag{5.23}$$

The integrals in Eq. (5.22) are calculated by the Monte Carlo method at fixed values of a_1 and a_2. The best values of the parameters are determined from the condition that the free energy have a minimum. The calculations were carried out for a system of $N = 32$ particles for 41 values of the parameters a_1 and a_2.

The parameters at which the energy of liquid He^4 reaches a minimum at 0 atm and 0 K ($\rho = 2.20 \times 10^{22}$ atoms/cm^3) are

$$a_1 = 2.6 \text{ Å}, \quad a_2 = 5,$$

and the energy is in this case $0.77 \pm 0.09 \times 10^{-15}$ erg/atom. The experimental value of the energy is 0.998×10^{-15} erg/atom. Theory and experiment differ thus by ~20%. The density increases next all the way to its value at the phase transition. The parameter a_2 is regarded here as fixed and equal to 5, while a_1 is chosen such as to minimize E_N for each value of the density. The result is a discontinuity of the plot of a_1 versus density, which can be interpreted as the consequence of a liquid–solid phase transition. While the agreement with experiment is good, some disparities are also present. Thus, although the plot of the calculated distribution function $\mu(r)$ (see Fig. 20) agrees well with the plot of the experimental function, both in the steep region near 26 Å and in the region of the maximum, the experimental curve has a higher peak at 3.5 Å and oscillates more strongly at larger r. No better agreement with experiment can be obtained in the considered two-parameter theory.

Calculations were performed also for the one-particle density matrix

$$\rho_1(q_1 - q_1') = N \int \psi^*(q_1,q_2,...,q_N)\psi(q_1',q_2,...,q_N)$$

$$\times dq_2 \cdots dq_N \Big/ \int \psi^*(q_1,...,q_N)\psi(q_1,...,q_N)dq_1 \cdots dq_N. \quad (5.24)$$

As $q \to 0$ we have $\rho_1 \to \rho$, where ρ is the particle-number density; as $q \to \infty$ we have $\rho \to \rho_0$, where ρ_0 is the density of the number of particles with zero momentum.

The approach considered can be used in practice only if $\theta = 0$. At nonzero temperatures, the volume of the calculations increases strongly and much less information is obtained as a rule concerning the equations of state.

26. Molecular-dynamics method

Compared with the Monte Carlo method, the molecular-dynamics method is based on a simpler principle and consists of solving a system of Newton's equations for an N-body system (similar quantum calculations for N on the order of ten are not realistic at the present level of computation technology).

Let the potential energy of the interaction between the system particles be

$$U_N = \sum_{1 \leqslant i < j \leqslant N} \Phi(|q_i - q_j|). \quad (5.25)$$

Confinement to a pair interaction potential is not obligatory, but greatly simplifies the calculations and is therefore customarily used in calculations for sufficiently large N.

Regarding the system as closed, we obtain the total force acting on the ith particle:

$$F_i = -\sum_{j \neq i} \nabla\Phi(|q_i - q_j|). \quad (5.26)$$

Let all the particles have one and the same mass m. We write for them the system of Newton's equations

$$m\frac{dv_i}{dt} = m\frac{d^2q_i}{dt^2} = F_i. \quad (5.27)$$

A finite-difference form of these equations is more suitable for computer calculations. To this end, we break up the time scale into finite intervals Δt. We assume that at the initial instant of time t_0 the coordinates q_i^0 and velocities v_i^0 are specified for all particles. We have then in our case the difference problem[141]

$$q_i^{n+1} - 2q_i^n + q_i^{n-1} = F_i^n(\Delta t)^2/m,$$
$$q_i^{n+1} - q_i^{n-1} = v_i^n \Delta t, \tag{5.28}$$
$$q_i^0 = q_i(t_0), \quad q_i^n = q_i(t_0 + \Delta tn), \quad v_i^n = v_i(t_0 + \Delta tn).$$

Solution of Newton's equations implies obtaining information on the behavior of the coordinate as a function of time. We therefore eliminate q^{n-1} from the first equation of Eq. (5.28) with the aid of the second equation. We have then

$$q_i^{n+1} = q_i^n + v_i^n \Delta t + \frac{1}{2m} F_i^n \Delta t^2. \tag{5.29}$$

Thus, from the initial values of tie coordinates and velocities we can determine the time dependence of q_i. The difference system approximates a second-order differential operator. From the coordinates determined from the second equation of Eq. (5.28) we get the velocities.

The solutions of the equations of motion are subject to periodic boundary conditions, viz., a multiple of $L = V^{1/3}$ is added to the coordinate of each particle repeatedly, until the cubic cell is reproduced not less than 26 times. As a result, if one particle leaves the cell, another particle having the same momentum enters through the opposite face. The density and energy of the system are thus conserved. To simplify the calculations, the size of the cell is chosen much larger than the effective radius of the potential. Special calculation methods are used for systems with long-range Coulomb potential.

Boundary conditions other than periodic are also imposed. Thus, if random boundary conditions are used, particles crossing the boundary leave the system without returning, whereas at random time intervals new particles having random velocities enter the system at random time intervals and at random points. The system density fluctuates in this case, i.e., this approach corresponds to a grand canonical ensemble.

At high densities, the cell is chosen such that the periodic boundary conditions generate an ideal crystal lattice. For argon, a cubic cell is therefore usually chosen with $N = 4n$, where $n = 1, 2, 3, 4, 5, 6, \ldots$.

Having determined the time dependences of the system coordinates $q_i(t)$ and momenta $p_i(t)$, we can determine the mean value of any function $f[p_1(t),\ldots,p_N(t),q_1(t),\ldots,q_N(t)]$ of the dynamic variables

$$\langle f \rangle = \lim_{T\to\infty} \frac{1}{T} \int_0^T f[p_1(t),\ldots,p_N(t),q_1(t),\ldots,q_N(t)]\,dp_1\cdots dp_N\,dq_1\cdots dq_N. \tag{5.30}$$

Calculations by the molecular-dynamics method have shown that the distribution in velocity in the system becomes Maxwellian quite rapidly (within a time in which several collisions take place). The temperature is therefore defined in the molecular-dynamics method by the relation

$$T = \frac{1}{3k} \left\langle \sum_{1 \leqslant i \leqslant N} m_i v_i^2 \right\rangle. \tag{5.31}$$

When isothermal processes are simulated, T is maintained constant. When the temperature deviates from an admissible value, all the particle velocities are raised or lowered to attain stability of T.

A number of equilibrium characteristics of the system (heat capacity, compressibility, thermal pressure coefficient) are calculated from the fluctuations of the energy and of the virial $[r d\Phi(r)/dr]$ for a smooth paired potential. The fluctuations become large in the vicinity of a phase transition. Particularly large computational difficulties arise near the critical point.

Modern computers can operate with intervals Δt on the order of 10^{-14} s. In a hard-sphere system, several spheres can overlap simultaneously, i.e., a multiparticle interaction sets in. If Δt is short enough, however, this interaction can be regarded as a sequence of paired collisions.

In the molecular-dynamics method, the initial values of the coordinates are not randomly specified, i.e., overlapping configurations are excluded. The initial configuration can also be chosen to have the structure of a periodic lattice. The initial velocities are usually chosen equal but randomly directed. The total kinetic energy should then accord with the specified temperature. Once the atoms begin to drop in sequence from the initial state, the systems begin to relax to an equilibrium state.

The molecular-dynamics method can be used in two ways: either to determine the form of the intermolecular potential from the experimental data, or to use simple enough potentials to develop and refine the theories. Since MDM calculations can be carried out at arbitrary densities, the interaction potential is determined from data on the entire phase diagram (in contrast to the previously employed determination of the interaction potential from the equation of state at low densities). The potential obtained can then be reconciled either with the data on neutron or x-ray scattering, or with theoretically determined potentials. For inert gases, the potential determined by the MDM does not differ greatly from the theoretical one. The difference for ionic compounds, however, is appreciable.

The second method of using the MDM is based on using data for sufficiently simple systems, with attempts to fit them to different theories. This obviates the use, for liquids, of effects that are difficult to describe theoretically, such as the deviation from a pair potential, the presence of impurities, and others. In addition, this yields the properties of the so-called "ideal liquid" that can be used next as the basic model for more accurate theory.

It was observed that at a constant density the structure of the liquid depends weakly on the temperature T (at large T). Since the principal role is played at high temperatures by the repulsive part of the potential, a particle system with such a potential can be regarded as a good approximation for the description of the liquid state. The attractive part of the potential is treated as a perturbation.

The molecular-dynamics method has made it possible to analyze the complicated character of particle motion in a dense medium. Calculations were made possible for a great variety of models with a spherically asymmetric interparticle interaction potential.

As already noted, investigations by the molecular-dynamics method were carried out mainly for systems with interaction potentials described by simple model potentials, viz., hard-sphere and hard-disk systems in the three- and two-dimensional cases, respectively. This has permitted a detailed study of particle motion in these systems, particularly of the nature of transport phenomena (Ref. 142).*

It was previously believed that the molecules are transported by a hopping mechanism in which the hop is rarely as long as the intermolecular distance.[22] A particle can hop out of the surround of its nearest neighbors at some limiting distances between them. According to this diffusion mechanism, there should exist two characteristic free path lengths, one corresponding to the vibration of the particle in the cell, and the other of the order of the intermolecular distance. Let the particle have a probability $P(r)dr$ of negotiating the distance $r, r + dr$ between two successive collisions. According to the hopping diffusion mechanism, the probability density $P(r)$ should have two maxima, but this is not obtained from MDM calculations for a hard-sphere system. The latter calculations have shown that $P(r)$ decreases monotonically with increase of r. Furthermore, $\overline{P}(x)$ is practically independent of the density, in view of the dependence of $\overline{P}(x) = \lambda P(r/\lambda)$ on $x = r/\lambda$ [λ is the mean free path, $\lambda = v/2\pi\sigma^2\mu(\sigma)$ according to Enskog[142]], the molecular flux in the system is due to a large number of small hops, and not to a small number of large ones. At densities corresponding to an ordered structure the behavior of $\overline{P}(x)$ is similar, except that the curve lies somewhat lower (this is most strongly manifested at large x). The transport mechanism in a solid is therefore analogous to that in a homogeneous phase, i.e., the diffusion is via a large number of small hops that are more probable than hops of length equal to the intermolecular distance.

We introduce the temporal velocity correlation function (TVCF) $\langle v_x(0)v_x(t)\rangle$ (it is also called the velocity autocorrelation function[17]), from which we can assess the role of the temporal and spatial correlations and investigate the nature of the collective motions in a system.

Plots of the TVCF for various densities are shown in Fig. 21. Enskog's theory [143] calls for an exponential plot (labeled E in the figure). For low densities [$V/V_0 = 3$, curve 1 (for 108 particles); $V/V_0 = 2$, curve 2 (for 500 particles); V_0 is the volume corresponding to close packing] the TVCF is a monotonically decreasing function. In this case the moving particle drags with it the surrounding particles.

The asymptotic behavior of TVCF for long times is described by a function $\sim t^{-3/2}$ for a hard-sphere system and by a function $\sim t^{-1}$ for a hard-disk system. Since these functions decrease very slowly, the motion has a collective character and can be described by the laws of hydrodynamics. This was confirmed also by a direct numerical solution of the Navier–Stokes equation for a volume element equal to the volume (per particle) displaced in the liquid (the initial velocity was assumed equal to the mean squared particle velocity).

*We follow hereafter the exposition of Ref. 142.

Figure 21.

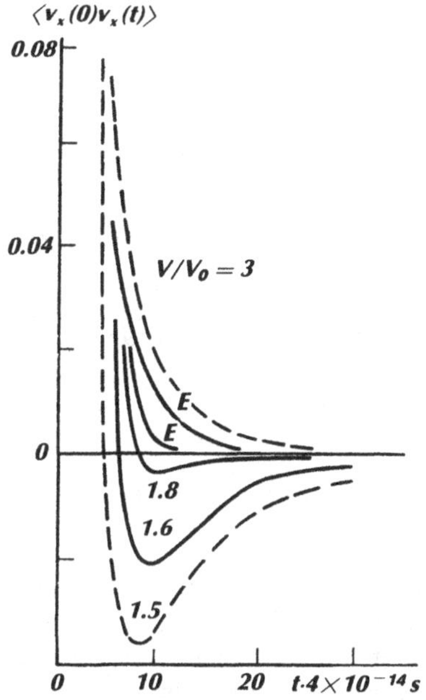

To investigate the long-time behavior of the system, the cell must be quite large, but not so large as to be crossed by the sound wave resulting from the periodic boundary conditions, i.e., $t < L/v_{ac}$ and $L = V^{1/3}$.

A theoretical analysis of the behavior of the TVCF at large t yielded the following asymptotic form[144]:

$$\langle v(0)v(t)\rangle \underset{t\to\infty}{\to} \frac{2\theta v}{m}\,[4\pi(D+\nu)t]^{-3/2}, \tag{5.32}$$

where D is the diffusion coefficient and ν the kinematic-viscosity coefficient. It was shown at the same time that the particle drifts together with its surrounding neighbors, so that D in Eq. (5.32) is the Lagrangian diffusion coefficient that determines the diffusion of a drop of characteristic dimension $R = \frac{8}{3}\sqrt{\tau\nu}$, where τ is the relaxation time of the viscous stresses.

With increase of the velocity, the TVCF acquires a negative part (at $V/V_0 = 1.8$, 1.6, and 1.5 in Fig. 21). Since the particle is surrounded at high densities by a barrier, the motion of the particles surrounding a given particle is on the average reversed, so that the TVCF now has a negative part. The role of the negative temporal correlation increases with increase of density.

In some transport phenomena, such as viscosity and heat conduction, potential terms take on a substantial role alongside the kinetic ones. Neglect of attraction forces leads then to a considerable distortion of the phenomenon in question. It is therefore impossible to stay within the framework of the hard-sphere approximation even in the zeroth approximation. In this simplest case these data are considered by using the Lennard-Jones interaction potential

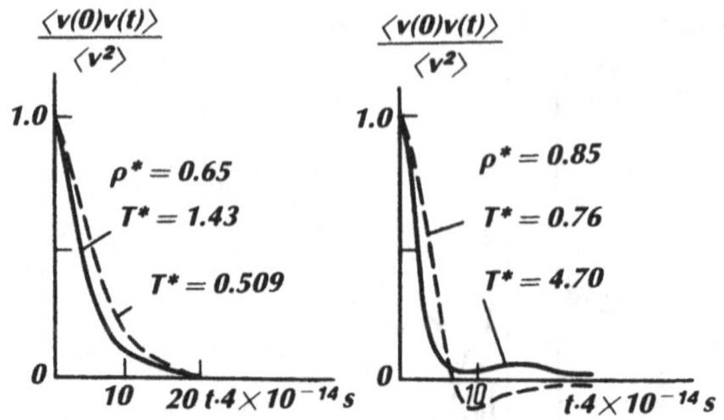

Figure 22.

$$\Phi(r) = 4\varepsilon\left[\left(\frac{\sigma}{r}\right)^{12} - \left(\frac{\sigma}{r}\right)^{6}\right].$$

A system of particles with an interaction potential of this type describes the principal chracteristic features of liquids. The molecular-dynamics method reveals many properties of these systems. Attraction between the particles leads to the formation of vacancies and clusters in the liquid phase. The disposition of the vacancies in space is irregular. Near the edge of a vacancy, the particles produce a microscopic surface tension that prevents them from entering the unoccupied region. The vacancy lifetime is $\sim 3\times 10^{-12}$ s. The most thoroughly investigated is the two-dimensional particle system, for which the positions of the particles at different instants of time and at various frequencies were determined. It was observed that clusterlike aggregates exist even above the critical temperature. Tracking the motion of a single particle reveals a "hopping" diffusion mechanism. The particle moves mainly in one direction together with its accompanying group. At higher density and longer time, the cluster diffusion is replaced by chain diffusion.[142]

Since the interaction potential between the particle has, besides a repulsive part, also an attractive one, the collision durations and the intervals between them become commensurate. Figure 22 shows the TVCF of particles with a Lennard-Jones interaction potential (here $\rho^* = \sigma^3/v$; $\theta^* = \theta/\varepsilon$; $\varepsilon = 119.8$ K; $\sigma = 3.405$ Å, Ref. 142). For short times, the TVCF has a Gaussian form, in contrast to the hard-sphere system, where it is exponential.

In addition to the model considered above, there is also another model[145,146] in which the liquid constitutes a system of hard spheres that move between collisions along Brownian trajectories that result from collisions caused by the attractive part of the potential. Since the latter collisions disturb the temporal correlation of the particle motion, this motion can be regarded as uncorrelated. On the basis of these assumptions, we can write kinetic equations for the distribution functions and obtain the kinetic coefficients from their collisions.

The molecular-dynamics method was used in Ref. 147 to investigate a system with a potential

$$\Phi(r)=\begin{cases} 0 & (r>b) \\ -\varepsilon & (\sigma<r<b) \\ \infty & (r<b) \end{cases}. \tag{5.33}$$

We have excluded from the total number of collisions the ones due to the hard core, and named the remainder "soft." Calculations have shown that the ratio of the number of "soft" collisions to the number of hard-sphere collisions is ~ 0.5. Doubts were therefore cast on the statement that the correlations are greatly weakened by "soft" collisions, since they produce a small change in the particle momentum.

It is assumed in this theory that the time evolution of the particle momentum between hard-sphere collisions obeys the equation

$$\frac{dp}{dt}=-\frac{\xi}{m}p+X(t), \tag{5.34}$$

where $X(t)$ is a random force and ξ is the friction constant. We introduce the autocorrelation function of the random forces[142]:

$$\zeta(\tau)=\langle X(t)X(t+\tau)\rangle. \tag{5.35}$$

It is assumed that $\zeta(\tau)=0$ if $\tau>\tau_1$, where τ_1 is much less than the velocity correlation time. It is assumed in addition that $\zeta(\tau)$ has a sharp peak at $\tau=0$ and $\tau \ll m/\xi$. In this case

$$\xi=\frac{1}{\theta}\int_0^\infty d\tau\,\zeta(\tau). \tag{5.36}$$

If the autocorrelation function of the total force F is considered, then

$$\xi=\frac{1}{3\theta}\int_0^{\tau_1} d\tau\langle F(t)F(t+\tau)\rangle, \tag{5.37}$$

where τ_1 is obtained from the condition

$$\langle F(t)F(t+\tau_1)\rangle=0. \tag{5.38}$$

Calculations carried out by the molecular-dynamics method have shown that appreciable correlations are present in the system. In addition, if the collision operators are to satisfy the assumptions made above, it is necessary that their eigenvalue spectra not overlap, and in this case the relaxation times in a hard-sphere system and in a system of particles with a van der Waals interaction potential must be substantially different. The MDM data do not confirm this. Calculations of the transport coefficients show a difference, by a factor 1.5 to 2, between the theoretical and experimental data. The main assumptions of the considered approach are therefore not satisfied. This example shows how the MDM makes it possible to analyze and verify the main tenets of the suggested theories.

The MDM can be used to interpret and to predict neutron-scattering spectra. We introduce to this end first an equilibrium density correlation function

$$G(r,t) = V\left\langle \sum_{i,j} \rho_i(r,t)\rho_j(0,0) \right\rangle, \tag{5.39}$$

where

$$\rho_i(r,t) = \delta[r - r_i(t)]. \tag{5.40}$$

We represent $G(r, t)$ in the form

$$G(r,t) = G_s(r,t) + G_d(r,t), \tag{5.41}$$

where

$$G_s(r,t) = V\langle \rho_i(r,t)\rho_i(0,0)\rangle. \tag{5.42}$$

Here $G_s(r,t)$ is the probability density of finding the particle at the instant of time t at the point r, if its position at $t = 0$ was $r = 0$. $G_s(r,t)$ can be calculated by the MDM, by introducing the relation

$$G_s(r,t) = \frac{1}{4\pi r^2 \Delta r}\langle \theta\,[r_i(t) - r_i(0) - r]\rangle, \tag{5.43}$$

where

$$\theta(x - r) = \begin{cases} 1 & (|r| < |x| < |r + \Delta r|) \\ 0 & (|x| < |r| \text{ or } |x| > |r + \Delta r|) \end{cases}$$

A relation can also be introduced for $G_d(r,t)$, the pair correlation function

$$G_d(r,t) = \frac{V_n(r,t)}{N 4\pi r^2 \Delta r}, \tag{5.44}$$

where $n(r,t)$ is the number of particles located at an instant of time at a distance $r + \Delta r$ from the position occupied by a definite atom at $t = 0$.[142]

We denote by $S_s(k,\omega)$ the Fourier transform of the function $G_s(r,t)$. This transform determines the differential cross section for incoherent neutron scattering. On the other hand, the structure factor $S(k,\omega)$ is the Fourier transform of $G(r,t)$ and is the differential cross section for coherent scattering. Although $G_s(r,t)$ and $G(r,t)$ can be directly measured, this cannot be done for all k and ω, since such measurements are complicated and consume much material. The use of the MDM is therefore effective. It cannot be used for investigations in the case of large wavelengths, so that the data are obtained here from experiments on light scattering in a liquid. The result is information on the function $S(k,\omega)$ for almost all values of k and ω.

In theoretical investigations it is customarily assumed that $G_s(r,t)$ have a Gaussian form, i.e.,

$$G_s(r,t) = [4\pi A(t)]^{-3/2} \exp[-r^2/4A(t)], \tag{5.45}$$

where $A(t) = \frac{1}{6}\langle r^2\rangle$.[142] It is therefore necessary to determine hence $\langle r^{2n}\rangle$:

$$\langle r^{2n}\rangle = C_n\langle r^2\rangle^n = \frac{(2n+1)!!}{3^n}\langle r^2\rangle, \quad n = 1,2,3,\ldots. \tag{5.46}$$

Analogous quantities can be determined also by the MDM method, and will be designated $\langle r^{2n}\rangle_{MD}$. The deviation of $\langle r^{2n}\rangle$ from $\langle r^{2n}\rangle_{MD}$ is characterized usually by the quantity

$$\alpha_n(t) = \frac{\langle r^{2n}\rangle_{MD}}{C_n\langle r^2\rangle^n} - 1. \tag{5.47}$$

Calculations yield $\alpha_n \geqslant 0$, i.e., the true $G_s(r,t)$ tends to zero more slowly than Eq. (5.45). As $t\to\infty$ we have $\alpha_n \to 0$ ($\alpha_n \sim 0$ already at $t \sim 10^{-11}$, and $\langle r^2\rangle^{1/2}$ is equal to the distance to the first neighboring particle).

Many other examples of productive utilization of the MDM for the analysis of various approaches to the analysis of many-particle systems can be cited. In addition, this method yielded fundamental results on the behavior of systems of hard disks and hard spheres and on phase transitions in these systems, thereby adding greatly to our knowledge of the behavior of statistical systems. In the next sections of this chapter we consider mainly the results obtained for different systems by numerical methods.

27. Hard-disk and hard-sphere systems

Hard-disk and hard-sphere systems are the most widely investigated by the molecular-dynamics and Monte Carlo methods. The particle-interaction potential is in this case of simplest form, so that the computer time is considerably shortened and a sufficiently large number of particles can be included. At an equal number of particles, the hard-disk system is much more effective than a system of hard spheres, and is considerably less influenced by the boundary conditions.

We have already dealt with the results of the molecular-dynamics system when we considered cluster expansions for hard-disk and hard-sphere systems, for which the maximum number of virial coefficients is known. Results of computer experiments for such very simple systems, albeit not encountered in nature, can shed light on the possibility of using the virial and other cluster expansions.

The results of investigations of the equations of state of a system of hard disks, both by the Monte Carlo method and by the MDM, are in good agreement, including those in phase-transition regions. The scatter of the points is due to various factors, such as errors due to the statistical spread [$\sim o(N^{-1})$], to ergodicity, to suppression of the momentum fluctuations in the molecular-dynamics method,[17] and others.

A correlation was found between the geometric configuration of the system and the pressure fluctuation in the region between the isotherms of the ordered and homogeneous states.[148,149]

The Monte Carlo method is used to reveal phase transitions from an ordered into a homogeneous state, but not transitions in reversed order. Treatment of this system in a gravitational field did not permit separation of the phases. Doubts are therefore cast on the premise that the resultant transi-

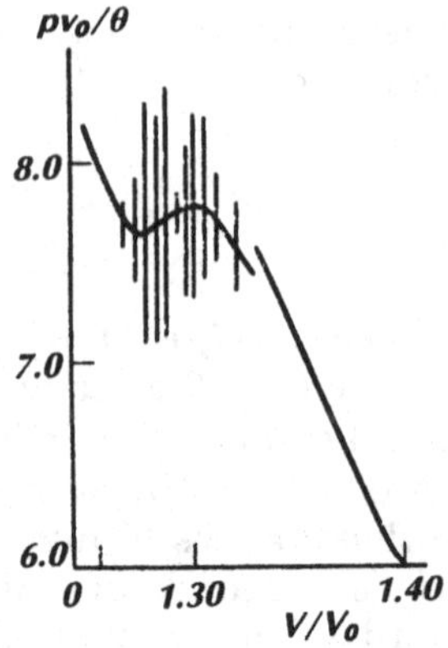

Figure 23.

tion is of first order, and were dispelled by the results of MDM investigations.[149] A study of a three-dimensional system of 500 particles indicates that the character of the particle motion changes fundamentally. Assume that the system was initially ordered and had an hcp structure, and that the particles moved about some equilibrium position. When the volume was decreased by 30% relative to close packing, the system became unstable and transitions betwen ordered and inhomogeneous phases were observed in it, but coexistence of the two phases could not be discerned. Two-dimensional systems of hard disks were therefore investigated, since fewer particles were needed to form a cluster of particles of one phase of arbitrary specified diameter, smaller than in the case of three-dimensional systems. Therefore, the considered system of 870 hard disks was much more effective than the system of 500 hard spheres.[140] If, however, a system of few (72) particles is considered in the two-dimensional case, its behavior is similar to that of a three-dimensional system, viz., there are two unconnected modes: in the range from $S^* = S/S^0 = 1.33$ to 1.35 the system fluctuates strongly between the high-pressure mode corresponding to the uniform phase and the mode corresponding to the ordered structure (S_0 is the area corresponding to close packing of the particles). At $S^* \leqslant 1$ the ordered phase is always stable. Coexistence of two phases was observed in the system of 870 hard spheres (see Fig. 23). It was investigated over a period of time corresponding to several million collisions. Large fluctuations took place in the phase-transition region, so that both phases could coexist with equal probability. To investigate the fluctuations, series of 10 million collisions each were broken up into segments of 50 000 collisions each, over which the averaging was carried out. The thin vertical lines in Fig. 23 denote the range of pressures found among these segments. As seen from the figure, the number of fluctuations is significantly larger in the phase transition region than in the region of fully ordered or perfectly uniform phase. Photographs of a system of hard disks at densities corresponding to the phase transition indicate clearly the presence of two coexisting phases. The existence of a phase transition in a system of hard disks was thus demonstrated. There is no attractive component of the potential in the hard-disk system, so that the phase transition obtained is purely geometric. It is usually attributed to the presence of a shear instability in the motion of the ordered-structure particles, since "it is impossible to shift one layer relative to another in the close packed state, but slippage becomes possible when the system volume is increased."[142]

The fact that results were obtained for a system of hard disks in the case of different numbers of particles makes it possible to extrapolate these results towards large N. The absence of coexistence of two phases in a system of 72 particles is attributed to the high interface energy of the two phases, which the fluctuations are not strong enough to produce. If, however, a system of 72 disks is observed long enough, averaging over all the states produces on the pressure plot a horizontal plateau which lies in this case 10% below the analogous plateau for a system of 870 particles.[142] For a system of 870 hard disks, the free surface energy per particle is small compared with its average kinetic energy, so that it became possible to observe phase coexistence in this system.

The MDM is used also to investigate metastable states of liquids.[150,151]

As already noted, for a system of hard spheres, in contrast to that of hard disks, the problems connected with the number of considered atoms, with the size of the cell, etc., become much more complicated. The hard-sphere systems considered at the present time are too small to observe coexistence of a uniform structure with a periodic one. It is usually possible to observe a transition from an ordered into a uniform phase, but the inverse process leads to a transition of the uniform phase into a metastable state. The metastable state of a macroscopic system is upset by large fluctuations that cannot arise in small systems with periodic boundary conditions. In small systems the metastable states can exist for a very long time. This makes the investigation of hard-sphere systems quite complicated.

The isotherms of the ordered and uniform phases differ by 10%. A transition between them is therefore possible.[17] To draw a coexistence line for the two phases it is necessary to use a thermodynamic analysis. Two coexisting phases should have equal chemical potentials as well as equal pressures. For the uniform phase, the expressions for the absolute value of the entropy, meaning also for the chemical potential as well as for the pressure, are known with high ($\sim 1\%$) accuracy. The entropy of a periodic structure, on the other hand, is determined by integration, accurate to an additive constant. A method for determining this entropy was proposed in Refs. 152 and 153. To this end a system is considered in which no phase transition is possible. It is assumed that the center of the particle cannot go outside of a unit cell of volume $v = V/N$ at any density. At sufficiently large v the particles will collide both with their neighbors and with the cell walls. At high densities, a particle collides mainly with neighboring particles, and at low ones mainly with the cell walls. The presence of walls prevents disorder of the structure at low densities. At low densities it is possible to calculate exactly the thermodynamic properties of an artificial cellular system, as well as of a uniform system. At high densities, introduction of cells plays no role, since it makes no additional contribution to the collective entropy. It is assumed at present that the earlier opinion that a collective entropy appears upon melting [17] is in error. Extrapolation of an ordered structure through the metastability region into the low-density region has made it possible to determine the absolute value of the entropy in the entire density range.

We measured the quantity

$$\left(\frac{\partial S}{\partial V}\right)_{N,\theta} = kP/\theta \tag{5.48}$$

up to 0.75 of the density for close packing, a value in the stability region of the periodic structure. To determine the additive constant, Eq. (5.48) was integrated from a known value at low density. This yielded for high-density entropy

$$S(\rho) = S_0(\rho_0) + N\int_{\rho}^{\rho_0} \frac{k\rho}{\theta\rho^2}\,d\rho, \tag{5.49}$$

where $\rho = 1/v$ and $S_0(\rho_0)$ is the calculated entropy of an artificial ordered structure at a certain low density [it is assumed that $S_0(\rho_0)$ is known]. The unknown entropy was thus determined accurate to $0.015Nk$. The pressure at which the two phases coexist was calculated for a hard-sphere system to be $(8.27 \pm 0.13)N\theta/V_0$, where V_0 is the closed-packed volume. The limiting relative densities are then $(0.736 \pm 0.003)\,v_0/v$ and (0.667 ± 0.003) for the ordered and uniform structures, respectively.

Considering the entropy difference for a system without constraints and for a system with single-particle occupancy of the cells, we obtain the so-called collective entropy:

$$\Delta S = (S - S_{03}). \tag{5.50}$$

The calculations showed that, up to the phase-transition region, the collective entropy is proportional to the density and depends little on the system dimensions. If the collective entropy is taken into account in the cell model, it is therefore necessary to add a term proportional to the density.

The molecular-dynamics method, and also the Monte Carlo method, have demonstrated the geometric character of the transition between the ordered and uniform phases, thereby confirming Lindeman's empirical law[154] that describes the melting of a large class of substances. In its initial formulation, Lindeman's law reduces to a statement that a substance begins to melt when the maximum displacement of the atom from the lattice site reaches 0.1 of the atom's radius, i.e., melting begins after the volume of the solid has increased by approximately 30% from its value in the close-packed state at 0 K. Lindeman's law is usually expressed in terms of the ratio of the potential energy for the maximum displacement of the atom to its kinetic energy, using a harmonic approximation for the atom motion, and expressing the elastic constant in terms of the Debye temperature. This approach, however, obscures the geometric nature of the phase transitions, since an impression may be gained that it can take place in a system with purely harmonic forces.[93]

Investigation of the properties of liquids and solids shows that when a solid melts it becomes unstable to the long-wave shear mode.[155] The calculations entail in this case a determination of the instability of nonlinear equations whose nonlinearity is due to anharmonicity.[156,157] The molecular-dynamics method demonstrates the validity of this approach. A simple model is considered, called the correlated lattice model, in which the central particle is correlated with some (usually nearest) neighbors, and the remaining particles are fixed in their sites, just as in the usual Einstein model of a solid. If a system of hard disks is considered in the correlated lattice model, an analytic solution can be obtained.[93] The equation of state has then a van der Waals loop close to that determined from the molecular-dynamics method. The model includes an interpretation of melting from the geometric standpoint and from the standpoint of shear instability: the minimal density at which one row of atoms is still unable to glide between two others determines the limit of existence of the ordered structure. In addition to its advantage, the model has a shortcoming. It is incapable of providing a qualitatively correct description of the single-particle distribution function. In the three-dimensional case, it does not include melting as a first-order phase transition.[93]

The single-particle distribution function is quite sensitive to low-frequency oscillations. This was first deduced from the elasticity theory of solids.[159] As $N\to\infty$ the half-width of the single-particle distribution function in a two-dimensional system tends to infinity. In the three-dimensional case, however, the half-width is bounded. Therefore, in contrast to the two-dimensional case, for three dimensions the form of the single-particle distribution function of an ordered phase differs in principle from that of uniform phase. The sufficient condition for the existence of a solid state in two-dimensional systems is only a relative ordering of the particles. Consider a high-density system of hard disks or spheres with $v\sim v_0$ and $v/v_0 - 1 \ll 1$. In this case we write the equation of state in the series form[93]

$$\frac{pv}{\theta} = D/\alpha + C_0 + C_1\alpha + C_2\alpha^2 + \cdots, \tag{5.51}$$

where $\alpha = (v - v_0)/v_0$, and D is the dimensionality of the system.

As shown above, the first term on the right-hand side of Eq. (5.51) is asymptotically exact as $\alpha\to 0$. There is no general approach to the derivation of the expansion (5.51). The coefficients $C_i\,(i = 0, 1, 2,...)$ can be determined by the molecular-dynamics method as well as on the basis of the lattice theory and the correlated latttice theory. Data for the system of hard disks are given in Table IV.[93]

We see that the correlated lattice theory describes correctly the ordered phase in a system of hard disks.

Hard-sphere systems were investigated for face-centered and hexagonal structures. Within the accuracy limits of a computer experiment ($\sim$0.01%), no differences were observed between the equations of state of these structures. Since these structures have equal arrangements of the first- and sec-

Table IV. Data for the system of hard disks.

	C_0	C_1	C_2
Numerical experiment	1.80	0.8	>0
Lattice theory	1.56	−0.1	>0
Correlated lattice theory	1.80	0.8	>0

ond-nearest neighbors, their spectra have identical low- and high-frequency parts and differ only in their central parts.[93]

Hard-sphere and hard-disk systems were investigated in detail at low densities by a computer experiment. The results were compared with the virial equation of state, for which seven coefficients are known at present. The agreement is improved by constructing a Padé approximant that permits a highly accurate description of the equation of state of a uniform phase. Some leeway in the choice of the form of the approximant is restricted by the required best fit to the computer-experiment data. Therefore, a Padé approximant expanded in a virial series yields, first, exact values of the known coefficients and, second, estimates of the higher virial coefficients B_{i0}, the accuracy decreasing as i is increased. Therefore, only the estimate of the eighth virial coefficient can be regarded as sufficiently reliable for a hard-sphere system. This is confirmed by the fact that if a Padé approximant is constructed using six virial coefficients, the seventh virial coefficients agrees well enough with its known numerical value.

The molecular-dynamics method and the Monte Carlo method provide thus a complete description of a system of hard disks and a system of hard sphere and yield their thermodynamic properties.

28. Particle systems with square potential wells and with Lennard-Jones potentials

Although the hard-disk and hard-sphere systems considered in the preceding section exhibit many characteristic properties of real systems, the absence of an attractive component in their potentials makes a description of the entire phase diagram impossible, and there is no distinction between a liquid and a gas in these model systems. Numerically investigated systems have therefore more complicated potential functions, the most widely used among which are

$$\Phi(r) = \begin{cases} \infty & (r<\sigma) \\ -\varepsilon & (\sigma \leqslant r \leqslant g\sigma) \\ 0 & (g\sigma < r < \infty) \end{cases}, \tag{5.52}$$

and also the Lennard-Jones potential

$$\Phi(r) = 4\varepsilon\left[\left(\frac{\sigma}{r}\right)^{12} - \left(\frac{\sigma}{r}\right)^{6}\right]. \tag{5.53}$$

In Eq. (5.52), $g\sigma$ is the action limit of the attractive potential, ε is the well depth, σ is the diameter of the hard core, and g is a numerical parameter. Putting $\varepsilon = 0$ transforms Eq. (5.52) into the hard-sphere potential. The same holds for the properties of the thermodynamic functions if $T \to \infty$. We consider first the results of calculations for a potential of the form (5.52). The calculations were performed by both the Monte Carlo and the molecular-dynamics methods. In the one-dimensional case, a system for $g = 150$ was investigated by the Monte Carlo method. Dense clusters were produced then when the temperature T was lowered. No equation of state was obtained in this case.[159] If $\varepsilon/\theta \ll 1$, clusters are produced even at $g < 2$ (Ref. 17). Calculations were performed also for a three-dimensional system with a potential (5.52), both by the Monte Carlo method ($g = 1.50$),[160] and by the MDM ($g = 185$).[161] The equation of state of such a system is written in the form [17]

$$\frac{pv}{\theta} = 1 + \frac{2\pi\sigma^3}{3v}\{\mu(\sigma) - [1 - \exp(-\varepsilon/\theta)]\mu(g\sigma)g^3\}, \tag{5.54}$$

with $\mu(\sigma)$ and $\mu(g\sigma)$ functions of the density. The second term in the curly brackets makes possible a van der Waals loop possible under certain conditions. The isotherm of a system of hard spheres is obtained as $\varepsilon/\theta \to 0$. Isotherms were calculated in Ref. 160 for $\varepsilon/\theta = 0, 0.33$, and 1.2 and in Ref. 161 for $\varepsilon/\theta = 0.58$ and 0.77, with the data in the latter case limited by the low densities. A second family of van der Waals loops is formed at low densities (at $v/v_0 \approx 3$). The onset of negative pressures at $\varepsilon/\theta = 1$ and 2 in Ref. 160 is attributed to the periodic boundaries, and the present of sections where $(\partial p/\partial v)_{N,\theta} < 0$ is attributed to the finite size of the considered systems. No such sections should be observed in large systems in view of the presence of spontaneous fluctuations in them. We note that the presence of two families of van der Waals loops was noted in an earlier study [162] of isotherms in the cell theory. The loops had in the high-density region much larger amplitudes than in Ref. 160, and were preserved down to the critical point $\theta_C \sim 5\varepsilon$. The presence of two families of van der Waals loops was attributed in Ref. 162 to a liquid–gas and liquid–solid transition. It is assumed at present that there is no critical point for the liquid–solid transition, so that the interpretation in Ref. 162 of the presence of two van der Waals families is apparently not valid.

Compared with the potential (5.52), the Lennard-Jones potential (5.53) is of great interest, since it describes well enough the interaction between particles of a number of real substances for which many experimental data are available. A system of particles with a Lennard-Jones interaction potential is not only of theoretical but also of practical interest. In one of the first applications of the MDM to a study of a particle system with a Lennard-Jones interaction potential, the results of a numerical experiment were compared with the data for argon.[163] The Lennard-Jones interaction potential has two parameters. The calculation results are given in relative units of ε for energy and σ for length. The calculation results are different for specific substances only because of their different ε and σ. Conversely, the experimental data can be used to determine ε and σ.

The equations of state calculated for argon by the MDM were in good agreement with experiment at all densities up to the triple point. This agreement, however, became worse as the pressure increased. This is usually regarded as an indication that the multiparticle interactions make a substantial contribution. To take them into account effectively, the two-particle potential is assumed to be density dependent. This raises the question of the validity of using a two-particle potential for a description of interactions in a real multiparticle system. It is shown in a number of papers that even a density-independent two-particle potential is effective and takes multiparticle interactions into account. In fact, for example, the parameters of the Lennard-Jones potential are determined on the basis of some or other experimental data that reflect all the interactions that take place in the system, and therefore these parameters also depend effectively on all the types of interaction in the system. The plot of a true (two-particle) interaction potential lies somewhat below that of the Lennard-Jones potential used in practice.*

Calculations for systems with Lennard-Jones potentials were made also by the MDM. In Ref. 139, in particular, the isotherm at 126 K is compared with the experimental data. The isotherm has the usual form, with a broad positive region with height ~ 10 atm relative to the equilibrium pressure, as well as a narrow negative peak ($T = 126$ K lies between the triple point 83.8 K and the critical point 150.9 K of argon). The region of the minimum cannot be accurately determined for several reasons, including the small number of the points determined and the difficulty of determining the equation of state of a liquid by using Eq. (5 14) (Ref.130). Experiment yielded for a liquid at equilibrium with its vapor at 126.7 K a value $pv/\theta \sim 0.06$. Since pv/θ is defined as the difference between unity and another quantity [see Eq. (5.14)] determined with a standard deviation 0.04–0.07, the accuracy of pv/θ is of the same order as pv/θ itself. The negative part of the van der Waals loop is due to the periodic boundary conditions, since this loop does not tend to become smoothened when N is increased. The possibility of drawing the liquid–vapor coexistence line in accordance with the Maxwell rule at high accuracy is therefore problematic.

The Monte Carlo method yielded a van der Waals loop corresponding to the liquid–solid transition. An investigation at $T = 126$ K and $\rho^* = 0.92$ has also shown that the solid structure is close to face-centered-cubic. To be sure, the particles are somewhat displaced in this case from their positions in an ideal face-centered lattice. At $\rho^* = 0.84$ the system acquires the structure of a liquid, where appreciable random displacements of the particles take place. Also calculated was the radial distribution function at $\rho^* = 0.92$ and $\rho^* = 0.84$, albeit for a small number of particles ($N = 32$). The distribution function obtained at $\rho^* = 0.92$ did not reflect the fact that a crystal structure is present. It is noted in Ref. 139, at the same time, that at $r > \sigma\sqrt{2}$ the crystal structure is determined directly by the periodic boundary conditions.

*In the classical treatment of a statistical system the potential takes into account effectively also the contributions due to quantum effects.

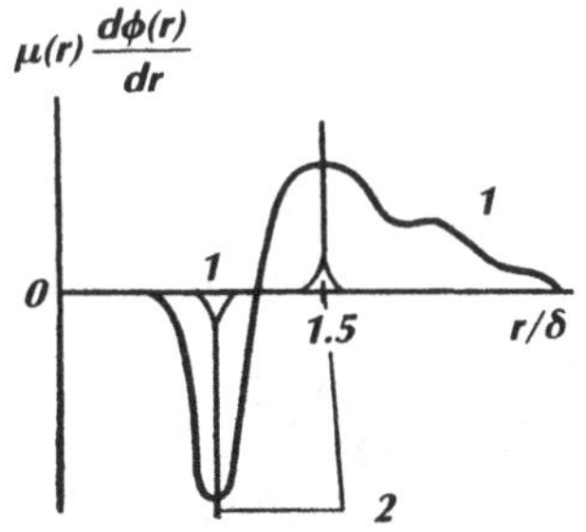

Figure 24.

To analyze the singularities introduced by the Lennard-Jones potential compared with the hard-sphere potential, and also with a potential of type (5.52), we rewrite Eq. (5.52) in the form[164]

$$\Phi(r) = \begin{cases} \infty & (r<\sigma) \\ -\varepsilon & (\sigma<r<1.5\sigma), \\ 0 & (r>1.5\sigma) \end{cases} \tag{5.55}$$

i.e., we put $g = 1.5$. The equation of state takes in this case the form[17]

$$\frac{pv}{\theta} = 1 + \frac{2\pi\sigma^3}{3}\rho\{\mu(\sigma) - [1 - \exp(-\varepsilon/\theta)](1.5)^3\mu(1.5\sigma)\}. \tag{5.56}$$

From the condition $p = 0$ we have a relation for the determination of ρ:

$$(1.5)^3[1 - \exp(-\varepsilon/\theta)]\mu(1.5\sigma) - \mu(\sigma) = 1.5/\pi\rho\sigma^3.$$

Figure 24 shows the expressions for the virials $\mu(r)[d\Phi(r)/dr]$ of a system of particles with a Lennard-Jones potential (curve 1) and for a system with a potential (5.55) (curve 2).[17]

The presence of a positive part in the expression for $\mu(r)d\Phi(r)/dr$ can cause the isotherm to have a region in which $pv/\theta < 1$. For a hard-sphere system, $\mu(r)d\Phi(r)/dr$ has only a negative δ function at $r = \sigma$. The potential (5.55) describes approximately correct the behavior of $\mu(r)d\Phi(r)/dr$ of a system with a Lennard-Jones interaction potential (when for the identical $T^* \sim 1.0$).

The Monte Carlo and molecular-dynamics methods were used to calculate various thermodynamic properties of particle systems with Lennard-Jones potentials. These were gathered in Ref. 17 and compared there with the experimental data.[194–196]

Computer-experiment methods are widely used to analyze solutions of various integral equations for distribution functions and also to verify the principal assumptions under which these equations were derived. It is then possible to eliminate effects due to an inaccurate choice of the potential, effects always encountered when solutions of integral equations are compared with experiment.

The MDM can be used to determine $\mu(r)$, meaning also $h(r) = \mu(r) - 1$ and also the direct correlation function $C(r)$ determined from the Ornstein–Zernike equation

$$h(|r|) = C(|r|) + \rho \int h(|r' - r|)C(|r'|)dr.$$

In this case one can verify the validity of the approximation

$$C(r) = -\mu(r)[\exp(\beta\Phi) - 1],$$

which leads to the Percus–Yevick equation, as a function of the volume and temperature. It was found that the Percus–Yevick equation describes well a statistical system at densities below critical.[165] In another check on the Percus–Yevick equation the latter is used to determine the intermolecular potential, with $\mu(r)$ and $C(r)$ calculated by the MDM. It was found that the Percus–Yevick equation is not effective at high densities and at temperatures lower than critical.

The MDM was also used to investigate the superposition approximation, and it was shown that it is not valid for a Lennard-Jones potential at short distances (see Ref. 17 for a detailed discussion of this question).

An important but much more complicated task is the investigation of surface phenomena. In this case, compared with a three-dimensional system, the number of considered atoms is decreased, and this leads to scantier statistical data. An investigation, by the MDM, of a transition layer in a particle system with a Lennard-Jones potential, nevertheless met with success.[17]

Computer experiments thus provide descriptions of models of many-particle systems with sufficiently realistic potentials, and with results that are in accord in the main with real systems. For an all-inclusive description of the phase diagrams, however, the computation accuracy is still insufficient.[166]

We consider now questions connected with the accuracy of the MDM, which becomes particularly important when the intermolecular-interaction potential has a more complicated form, since the computation time is significantly increased in this case. Modern computers can cope with calculations for systems consisting of several hundred particles. It is therefore important to analyze the effectiveness of the employed difference schemes. For a hard-sphere system, the difference schemes converge quite well, but for particle systems with a Lennard-Jones interaction potential the convergence is much worse, since the interaction potential depends strongly on the distance. Difference-iteration schemes were therefore used in the initial investigations. For sufficiently small chosen difference steps, however, as was subsequently shown, the convergence is satisfactory, and the quality criteria of a difference scheme are a satisfaction of the energy conservation law and the dependence of the coordinate as a function of time on the difference step.[142]

Besides the optimal choice of the difference scheme, the number of calculations can also be decreased by taking into account other factors governed by the large number of the considered particles ($\sim 10^3$) and by the form of the interaction potential, for in this case it is not necessary to calculate all the distances between the particles and all the interaction forces. The interaction between particles in a van der Waals potential and separated by large distances $\Gamma \geqslant 3.3$ can be set equal to zero within the limits of the calculation

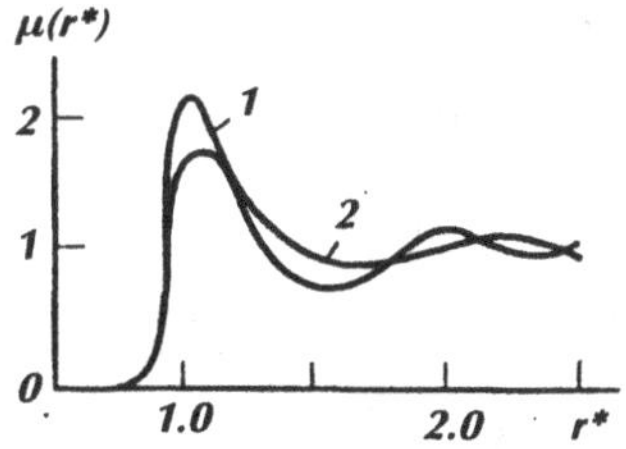

Figure 25.

accuracy. At particle distances $2.5. < r < 3.5\sigma$, where the Lennard-Jones potential varies slowly, the interaction forces can be calculated not for each step but for groups of several steps. These simplifications reduce appreciably the number of calculations, and the error is less than that due to introduction of periodic boundary conditions.

Estimates of the errors due to introduction of periodic boundary conditions have shown them to be of the order of $o(1/N)$ in the second and third virial coefficients. An investigation of the dependence of the diffusion coefficient D on the number of particles led to the equation [167]

$$D_N = D_\infty (1 - 2/N), \tag{5.57}$$

where D_∞ is the diffusion coefficient as $N \to \infty$. Hence we see that for the system with $N > 40$ particles the result is comparable in accuracy with that of the experiment. The asymptotic behavior of some quantity $p_N(1/N)$ as $N \to \infty$ is usually determined by the following interpolation method: the abscissa (x) is $1/N$ and the ordinate (y) the quantity $p_N(1/N)$ approximated by a linear function to determine its value at $N^{-1} = 0$.

To conclude this section, we discuss the results obtained for the pair distribution function of a system of particles with Lennard-Jones interaction potentials. Figure 25 shows a plot of $\mu(r^*)$ $(r^* = r/\sigma)$ for $\theta^* = \theta/\varepsilon = 2.89$ at a density $\rho^* = \rho\sigma^3 = 0.85$ (curve 1) and for $\theta = 2.64$ and $\rho^* = 0.55$ (curve 2). It can be seen from the figure that the plots do not differ in principle from the similar ones obtained for a system of particles with a hard-sphere interaction potential. With increase of density, the heights of the peaks and the slope of the first rise increase, while the maximum shifts to the left, i.e., the structure becomes more pronounced. Figure 26 shows a plot of $\mu(r^*)$ for one density $\rho^* = 0.85$ and at different temperatures, $T = 2.89$ (curve 1) and $T^* = 0.66$ (curve 2).[165] We see that the ordering of the structure becomes more pronounced when the temperature is lowered.

As noted above, the Percus–Yevick equation cannot be used for a system of particles with a Lennard-Jones potential at low temperatures. From the viewpoint of the behavior of $\mu(r^*)$, this is manifested by the fact that the first maximum turns out to be high and is shifted to the left relative to the one actually obtained by the MMD (at high densities). On the other hand, however, at very high densities $\mu(r^*)$ is very well described by the pair distribution function of a hard-sphere system. This has made it possible to use a system of

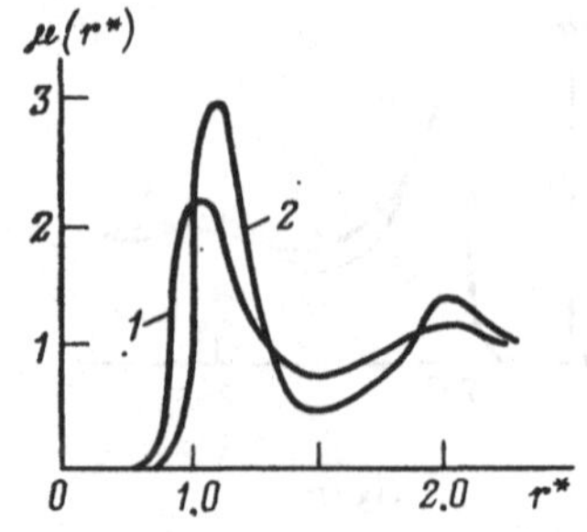

Figure 26.

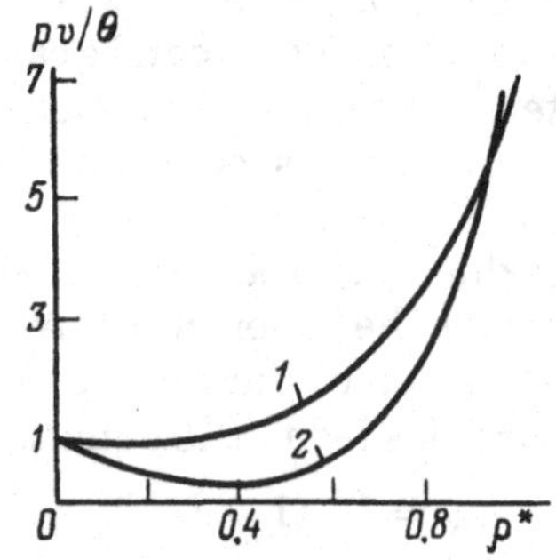

Figure 27.

hard spheres with a diameter that depends on density and temperature as the basic one for the description of liquids, with the attracting part of the potential treated as a perturbation.

Figure 27 shows the equations of state for a system of particles with a Lennard-Jones interaction potential at $T^* = 2.74$ (curve 1) and $T^* = 1.35$ (curve 2),[168] i.e., at a temperature above critical. For ρ^*, the curves start out from the value $pv/\theta = 1$, after which pv/θ decreases, inasmuch as at low densities a large influence is exerted by attractive interactions that cause a decrease of the pressure. With increase of density, pv/θ (the compressibility) first reaches a minimum, and then increases rapidly to values greater than unity, indicating thus that the repulsive part of the potential assumes a larger role. As the temperature is lowered, the influence of the attracting part of the potential increases.

Thus, an investigation of a system of particles with a Lennard-Jones potential has yielded information that adds to our knowledge of the behavior of real statistical systems.

Chapter 6
Principles of the theory of the crystalline state

We study in this chapter the crystalline state of a system of many particles and their ordered arrangement in space. The physics of solid crystalline bodies covers at present a large field. We begin with an exposition of the main tenets of the structure of solids and with an examination of its initial models that serve as the basis for the development of the theory of the crystalline state.

29. Crystal classification by type of structure

A characteristic feature of single crystals is their anisotropy, which is due to the ordered arrangement of the particles. This arrangment is manifested in the crystals by their regular form. In nature, however, one encounters more frequently polycrystalline solids, consisting of a large number of single crystals that have become fused together in disorderly fashion. No anisotropy is therefore observed in polycrystals. We consider here single crystals and take the crystalline state to mean a fully ordered structure.

We proceed to classify the crystalline state by type of structure. We consider a regular three-dimensional lattice that can be obtained by repetition, in three different noncoplanar directions, of a unit crystal lattice (see Fig. 28), called the Bravais lattice. The analytic expression for the coordinates of the lattice points is of the form

$$R = n_1 a_1 + n_2 a_2 + n_3 a_3 \,, \tag{6.1}$$

where n_1, n_2, and n_3 are integers, while a_1, a_2, and a_3 are three arbitrary noncoplanar vectors. The lengths of the parallelepiped edges $|a_1|$, $|a_2|$, and $|a_3|$ are called the lattice identity periods. The vectors a_1, a_2, and a_3 thus generate the lattice. The unit cell can be uniquely specified by the quantities $|a_i|$ $(i = 1, 2, 3)$ and by the angles α, β, and γ.

Since all its points are equivalent, the lattice should be of infinite length. A real crystal, however, has finite dimensions. In the overwhelming number of cases, however, the boundary effects can be neglected. When the finite size of the crystal must be taken into account, it is regarded as occupying only a part of the Bravais lattice. In some cases the finite size of the crystal is taken into account for convenience. In that case one chooses a part of a Bravais lattice that has the simplest form, for example,

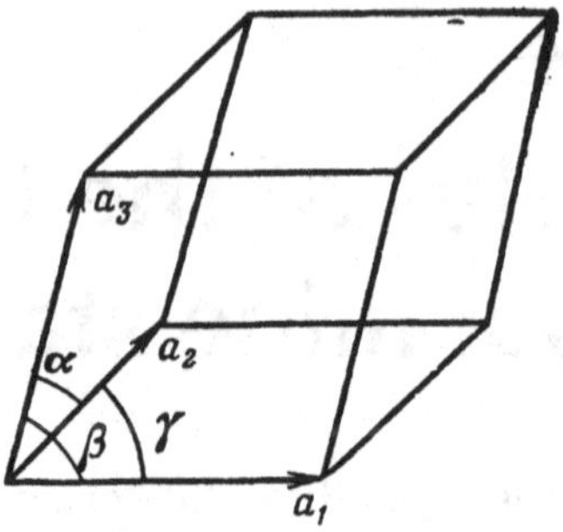

Figure 28.

$$\begin{aligned} R &= n_1a_1 + n_2a_2 + n_3a_3 , \\ n_1 &= 0, 1,\ldots,N_1 , \\ n_2 &= 0, 1,\ldots,N_2 , \\ n_3 &= 0, 1,\ldots,N_3 . \end{aligned} \tag{6.2}$$

Therefore $N = N_1N_2N_3$.

The choice of the vector triad a_1, a_2, and a_3 is not unique, so that in addition to Eq. (6.1) the definition of the Bravais lattice is formulated differently: The "Bravais lattice is an infinite periodic structure made up of discrete points and having an absolutely identical order and orientation regardless of the point from which we start." [168] From this definition it is easily seen that the body-centered-cubic (bcc) and face-centered-cubic (fcc) lattices are Bravais lattices. The principal vectors a_1, a_2, and a_3 for a bcc lattice can be chosen to be

$$a_1 = a_i, \quad a_2 = a_j, \quad a_3 = \frac{a}{2}(i + j + k) , \tag{6.3}$$

where i, j, and k are the unit vectors of a Cartesian coordinate frame, and a is a constant (see Fig. 29). A more symmetric set of the basic vectors is

$$a_1 = \frac{a}{2}(j + k - i), \quad a_2 = \frac{a}{2}(k + i - j) ,$$

$$a_3 = \frac{a}{2}(i + j - k) . \tag{6.4}$$

In the case of an fcc lattice, the symmetric set of basic vectors is written in the form

$$a_1 = \frac{a}{2}(j + k), \quad a_2 = \frac{a}{2}(k + i), \quad a_3 = \frac{a}{2}(i + j) \tag{6.5}$$

(see Fig. 30). The bcc and fcc lattices are important because their structures are possessed by many bodies encountered in nature.

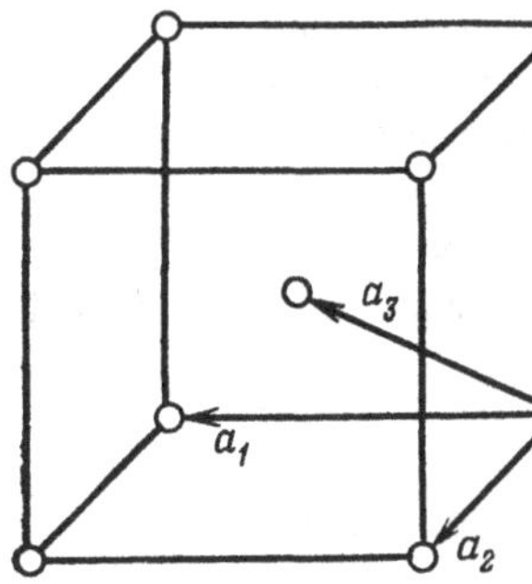

Figure 29.

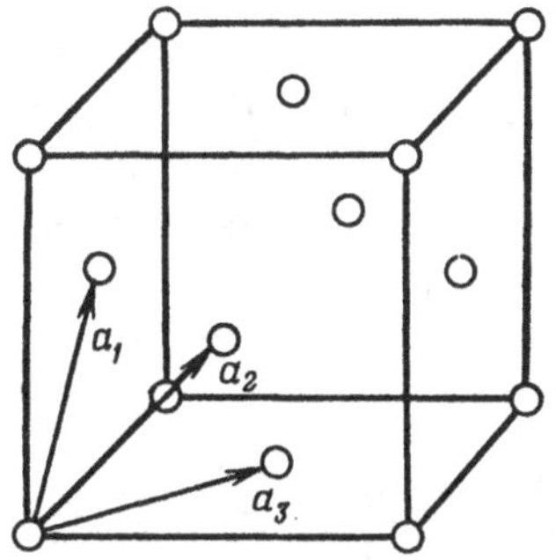

Figure 30.

For each point of a Bravais lattice, the number of nearest neighbors, i.e., of points lying closest to a given point, is the same. It is called the first coordination number (if second, etc., neighbors are considered), or simply the coordination number z_1. It is equal to 6 for a simple cubic (sc) lattice, to 8 for a bcc lattice, and to 12 for an fcc.

Let us consider a volume of space, called the primitive cell, which fills, following the translations that make up the Bravais lattice, all of space without gaps and is not self-intersecting (intersection with a zero volume is disregarded). The primitive cell contains one point of the Bravais lattice. The analytic expression for the primitive cell is

$$r = \sum_{i=1}^{3} x_i a_i \,, \tag{6.6}$$

where $0 \leqslant x_i \leqslant 1$ $(i = 1, 2, 3)$, i.e., the set of all values of r. In place of the primitive cell one usually chooses the elementary unit cell, which contains a large number of points, but has the lattice symmetry. An arbitrary cubic cell double the primitive cell is used to describe a bcc lattice, and one four times larger is used for the fcc lattice. Another approach to the solution of this problem is to introduce a Wigner–Seitz cell (see Sec. 10).

A real crystal structure is described by a lattice with a basis: a Bravais lattice is specified and the locations of the atoms, ions, etc., in an individual unit cell is indicated. If the basis consists of one atom or ion, the crystal lattice is called a monatomic Bravais lattice.

If a nonprimitive arbitrary cell is chosen, the Bravais lattice can be specified as a lattice with a basis. For a bcc lattice we introduce the basis

$$0, \quad \frac{a}{2}(i+j+k), \tag{6.7}$$

and for an fcc lattice the four-point basis

$$0, \quad \frac{a}{2}(i+j), \quad \frac{a}{2}(j+k), \quad \frac{a}{2}(k+i). \tag{6.8}$$

In this case they can be described as simple cubic lattices generated by the vectors a_i, a_j, and a_k.

By way of example of a lattice with a basis we consider a lattice of the diamond type, which consists of two fcc Bravais lattices relatively displaced along the body diagonal of a cubic lattice by $\frac{1}{4}$ the length of this diagonal. This lattice can be regarded as an fcc lattice with a basis consisting of two points: 0 and $a/4(i+j+k)$. In this case $z_1 = 4$.

The hexagonal close-packed (hcp) structure is not a Bravais lattice. Its triad of basis vectors is given by

$$a_1 = a_i, \quad a_2 = \frac{a}{2}i + \frac{\sqrt{3}}{2}a_j, \quad a_3 = Ck. \tag{6.9}$$

The first two vectors generate a triangular lattice in the (x, y) plane, and a_3 corresponds to placing one plane above the other at a distance C. The hcp structure consists of two simple hexagonal Bravais lattices, which are shifted relative to one another by a vector $(a_1 + a_2)/3 + a_3/2$.

A lattice with a basis is needed in the presence of two species of ions or atoms located at Bravais-lattice points, and there is no complete translational symmetry of the Bravais lattice. An example is the sodium chloride crystal (see Fig. 31), in which the sodium and chlorine ions are located at the sites of a simple cubic lattice in such a way that the nearest neighbors of the ions of one kind are six ions of the other kind. This structure can be described as an fcc Bravais lattice with a two-point basis consisting of a sodium ion at the point 0 and a chlorine ion at the point $(a/2)(i+j+k)$ (see Fig. 32).

The ions in a compound of the zinc-blende type are arranged in a diamond-type lattice in such a way that an ion of one type has four ions of the other type as neighbors. A lattice with a basis must be introduced in this case, both because of the geometric arrangement of the ions and because of the presence of two ion types.

Let us consider all possible symmetry types of Bravais lattices. The aggregate of all the rigid operations (i.e., operations that preserve the distances between all lattice points) transforming the lattice into itself makes up the symmetry group or the space group of the lattice. The operations of the space

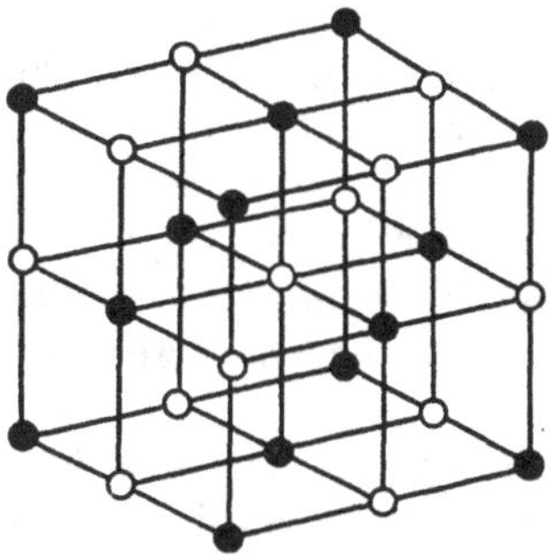

Figure 31.

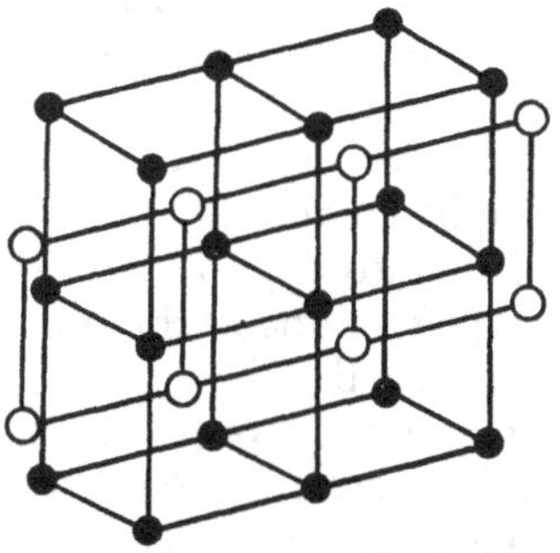

Figure 32.

group include all the translations by the lattice vectors, rotations, reflections, and inversions that transform the lattice into itself. If the lattice coincides with itself upon rotation about a certain axis through an angle $2\pi/n$, this axis is an n-fold symmetry axis. In addition to the trivial onefold axis, only twofold, threefold, fourfold, and sixfold symmetry axes are possible.

We examine now the possible types of Bravais-lattice symmetry, called crystal systems or syngonies.[18] With respect to rotations and reflections, there are seven such (point) symmetry groups.

1. *Triclinic system.* This system corresponds to symmetry group C_i. The sites are vertices of identical parallelepipeds with $|a_1| \neq |a_2| \neq |a_3|$, $\alpha \neq \beta \neq \gamma$. This is the least symmetric system. The triclinic-system Bravais lattice is designated Γ_i.

2. *Monoclinic system.* Its symmetry group is C_2h: the symmetry elements are a twofold axis and a symmetry plane perpendicular to it. In this case $|a_1| \neq |a_2| \neq |a_3|, \alpha = \gamma = 90°, \beta \neq 90°$ the unit cell is a right parallelepiped with a parallelogram as the base. There exists a simple monoclinic Bravais lattice Γ_m, where the sites are the corners of right parallelepipeds, as well as a lattice Γ_m^0 with a centered base, where the sites are corners as well as the centers of the opposite rectangular faces of the parallelepipeds.

3. *Rhombic system.* The system corresponds to the point group D_{2h}. In the case considered, $|a_1| \neq |a_2| \neq |a_3|$, $\alpha = \beta = \gamma = 90°$. There are four types of Bravais lattices of the rhombic system: Γ_0, the sites are at the corners of the rectangular parallelepipeds; Γ_0^b, the sites are both at corners of rectangular parallelepipeds and at the centers of two opposite faces of each parallelepiped; Γ_0^v, the sites are both at the corners and at the centers of the parallelepipeds; Γ_0^f, the sites are both at the corners and at the centers of the faces of the parallelepipeds.

4. *Tetragonal system.* The system corresponds to point group D_{4h}. In this case $|a_1| = |a_2| \neq |a_3|$. There are two types of such a Bravais lattice: Γ_q, simple tetragonal lattice (the sites at the vertices of right prisms); Γ_q^v, body-centered tetragonal lattice (sites located at the vertices as well as at the centers of right prisms).

5. *Rhombohedral (or trigonal) system.* Point group D_{2d}, $|a_1| = |a_2| = |a_3|$, $\alpha = \beta = \gamma \neq 90°$. There is only one Bravais lattice of this system, $\Gamma_{\rm rh}$ (the sites are at the corners of a rhombohedron).

6. *Hexagonal system.* Point group D_{6h} with $|a_1| = |a_2| \neq |a_3|$; $\alpha = \beta = 90°$, $\gamma = 120°$. Three unit cells can be combined into a regular right-angle prism. Only one Bravais lattice of this group, Γ_h, is realized (the sites are at the corners of regular hexagonal prisms and at the centers of their bases, in other words, at the corners of the unit cell).

7. *Cubic system.* The point group is 0_h, and $|a_1| = |a_2| = |a_3|$, $\alpha = \beta = \gamma = 90°$. The unit cell is a cube. Three Bravais lattices of this system are realized: Γ_C, simple cubic; Γ_C^v, bcc; Γ_C^f, fcc.

Fourteen types of Bravais lattice are thus possible. The set of symmetry elements defined by the set of axes and symmetry planes (screw axes and slip planes are regarded as simple axes and planes) are called crystal classes. To determine the classes pertaining to a given system it is necessary to determine all the point groups that contain all or some of the symmetry elements of the system. For this procedure to be unique, each class pertains to the least symmetric of all the systems in which it is contained. There are altogether 32 classes, distributed over the systems in the following manner:

System	Classes
Triclinic	...... C_1, C_i
Monoclinic	.... C_s, C_2, C_{2h}
Rhombic	.... C_{2v}, D_2, D_{2h}
Tetragonal	.... $S_4, D_{2d}, D_{2h}, C_4, C_{4h}, D_{4v}, D_4, D_{4h}$
Rhombohedral	...... $C_3, S_6, C_{2v}, D_3, D_{3d}$
Hexagonal	.... $C_{3h}, D_{3h}, C_6, C_{6h}, C_{6v}, D_6, D_{6h}$
Cubic	..... T, T_h, T_d, O, O_h

The space group of a crystal lattice is the aggregate of all its symmetry elements. The possible space groups are distributed over the crystal classes. Altogether, 230 different space groups are possible, as first demonstrated by Fedorov. The number of space groups in each class is listed in Table V.

Table V. Number of space groups.

Class	C_1	C_i	C_s	C_2	C_{2h}	C_{2v}	D_2	D_{2h}	S_4	C_4	
Number of groups	1	1	4	3	6	22	9	28	2	6	
Class	C_{4h}	D_{2d}	C_{4v}	D_4	D_{4h}	C_3	S_6	C_{3v}	D_3	D_{3d}	C_{3h}
Number of groups	6	12	12	10	20	4	2	6	7	6	1
Class	C_6	C_{6h}	D_{3h}	C_{6v}	D_6	D_{6h}	T	T_h	T_d	O	O_h
Number of groups	6	2	4	4	6	4	5	7	6	8	10

Table VI. Examples of elements with rhombohedral lattices.

Element	a (Å)	θ	Atoms per unit cell	Basis
Hg (5 K)	2.99	70°45′	1	$x=0$
As	4.13	54°10′	2	$x=\pm 0.226$
Sb	4.51	57°6′	2	$x=\pm 0.233$
Sm	9.00	23°13′	3	$x=0\pm 0.222$

Table VII. Examples of elements with tetragonal lattices.

Element	a (Å)	c (Å)	Basis
In	4.59	4.94	In face-centered positions of arbitrary cell
Sn (white)	5.82	3.17	At points with coordinates 000, 0 1/2 1/4, 1/2 0 3/4, 1/2 1/2 1/2 relative to the axis of the arbitrary cell

In nature, 70% of all elements have in the crystalline state a structure of type fcc, bcc, hcp, and diamond. The remaining elements have a great variety of structures. In Tables VI, VII, and VIII are shown examples of elements with rhombohedral, tetragonal, and rhombic Bravais lattices, respectively.[168,170] In Table VI, a is the length of the basis vectors, and θ is the angle between any two of them. The coordinates of the basis points are connected with the basis vectors by the relation $x(a_1+a_2+a_3)$. In Tables VII and VIII we have, respectively, $|a_1|=|a_2|=a$, $|a_3|=c$ and $|a_1|=a$, $|a_2|=b$, $||a_3||=c$.

30. Crystal classification by type of bond

We have considered above a classification of crystalline solids on the basis of the symmetry properties of their crystal structure. This classification, however, is incapable of accounting for many important properties. In this section we classify the crystals by the type of the valence–electron distribution, which determines the type of the bond.

Table VIII. Examples of elements with rhombic Bravais lattices.

Element	*a* (Å)	*b* (Å)	*c* (Å)
Ga	4.511	4.514	7.645
P (black)	3.31	4.38	10.50
Cl (113 K)	6.24	8.26	4.48
Br (123 K)	6.67	8.72	4.48
I	7.27	9.79	4.79
S (rhombic)	10.47	12.87	24.49

1. *Ionic crystals.* Ions of both signs are located on the lattice sites of these crystals. The electron charge is well localized near the ions. The charge distribution of the ions that make up the solid does not differ greatly from the charge distribution of free ions. The most important role is played here by the strong electric interaction forces between the ions.

The simplest model proposed for an ionic crystal is that its constituent ions are impenetrable charged spheres. The impenetrability is due to the Pauli principle and to the fact that the ions possess, in this case, filled electron shells. On the other hand, the notion that the ions are absolutely impenetrable, so that the repulsive part of the potential is approximated by a hard-sphere potential, is not accurate. A more realistic repulsive potential is used for the calculations.

The compounds most closely represented by the simplified model of the ionic crystal are alkali halides. All are cubic at normal pressure. The cation is one of the alkali metals (Li^+, Na^+, K^+, Rb^+, Cs^+), and the anion is the halide (F^-, Cl^-, Br^-, I^-). Under normal conditions alkali-halide crystals have the sodium–chloride structure (see Fig. 31). The fact that alkali-halide compounds are made up of spherical ions is confirmed by the electron charge-distribution data obtained by x-ray diffraction and from band-structure calculations.

The distance d between the nearest neighbors is connected with the side a of the relative cubic cell by the relation $d = a/2$ for the sodium–chloride structure and $d = \sqrt{3}a/2$ for cesium chloride. The values obtained from x-ray data for the a sides of the cubic cells of 12 alkali-halide crystals agree with the elementary model in which the ions are regarded as impenetrable spheres.

If the ions are in contact with one another, the distance between the nearest neighbors is defined as the sum of the radii of these ions:

$$d_{ij} = \Gamma_i + \Gamma_j . \tag{6.10}$$

This equation is valid if the radius $r^>$ of the larger ion does not exceed greatly the radius $r^<$ of the smaller one. In the opposite case $r^> \gg r^<$ we have in place of Eq. (6.10),

$$d = \sqrt{2} r^> , \tag{6.11}$$

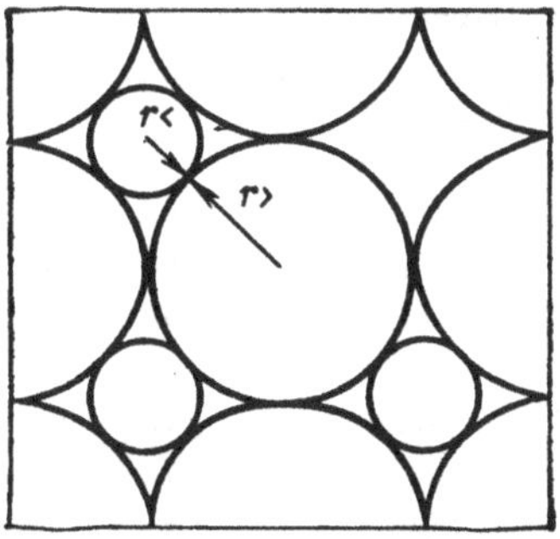

Figure 33.

i.e., d is independent of the radius of the smaller ion. Let us determine the ratio of the critical radii when the ion with the smaller radius comes in contact with the ion having the larger radius. In this case a large ion is in contact with a small one as well as a large one (see Fig. 33). It can be seen directly from the figure that

$$\frac{r^{>} + r^{<}}{r^{>}} = \sqrt{2}\,,$$

hence

$$\frac{r^{>}}{r^{<}} = \frac{1}{\sqrt{2}-1} = \sqrt{2}+1 \cong 2.41. \qquad (6.12)$$

This reasoning is valid for the sodium–chloride structure. For a structure of the cesium–chloride type we have

$$\frac{r^{>}}{r^{<}} = \frac{1}{2}\,(\sqrt{3}+1) = 1.37. \qquad (6.13)$$

Table IX (Refs. 171 and 169) lists the ion radii for alkali-halide components. The numbers in the parentheses are the radii in angstroms. The table lists also the sums $r^{+} + r^{-}$ and the ratio $r^{>}/r^{<}$. If this ratio exceeds the critical value, the value of d determined from Eq. (6.11) is given in square brackets.

Ionic crystals can also be compounds of doubly ionized elements of groups II and VI of the Periodic Table. All (except beryllium compounds and MgTe) have the sodium–chloride structure.

2. *Covalent* (*atomic*) *crystals*. The atoms in the lattice sites of these crystals are neutral. The bond between the atoms in the crystal is called homopolar (or covalent). In principle, the spatial distribution of the electrons in covalent crystals is of the same form as in metals, but in this case there are no partly filled bands in k space. The electrons are not strictly in the vicinities of the atoms, but neither is their density uniformly distributed between the

Table IX. Ion radii for alkali-halide components.

	Li^+ (0.60)	Na^+ (0.95)	K^+ (1.33)	Rb^+ (1.48)	Cs^+ (1.69)
F^- (1.36)					
d	2.01	2.31	2.67	2.82	3.00
$r^- + r^+$	1.96	2.31	2.69	2.84	3.05
$r^> / r^<$	2.27	1.43	1.02	1.09	1.24
Cl^- (1.81)					
d	2.57	2.82	3.15	3.29	3.57
$r^- + r^+$	2.41	2.76	3.14	3.29	3.50
$r^> / r^<$	3.02[2.56]	1.91	1.36	1.22	1.07
Br^- (1.95)					
d	2.75	2.99	3.30	3.43	3.71
$r^- + r^+$	2.55	2.90	3.28	3.43	3.64
$r^> / r^<$	3.25[2.76]	2.05	1.47	1.32	1.15
I^- (2.16)					
d	3.00	3.24	3.53	3.67	3.05
$r^- + r^+$	2.76	3.11	3.49	3.64	3.85
$r^> / r^<$	3.60[3.05]	2.27	1.62	1.46	1.28

ions: the electrons are distributed along certain preferred directions. In the covalent diamond crystal, for example, the electron density is highest in the interstices; the electrons are concentrated mainly along directions joining a carbon ion with its four nearest neighbors. Thus, whereas in NaCl the electron density along the line joining nearest neighbors does not exceed 0.1 electron per angstrom, in diamond it is not lower than 5 electrons per angstrom. A structure similar to that of diamond is formed by elements of group IV of the periodic systems (silicon, germanium, and grey tin). A carbon atom in a diamond is part of four covalent bonds. Notwithstanding the electrostatic origin of the bonds, it is impossible to use in this case the simple model of positively and negatively charged spheres. Since their electrons are delocalized, covalent crystals are not as good insulators as ionic crystals. All semiconductors are covalent-bond crystals, whereas III–V compounds have certain ionic-bond attributes. Almost all have the zinc-blende structure typical of covalent crystals. They are usually described as covalent crystals having an excess charge density around the ionic residues.

3. *Molecular crystals.* In this case the molecules in the lattice sites are oriented in a definite manner. The most characteristic examples of a molecular crystal are those of solidified inert gases (with the exception of helium). They have completely filled electron shells that are insignificantly distorted when the solid is formed. The electrons have a low density in the spaces between the ions and are localized around the ions.

Solidified inert gases (except helium) form crystals with a single-atom fcc Bravais lattice. The atoms are bound by van der Waals forces. To explain their character, we determine the energy of the interaction between two hy-

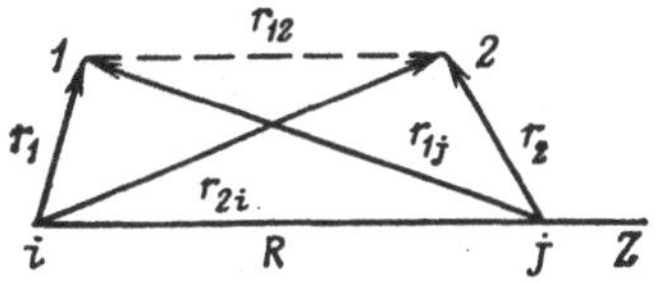

Figure 34.

drogen atoms.* We assume that the position of electron 1 relative to the nucleus i is determined by a radius vector r_1 with components x_1, y_1, and z_1, and the position of electron 2 relative to nucleus j is determined by radius vector r_2 with components x_2, y_2, and z_2, and direct the z axis along the line joining the nuclei of the atoms (see Fig. 34). In the Born–Oppenheimer approximation (we assume that the nuclei are at rest) the Hamiltonian of this system is of the form

$$H = H_0 + V\,, \tag{6.13'}$$

where

$$H_0 = -\frac{\hbar^2}{2m}(\nabla_1^2 + \nabla_2^2) - \frac{e^2}{r_1} - \frac{e^2}{r_2} \tag{6.14}$$

is the Hamiltonian of two independent atoms, and

$$V = \frac{e^2}{R} + \frac{e^2}{r_{12}} - \frac{e^2}{r_{1j}} - \frac{e^2}{r_{2i}} \tag{6.15}$$

is the potential describing the interaction. We regard V as a perturbation.

We expand V in powers of R^{-1}, assuming $r_1 \ll R$. The main term of the expansion corresponds then to the interaction of two dipoles $i1$ and $j2$ with moments $\mathbf{p}_1 = -e\mathbf{r}_1$ and $\mathbf{p}_2 = -e\mathbf{r}_2$. Retaining in Eq. (6.15) only this principal term, we have

$$V = \frac{\mathbf{p}_1\mathbf{p}_2}{R^3} - 3\,\frac{(\mathbf{p}_1\mathbf{R})(\mathbf{p}_2\mathbf{R})}{R^5}\,. \tag{6.16}$$

In our coordinate frame this expression takes the form

$$V = \frac{e^2}{R^3}(x_1x_2 + y_1y_2 - 2z_1z_2)\,. \tag{6.17}$$

If $\psi_0(r)$ is the atom's wave function in the ground state, the wave function of the system in zeroth-order perturbation theory can be represented in the form

$$\psi(1,2) = \psi_0(r_1)\psi_0(r_2)\,. \tag{6.18}$$

We have neglected here exchange effects which are small when R is large.

*We assume that all the hydrogen atoms are spaced widely enough to be able to neglect the overlap of the electron shells, for otherwise the character of the interaction would change. This model was chosen solely to simplify the analysis.

In the zeroth approximation, the total system energy is the sum of the energies of the noninteracting atoms. The first-order energy correction E' is equal to

$$E' = \langle \psi(1,2)|V|\psi(1,2)\rangle$$
$$= \frac{e^2}{R^3}\langle \psi_0(r_1)\psi_0(r_2)|x_1x_2 + y_1y_2 - 2z_1z_2|\psi_0(r_1)\psi_0(r_2)\rangle .$$

The first term of the sum is

$$\frac{e^2}{R^3}\langle \psi(1,2)|x_1x_2|\psi(1,2)\rangle = \frac{e^2}{R^3}\langle \psi_0(r_1)|x_1|\psi_0(r_1)\rangle^2$$
$$= \frac{e^2}{R^3}\left[4\pi \int \psi_0^2(r)xr^2\,dr\right]^2 = 0$$

by virtue of the spherical symmetry of $\psi_0(r)$. It can similarly be shown that the remaining terms of the sum are also zero. Therefore,

$$E' = 0. \tag{6.19}$$

The second-order perturbation-theory correction to the energy is given by

$$E'' = \sum_n{}' \frac{|\langle 0|V|n\rangle|^2}{E_0 - E_n} . \tag{6.20}$$

The summation here is over all the excited states, with 0 corresponding to the ground state. Hence $E_n > E_0$ and

$$E'' = -\frac{C}{R^6}, \tag{6.21}$$

where C is a positive constant independent of R. The total energy of the two hydrogen atoms is thus

$$E = E^0 + E' + E'' = E^0 - \frac{C}{R^6}, \tag{6.22}$$

where E^0 is the energy of two unbound hydrogen atoms. We see thus that the attraction force between the atoms is $\sim 1/R^7$, i.e., the van der Waals value.

A similar analysis can be carried out also for inert-gas atoms. In this case we get

$$H_0 + H_0^1 + H_0^2 , \tag{6.23}$$

where H_0^1 and H_0^2 are the Hamiltonians of the isolated atoms 1 and 2, respectively, and

$$V = e^2\left[\frac{z^2}{R} - \sum_{i=1}^{z}\left(\frac{z}{|R - r_i^{(1)}|} + \frac{z}{r_i^{(2)}}\right) + \sum_{i,j=1}^{z}\frac{1}{|r_i^{(1)} - r_j^{(2)}|}\right], \tag{6.24}$$

where R is the distance between the centers of the inert atoms. It is assumed that the nuclei are immobile and are located at the points 0 and $\mathbf{R}$, and the charge of each nucleus is zero. The electrons bound to the nucleus located at the point 0 have coordinates $r_i^{(1)}$, and those bound to the nucleus at R have

Table X. Comparison of the metal ion radii.

Metal	r_{ion} (Å)	r_{met} (Å)	r_{met}/r_{ion}
Li	0.60	1.51	2.52
Na	0.05	1.83	1.03
K	1.33	2.26	1.70
Rb	1.48	2.42	1.64
Cs	1.60	2.62	1.55
Cu	0.06	1.28	1.33
Ag	1.26	1.45	1.15
Au	1.37	1.44	1.05

coordinates $r_i^{(2)}$ ($i = 1, 2,...,z$). The distance R is assumed large enough to be able to neglect the overlap of the electron shells. The calculations show that the interaction energy is given by Eq. (6.22) in this case, too.

For a more accurate analysis of the intermolecular interactions, account must be taken of the interaction in groups of three and more atoms. This interaction cannot be represented as a sum of pair interactions and falls off with distance more rapidly than $1/R^6$. Thus, taking into account the dipole–dipole interaction energy of a triplet of inert gas atoms, Axelrod and Teller obtained, besides the van der Waals part $-C(r_{12}^{-6} + r_{13}^{-6} + r_{23}^{-6})$ of the interaction potential of a system of three particles, a term corresponding to three-particle interaction:

$$\Phi(r_1, r_2, r_3) + C_3\frac{1 + 3(e_{12}e_{13})(e_{23}e_{21})(e_{31}e_{32})}{r_{12}^3 r_{23}^3 r_{31}^3}, \tag{6.25}$$

where $e_{ij}(r_i - r_j)/|r_{ij}|$ is a unit vector along r_{ij}.[172–174]

4. *Metallic crystals.* In this case the positive metal ions are located at the corners of the crystal lattice, and the electrons move randomly between them. The most typical metallic crystals are the alkali metals of the first group, many properties of which fit the Sommerfeld free-electron model. The valence electrons are then completely separated from the ions and form a gas of constant density. In Table X are compared the metal ion radii determined from the data on the structure of ionic crystals and those determined from the relation[169]

$$r_{met} = d/2, \tag{6.26}$$

where d is the distance between the nearest neighbors in the metal. We see from the table that the ion-radius concept is not helpful when it comes to determine the lattice constant. The compressibility is determined in alkali metals mainly by the electron gas. In noble metals, closed d shells play an important role in the determination of a number of mechanical properties. Both in metals and in ionic crystals, the ion sizes are governed to a great degree by the d shells.

Most metals have bcc, fcc, or hcp lattices.

Since hydrogen-bond crystals have a number of distinctive properties, they are set apart as a separate group.[169]

31. Static model of a crystal

We consider the simplest crystal model based on the representation of its structure in classical mechanics. At absolute zero temperature, according to classical mechanics, all the crystal particles should be at rest at the crystal-lattice sites (we are considering here an ideal defect-free crystal structure). This is of course an approximation: In a real crystal, in view of the Heisenberg uncertainty principle, the particles cannot be strictly localized at the crystal lattice sites even at $T = 0$ K. For a van der Waals crystal, for example, the dimensionless parameters of the zero-point vibrations are[175]

$$\Lambda = h/(\sigma\sqrt{m\varepsilon}) . \tag{6.27}$$

For small zero-point vibrations we have $\Lambda \ll 1$. The smaller the atom mass and the weaker its interaction with the neighbors, the larger the parameter Λ, viz., $\Lambda = 0.6$, 0.18, 0.1, 0.06, 3.1, and 2.7 for Ne, Ar, Kr, Xe, ^{3}He, and ^{4}He, respectively. We see therefore that the static-lattice model cannot be used for helium even at $T = 0$ K.

We begin our analysis with solidified inert-gas crystals. In this case the interaction has a Lennard-Jones potential (we disregard helium for the reasons cited above, since none of its isotopes can exist in the solid state at $T = 0$ and $p = 0$; ^{4}He crystallizes only at 25 atm, and ^{3}He at 33 atm).

At $T = 0$ K the free energy is equal to the internal energy of the system, given by

$$U = \frac{N}{2} \sum_k z_k \Phi(r_k) = \frac{N}{2} \sum_k z_k \Phi(\nu_k a) , \tag{6.28}$$

where a is the distance between nearest neighbors, z_k the number of particles on the kth coordination sphere, and $r_k = \nu_k a$ the radius of the kth coordination sphere. In the case of the fcc lattices of the solidified inert-gas crystals we have

$$\begin{aligned} &z_1 = 12, \quad \nu_1 = 1; \quad z_4 = 12, \quad \nu_4 = 2;\\ &z_2 = 6, \quad \nu_2 = \sqrt{2}; \quad z_5 = 24, \quad \nu_5 = \sqrt{5};\\ &z_3 = 24, \quad \nu_3 = \sqrt{3}; \quad \text{etc.} \end{aligned}$$

The sum in Eq. (6.28) is over all the coordination spheres. This expression can be rewritten in the form

$$U = 2\varepsilon N \left[A_{12} \left(\frac{\sigma}{a} \right)^{12} - A_6 \left(\frac{\sigma}{a} \right)^{6} \right] , \tag{6.29}$$

where

$$A_n = \sum_i \frac{1}{\nu_i^n} . \tag{6.30}$$

The values of A_n for a number of structures are listed in Table XI (Refs. 176 and 177). The value of A_n depends on the type of crystal structure. For large n we can put, with good accuracy, $A_n = z_1/\nu_1^n = z_1$, since $\nu_1 = 1$ by definition.

Table XI. Values of A_n for structures.

n	pc	bcc	fcc
3			
4	16.53	22.64	25.34
5	10.38	14.76	16.07
6	8.40	12.25	14.45
7	7.47	11.05	13.36
8	8.05	10.36	12.80
9	6.63	9.80	12.40
10	6.43	9.56	12.31
11	6.20	9.31	12.20
12	6.20	9.11	12.13
13	6.14	8.05	12.00
14	6.10	8.82	12.06
15	6.07	8.70	12.04
16	6.05	8.61	12.03
$n \geqslant 17$	$6+12(1/2)^{n/2}$	$8+6(3/4)^{n/2}$	$12+6(1/2)^{n/2}$

The pressure is given by

$$p = -\frac{dU}{dV} \tag{6.31}$$

(at $T = 0$ K). On the other hand $dV/V = dv/v = 3da/a$, therefore,

$$p = -\frac{a}{3v}\frac{d(U/N)}{da} = \frac{4\sqrt{2}\,a}{\sigma^3}\left[2A_{12}\left(\frac{\sigma}{a}\right)^{15} - A_6\left(\frac{\sigma}{a}\right)^{9}\right], \tag{6.32}$$

since $v = (\sqrt{2}/2)a^3$ for an fcc crystal. At pressures on the order of atmospheric:

$$\frac{p\sigma^3}{\varepsilon} \ll 1,$$

and we have therefore from Eq. (6.32),

$$a = \left(\frac{2A_{12}}{A_6}\right)^{1/6}\sigma = 1.09\sigma\,. \tag{6.33}$$

Substituting Eq. (6.33) in Eq. (6.29) we get

$$U/N = -\frac{\varepsilon A_6^2}{2A_{12}} = -8.6\varepsilon\,. \tag{6.34}$$

Next, we obtain the isothermal elastic modulus

$$\varepsilon_T = -V\left(\frac{\partial p}{\partial V}\right)_\theta = -\frac{a}{3}\left(\frac{\partial p}{\partial a}\right)_\theta .$$

From Eq. (6.32) we get

$$\varepsilon_T = \frac{4\sqrt{2}\,\varepsilon}{\sigma^3}\left[10A_{12}\left(\frac{\sigma}{a}\right)^{15} - 3A_6\left(\frac{\sigma}{a}\right)^{9}\right]. \tag{6.35}$$

Table XII. Calculated values of a, U/N, and ε for $T=0$ K and $p=0$, as well as corresponding experimental data.

		Ne	Ar	Kr	Xe
a (Å)	Theory	2.00	3.71	3.08	4.34
	Experiment	3.13	3.75	3.09	4.34
U/N (eV/atom)	Theory	−0.027	−0.080	−0.120	−0.172
	Experiment	−0.02	−0.08	−0.11	−0.17
$\varepsilon\times10^{10}$ (dyn/cm²)	Theory	1.81	3.18	3.46	3.81
	Experiment	1.1	2.7	3.5	3.6

For a defined by Eq. (6.33), ε_T is equal to

$$\varepsilon_T=\frac{4\varepsilon}{\sigma^3}A_{12}\left(\frac{A_6}{A_{12}}\right)^{5/2}=\frac{75\varepsilon}{\sigma^3}. \tag{6.36}$$

Table XII lists the calculated values of a, U/N, and ε for $T=0$ K and $p=0$, as well as the corresponding experimental data (Refs. 170, 179, 180, and 160).

The theoretically calculated values of the lattice constant a agree well with experiment for all the substances. The agreement worsens with decrease of atomic weight, in view of the increased contribution of the zero-point vibrations, which were not taken into account in this case. For U/N, similarly, the agreement between the theoretical and experimental values is good, the theoretical values being underestimated and the difference increasing with decrease of the mass of the material (ε and σ also decrease, meaning that Λ increases). This fact can also be attributed to neglect of the positive zero-point vibration energy. The agreement between the theoretical and experimental data for ε is not as good (the difference is 60% for neon and 20% for argon). The agreement with experiment improves also for ε with increase of mass.

We turn now to ionic crystals. The main contribution to the potential energy is made in this case by the Coulomb interaction between the ions. We write the pair-interaction potential $\Phi(r)$ in the form

$$\Phi(r)=\Phi_0(r)+\Phi_k(r), \tag{6.37}$$

where Φ_k and Φ_0 are, respectively, the Coulomb and repulsive parts of the potential. The interaction potential can also take on more complicated forms. Let, as above, a be the distance between neighbors. In view of the large effective radius of the Coulomb potential, it is not a simple matter to calculate its contribution to the internal energy. By definition, this fraction is equal to

$$U_k=\tfrac{1}{2}\sum_{k\neq0}\Phi(r_k), \tag{6.37$'$}$$

where the summation is over all but the central lattice points. Let, for the sake of argument, the ionic-crystal system consist of two types of ion, with the negative anions arranged in a lattice r_k and the positive cations in second lattice imbedded in the first and displaced by a vector d of absolute value a. We introduce a function $\alpha(\bar{r}_k)$ such that

$$|\bar{r}_k| = \alpha(r_k)a\,,$$
$$|r_k + d| = \alpha(r_k + d)a\,. \tag{6.38}$$

Then

$$U_k = -\frac{N}{2}\frac{e^2}{a}\left\{\frac{1}{\alpha(d)} + \sum_{k\neq 0}\left(\frac{1}{\alpha(\bar{r}_k + d)} - \frac{1}{\alpha(\bar{r}_k)}\right)\right\}. \tag{6.39}$$

The series on the right-hand side of this expression is conditionally convergent: its sum depends on the sequence of the summation. This is a reflection of the fact that, owing to the long-range character of the interactions, the particle-system energy can depend strongly on the configuration of a small number of particles on the surface. To eliminate this dependence, we shall sum in such a way that at each step the surface charges make no substantial contribution to the energy. We therefore subdivide the entire crystal into electrically neutral cells with a cubic-symmetry distribution of the charge in each, so that the energy of the interaction between the cells decreases as the fifth power of the distance between them, and the internal energy of each cell is easily calculated (it does not contain as many particles). Ewald's method[181] can also be used.

The resultant expression for U_k is

$$U_k = -\frac{N}{2}\frac{\alpha e^2}{a}, \tag{6.40}$$

where α is the Madelung constant and depends on the type of crystal structure,

$$\alpha = \sum_k \frac{(-1)^{k-1}z_k}{\nu_k}. \tag{6.41}$$

For a sodium–chloride structure we have

$$\alpha = \frac{6}{1} - \frac{12}{\sqrt{2}} + \frac{8}{\sqrt{3}} - \frac{6}{2} + \cdots = 1.747\,558,$$

for cubic-crystal structures such as cesium chloride

$$\alpha = 1.7627,$$

and for zinc blende

$$\alpha = 1.6381.$$

We see that α increases with an increase of the coordination number z_1.

The energies for ionic crystals are usually calculated per ion pair

$$\bar{U}_k = \frac{2U_k}{N} = -\alpha\frac{e^2}{a}. \tag{6.42}$$

The calculations show that the Coulomb interaction makes the main contribution to the energy U. At the same time, account must be taken of the repulsive part of the potential, assuming an inverse-power repulsion law:

$$\Phi_0(r) = \gamma / r^m . \tag{6.43}$$

This yields

$$U_0 = \frac{NC}{2a^m}, \tag{6.44}$$

where

$$C = \gamma \sum_k \frac{z_k}{\nu_k^m} . \tag{6.45}$$

We determine the distance between nearest neighbors from the energy-minimization condition

$$\bar{U} = \frac{C}{a^m} - \frac{\alpha e^2}{a}, \tag{6.46}$$

which yields

$$a^{m-1} = \frac{mC}{e^2 \alpha} . \tag{6.47}$$

In contrast to inert gases, we do not know the gas constant γ in Eq. (6.43). It is therefore easier to determine C from Eq. (6.47):

$$C = \frac{\alpha e^2 a^{m-1}}{m} . \tag{6.48}$$

Substituting this expression in Eq. (6.46), we get

$$\bar{U} = -\frac{\alpha e^2}{a} \frac{m-1}{m} . \tag{6.49}$$

At large m the value of U depends little on m.

We must obtain an equation for m. We cannot use for this purpose the expression for the energy, since small energy-measurement errors lead to large changes of m. It is better to use for this purpose an expression for the isothermal elastic modulus. We find first an expression for the pressure. For the sodium–chloride structure its value is

$$p = -\frac{dU}{dV} = -\frac{a}{6v}\frac{d\bar{U}}{da} = \frac{1}{6}\left[\frac{mC}{a^{m+3}} - \frac{\alpha e^2}{a^4}\right] . \tag{6.49$'$}$$

From this we can easily find also an expression for the isothermal elastic modulus

$$\varepsilon_T = -V\frac{\partial p}{\partial V} = -\frac{a}{3}\frac{dp}{da} = \frac{1}{18}\left[\frac{m(m+3)C}{a^{m+3}} - \frac{4\alpha e^2}{a^4}\right] . \tag{6.50}$$

Substituting here C from Eq. (6.48), we get

$$\varepsilon_T = \frac{1}{18}\left[\frac{m(m+3)}{a^{m+3}}\cdot\frac{\alpha e^2 a^{m-1}}{m} - \frac{4\alpha e^2}{a^4}\right]$$

$$= \frac{1}{18}\left[\frac{\alpha e^2(m+3)}{a^4} - \frac{4\alpha e^2}{a^4}\right] = \frac{(m-1)\alpha e^2}{18e^4}, \tag{6.51}$$

from which we determine m:

$$m = 1 + \frac{18\varepsilon_T a^4}{\alpha e^2}. \tag{6.52}$$

The values obtained for m from the experimental data using Eq. (6.52) range from 6 to 10.

Construction of a sufficiently simple model of a static lattice for crystals with another type of bond is not as illustrative and will not be considered here.

32. Einstein model of a crystal

The static model of a crystal considered in the preceding section is actually a rather crude approximation. The point is that at any temperature, including $T = 0$, the particles in the crystal execute small vibrations about equilibrium positions. It is expedient therefore to expand the potential energy in powers of the amplitudes of these small vibrations. If we confine ourselves only to terms up to quadratic inclusive, we obtain the potential energy in the harmonic approximation. Determination of the partition function for a crystal in the harmonic approximation yields much more information on the behavior of the crystal than in the static model. Even in this case, however, difficulties are encountered in the determination of the system oscillation spectrum, especially for crystals with complicated structures. On the other hand, inclusion of terms higher than quadratic in the expression for the potential leads to even greater difficulties.

The problem is greatly simplified by assuming that the atoms vibrate independently of one another. This approximation was used by Einstein in his treatment of a crystal in the harmonic approximation.[182] A crystal consisting of N particles is regarded as a system of $3N$ independent oscillators having one and the same frequency ω. The system energy was determined starting from Planck's formula for the average energy of a harmonic oscillator. The expression obtained for the energy was used to obtain for the crystal a heat capacity that, in contrast to the law of Dulong and Petit, agreed qualitatively with the experimental data.

In a wider sense, the Einstein approximation corresponds to a particle moving in the field produced by the remaining particles of the crystal, under the condition that they are immobile and fixed in the corresponding lattice sites. The potential energy of such a particle is easiest to determine if only two-particle interactions are present. Its value, by definition, is

$$U(q) = \sum_i{}' \Phi(|q - a_i|) - \tfrac{1}{2} \sum_i{}' \Phi(a_i)\,, \tag{6.53}$$

where the primes mean that the summations are over all sites but the one occupied by the considered particles. In the classical case, the thermodynamics of the crystal can be derived here directly. In fact, the free energy is

$$F = F_0 - \theta N \ln\left\{\frac{1}{ev} \int \exp[-u(q)/\theta]\,dq\right\}, \tag{6.54}$$

where F_0 is the free energy of an ideal gas. An undisputed advantage of this approach is that we take into account on its basis the anharmonicities of all order, without a series expansion of the potential function. This is particularly important for the determination of the equation of state near a phase transition. The correlation terms can be taken into account by perturbation theory or by using a cell-cluster expansion.

In the quantum case the problem becomes much more complicated in this approximation. It is convenient then to use the equation for the non-normalized single-particle density matrix

$$-\frac{\partial \rho}{\partial \beta} = H\rho\,, \tag{6.55}$$

where

$$H = -\frac{\hbar^2}{2m}\nabla^2 + u(q)\,. \tag{6.56}$$

The formal solution of Eq. (6.55) is

$$\rho = \exp[-\beta H]\,, \tag{6.57}$$

and the determination of the thermodynamic properties of the system reduces to a determination of the expression

$$\mathrm{Tr}\exp[-\beta H]\,. \tag{6.58}$$

The initial condition for Eq. (6.55) is

$$\rho(0) = 1. \tag{6.59}$$

In the coordinate representation, Eq. (6.55) takes the form

$$-\frac{\partial \rho(q, q'; \beta)}{\partial \beta} = Hq\rho(q, q'; \beta)\,. \tag{6.60}$$

[H_q is an operator acting on the variable $q = (x, y, z)$.] The initial condition is in this case

$$\rho = (q, q'; 0) = \delta(q - q')\,. \tag{6.61}$$

We consider $u(q)$ in the harmonic approximation. A Taylor expansion yields

$$u = u_0 + u_1 + u_2\,,$$

where u_i corresponds to the expansion term proportional to the ith power of the displacement, and u_0 is defined as

$$u_0 = \tfrac{1}{2} \sum_i{}' z_i \Phi(\nu_i a) , \tag{6.62}$$

and in the particular case of a Lennard-Jones potential as

$$u_0 = 2\varepsilon\left(\frac{\sigma}{a}\right)^6\left[A_{12}\left(\frac{\sigma}{a}\right)^6 - A_6\right] .$$

Let us show that

$$u_1 \equiv 0 . \tag{6.63}$$

To this end, we consider the derivative

$$\left.\frac{\partial u}{\partial x}\right|_{q=0} = \sum_i \frac{x - a_{ix}}{|q - a_i|}\Phi'(|q - a_i|)|_{q=0} = -\sum_i \frac{a_{ix}}{a_i}\Phi'(a_i) = 0 .$$

It can be shown similarly that

$$\left.\frac{\partial u}{\partial y}\right|_{q=0} = \left.\frac{\partial u}{\partial z}\right|_{q=0} = 0 .$$

Next,

$$\begin{aligned}\left.\left(\frac{\partial^2 u}{\partial x^2}\right)\right|_{q=0} &= \left.\left(\frac{\partial^2 u}{\partial y^2}\right)\right|_{q=0} = \left.\left(\frac{\partial^2 u}{\partial z^2}\right)\right|_{q=0} \\ &= \sum_i \left\{\frac{a_{ix}^2}{a_i^2}\Phi''(a_i) + \frac{1}{a_i}\Phi'(a_i) - \frac{a_{ix}^2}{a_i^2}\Phi'(a_i)\right\} ,\end{aligned}$$

$$\left.\left(\frac{\partial^2 u}{\partial x \partial y}\right)\right|_{q=0} = \left.\left(\frac{\partial^2 u}{\partial x \partial z}\right)\right|_{q=0} = \left.\left(\frac{\partial^2 u}{\partial y \partial z}\right)\right|_{q=0} = 0 . \tag{6.64}$$

For an fcc lattice

$$\sum_i \frac{a_{ix}^2}{a_i^2}\Phi''(a_i) = \frac{1}{3}\sum_i z_i \Phi''(a_i) ,$$

and we have from Eq. (6.64),

$$\left.\left(\frac{\partial^2 u}{\partial x^2}\right)\right|_{q=0} = \frac{1}{3}\sum_i z_i \left[\Phi''(a_i) + \frac{2}{a_i}\Phi'(a_i)\right] .$$

For a Lennard-Jones potential this expression takes the form

$$\left.\left(\frac{\partial^2 u}{\partial x^2}\right)\right|_{q=0} = \frac{8\varepsilon}{\sigma^2}\left(\frac{\sigma}{a}\right)^8\left[22\,A_{14}\left(\frac{\sigma}{a}\right)^6 - 5A_8\right] . \tag{6.65}$$

In the harmonic approximation, $u(q)$ is given by

$$u = u_0 + \frac{m\omega^2}{2}(x^2 + y^2 + z^2) , \tag{6.66}$$

where

$$\omega = \frac{2}{\sigma}\sqrt{\frac{2\varepsilon}{m}\left(\frac{\sigma}{a}\right)^4}\sqrt{22\,A_{14}\left(\frac{\sigma}{a}\right)^6 - 5A_8}$$

$$= \frac{2\sqrt{2}}{h}\varepsilon\Lambda\left(\frac{\sigma}{a}\right)^4\sqrt{22A_{14}\left(\frac{\sigma}{a}\right)^6 - 5A_8}\,. \tag{6.67}$$

By virtue of Eq. (6.66), the Hamiltonian is of the form

$$H_q = H_x + H_y + H_z\,. \tag{6.68}$$

The solution [Eq. (6.10)] can therefore be expressed as

$$\rho(q, q'; \beta) = \rho(x, x'; \beta)\rho(y, y'; \beta)\rho(z, z'; \beta)\,, \tag{6.69}$$

where $\rho(t, t'; \beta)$ $(t = x, y, z)$ satisfies the equation

$$-\frac{\partial\rho(t, t'; \beta)}{\partial\beta} = H_t\rho(t, t'; \beta) \tag{6.70}$$

with the initial condition

$$\rho(t, t'; 0) = \delta(t - t')\,. \tag{6.71}$$

Here

$$H_t = -\frac{\hbar^2}{2m}\frac{d^2}{dt^2} + \frac{m\omega^2}{2}t^2\,. \tag{6.72}$$

We thus write Eq. (6.70) in the form

$$-\frac{\partial\rho}{\partial\beta} = -\frac{\hbar^2}{2m}\frac{\partial^2\rho}{\partial t^2} + \frac{m\omega^2}{2}t^2\rho\,, \tag{6.73}$$

or, introducing new variables,

$$\eta = \sqrt{\frac{m\omega}{\hbar}}t,\quad f = \frac{\hbar\omega}{2}\beta = \frac{\hbar\omega}{2\theta} \tag{6.74}$$

in the form

$$-\frac{\partial\rho}{\partial f} = -\frac{\partial^2\rho}{\partial\eta^2} + \eta^2\rho \tag{6.75}$$

with the initial condition

$$\rho = \sqrt{\frac{m\omega}{\hbar}}\delta(\eta - \eta') \text{ as } f = 0. \tag{6.76}$$

As $j \to 0$, the oscillator will tend to behave as a free particle, for which the equation is of the form

$$\frac{\partial\rho}{\partial f} = -\frac{\partial^2\rho}{\partial\eta^2}\,. \tag{6.77}$$

The solution of this equation can be written down directly[141]:

$$\rho(\eta,\sigma';f)=\sqrt{\frac{4f}{\pi}}\exp\left[-\frac{1}{4f}(\eta-\eta')^2\right]. \tag{6.78}$$

We seek on this basis a solution of Eq. (6.75) in the form

$$\rho=\exp\{-[\alpha_1(f)\eta^2+\alpha_2(f)\eta+\alpha_3(f)]\}. \tag{6.79}$$

Substitution of Eq. (6.79) in Eq. (6.75) yields

$$\alpha_1'\eta^2+\alpha_2'\eta+\alpha_3'=(1-4\alpha_1^2)\eta^2-4\alpha_1\alpha_2\eta+2\alpha_1-\alpha_2^2 \tag{6.80}$$

($\alpha_i'=d\alpha_i/df$). From this we have

$$\alpha_1'=1-4\alpha_1^2\,, \tag{6.81}$$

$$\alpha_2'=-4\alpha_1\alpha_2\,, \tag{6.82}$$

$$\alpha_3'=2\alpha_1-\alpha_2^2\,. \tag{6.83}$$

Integrating Eq. (6.81) and using the initial condition, we get

$$\alpha_1=\tfrac{1}{2}\coth 2f. \tag{6.84}$$

Integrating now Eq. (6.82), we obtain

$$\alpha_2=\frac{A}{\sinh 2f},$$

$$\alpha_3=\frac{1}{2}\ln(\sinh 2f)+\frac{A^2}{2}\coth 2f-\ln B. \tag{6.85}$$

The integration constants A and B are determined from the condition of matching Eq. (6.79) with Eq. (6.78) as $f\to 0$. As a result we have

$$A=-\eta',\quad B=\sqrt{\frac{m\omega}{2\pi\hbar}}. \tag{6.86}$$

We ultimately have for ρ as a function of t, t', and β:

$$\rho(t,t';\beta)=\sqrt{\frac{m\omega}{2\pi\hbar\sinh 2f}}\exp\left\{-\frac{m\omega}{2\hbar\sinh 2f}[(t^2+t'^2)\sinh 2f-2tt']\right\}. \tag{6.87}$$

If $t=t'$, then

$$\rho(t,t;\beta)=\sqrt{\frac{m\omega}{2\pi\hbar\sinh 2f}}\exp\left[-\frac{m\omega}{\hbar}x^2\tanh f\right]. \tag{6.88}$$

We have deduced in Sec. 5 the thermodynamics of a system of harmonic oscillators by determining the partition function. We shall do this here by using the obtained density matrix. The three-dimensional density matrix of a harmonic oscillator is

$$\rho(q,q';\beta)=\left(\frac{m\omega}{2\pi\hbar\sinh 2f}\right)^{3/2}\prod_{t=x,y,z}\exp\left\{-\frac{m\omega}{2\hbar\sinh 2f}[(t^2+t'^2)\sinh 2f-2tt']\right\}. \tag{6.89}$$

We determine with the aid of this expression the free energy per harmonic oscillator:

$$F/N = -\theta \ln \int \rho(q,q;\beta)dq = \frac{3\hbar\omega}{2} + 3\theta \ln(1 - \exp[-\hbar\omega/\theta]) . \quad (6.90)$$

The density matrix enables us to find the mean value of any operator, including the kinetic-energy operator

$$E_k = \left\langle -\frac{\hbar^2}{2m}\left(\frac{d^2}{dx^2} + \frac{d^2}{dy^2} + \frac{d^2}{dz^2}\right)\right\rangle ,$$

$\langle ... \rangle$ denotes averaging with a normalized density matrix. This can be done, however, in a somewhat different manner. We determine from Eq. (6.90) the internal energy:

$$E = \frac{3\hbar\omega}{2} \coth f . \quad (6.91)$$

The potential energy is determined by the relation

$$E_n = \frac{1}{2} m\omega^2 \langle x^2 + y^2 + z^2 \rangle = \frac{3\hbar\omega}{4} \coth f . \quad (6.92)$$

Thus,

$$E_k = E - E_n = E_n = \frac{3\hbar\omega}{4} \coth f . \quad (6.93)$$

The Einstein approximation has made it possible to construct a crystal model and calculate all its thermodynamic properties, which agreed for the most part with the actually observed ones. Also observed was the crude character of this approximation, viz., the heat-capacity behavior at low temperatures did not agree with experiment, nor did allowance for the anharmonic terms lead in this approximation to good agreement between a number of thermodynamic characteristics and experiment at high temperatures. This indicates that the single-particle approximation is not always effective and account must be taken of the correlations of two, three, etc., particles, while a correct description of the behavior of the heat capacity at low temperatures calls for consideration of the collective behavior of all the particles.

33. Classical model of a crystal in the harmonic approximation. General treatment

As already noted, in a crystal the atoms execute small vibrations about equilibrium positions that coincide with the corresponding crystal-lattice sites. We denote by q_i the instantaneous position of an atom vibrating about the ith site defined by the vector a_i. If $u(a_i)$ is the vector of the atom displacement from site a_i, then

$$q_i = a_i + u(a_i)\,. \tag{6.94}$$

Thus, in contrast to the static lattice approximation, in which the potential energy was given by

$$U = \sum_{i<j} \Phi(|a_i - a_j|)$$

in the case of a pair interaction potential between particles we now have

$$U = \sum_{i<j} \Phi(|q_i - q_j|) = \sum_{i<j} \Phi[\,|a_i - a_j + u(a_i) - u(a_j)|\,]\,. \tag{6.95}$$

The Hamiltonian of the system considered is given by

$$H = T + U\,, \tag{6.96}$$

where T is the kinetic energy, equal in the simplest case to

$$T = \sum_i \frac{p_i^2}{2m}\,. \tag{6.97}$$

To determine the form of the Hamiltonian in the harmonic approximation, we use the expansion of the function U in a Taylor series, which takes for an arbitrary function f in the three-dimensional case the form

$$f(a+u) = f(a) + u\nabla f(a) + \frac{1}{2}(u\nabla)^2 f(a) + \frac{1}{3!}(u\nabla)^3 f(a) + \cdots\,. \tag{6.98}$$

We put in our case

$$a = a_i - a_j\,,$$
$$u = u(a_i) - u(a_j)\,.$$

We have then from Eq. (6.98),

$$U = \frac{N}{2}\sum_i \Phi(a_i) + \frac{1}{2}\sum_{i,j}[u(a_i) - u(a_j)]\nabla\Phi(a_i - a_j)$$
$$+ \frac{1}{4}\sum_{i,j}\{[u(a_i) - u(a_j)]\nabla\}^2\Phi(a_i - a_j) + O(u^3)\,. \tag{6.99}$$

Consider the quantity

$$-\sum_j \nabla\Phi(a_i - a_j)\,. \tag{6.100}$$

It is equal to the force applied to the atom in site a_i by the remaining atoms that are in equilibrium. This quantity should therefore be zero. We see hence that the linear term in Eq. (6.99) is zero. Expression (6.99) reduces then to the form[184]

$$U = U_0 + U_2\,, \tag{6.101}$$

where

$$U_0 = \sum_{i<j} \Phi(|a_i - a_j|)\,, \tag{6.102}$$

$$U_2 = \tfrac{1}{4} \sum_{\substack{i,j\\ \mu,\nu}} [u_\mu(a_i) - u_\mu(a_j)]\Phi_{\mu\nu}[u_\nu(a_i) - u_\nu(a_j)]\,, \tag{6.103}$$

$$\Phi_{\mu\nu} = \frac{\partial^2 \Phi(q)}{\partial q_\mu \partial q_\nu}\,, \tag{6.104}$$

and $\nu, \mu = x, y, z$. It is easy to take into account the static part of the potential energy when the thermodynamic quantities are obtained. The Hamiltonian for the analysis of a crystal in the harmonic approximation is therefore usually chosen in the form

$$H = T + U_2\,, \tag{6.105}$$

with a more expedient expression for U_2:

$$U_2 = \tfrac{1}{2} \sum_{\substack{i,j\\ \mu,\nu}} u_\mu(a_i) D_{\mu\nu}(a_i - a_j) u_\nu(a_j)\,. \tag{6.106}$$

In our case

$$D_{\mu\nu}(a_i - a_j) = \delta_{ij} \sum_k \Phi_{\mu\nu}(a_i - a_k) - \Phi_{\mu\nu}(a_i - a_j)\,. \tag{6.107}$$

The notation used above and the expression obtained for the potential energy are valid for one atom per unit cell. If the unit cell contains r atoms, it is more convenient to use a different notation.[50] We characterize the position of the lth unit cell by the vector

$$R(l) = l_1 a_1 + l_2 a_2 + l_3 a_3\,, \tag{6.108}$$

where l stands for the group of three numbers l_1, l_2, and l_3 (all integers), while a_1, a_2, and a_3 are three noncoplanar vectors making up the unit cell. We choose the origin in the unit cell such that the vector $x(\varkappa)$ that defines the position of the atom in the cell satisfies the condition $x(0) = 0$. The position of the $\varkappa$th atom in the lth cell ($\varkappa = 0, 1, 2, \ldots, r-1$) is therefore determined by the vector

$$x\binom{l}{\varkappa} = R(l) + x(\varkappa)\,. \tag{6.109}$$

We define the displacement of the atom from the equilibrium position as a result of the thermal fluctuations by the vector $u\binom{l}{\varkappa}$. If the atom has no internal degrees of freedom, the kinetic energy is given by

$$T = \frac{1}{2} \sum_{l,\varkappa} m_\varkappa \dot{u}_\alpha^2\binom{l}{\varkappa}\,, \tag{6.110}$$

$m_\varkappa$ is the mass of the atom of species $\varkappa$, and $u\begin{pmatrix} l \\ \varkappa \end{pmatrix}$ is the α component of the vector $u\begin{pmatrix} l \\ \varkappa \end{pmatrix}$ ($\alpha = x, y, z$). We expand the total potential energy of the system in a Taylor series in powers of $u\begin{pmatrix} l \\ \varkappa \end{pmatrix}$:

$$U = U_0 + \sum_{l,\varkappa,\alpha} \Phi_\alpha \begin{pmatrix} l \\ \varkappa \end{pmatrix} u_\alpha \begin{pmatrix} l \\ \varkappa \end{pmatrix} + \frac{1}{2} \sum_{\substack{l,\varkappa,\alpha \\ l',\varkappa',\beta}} \Phi_{\alpha\beta} \begin{pmatrix} l & l' \\ \varkappa & \varkappa' \end{pmatrix} u_\alpha \begin{pmatrix} l \\ \varkappa \end{pmatrix} u_\beta \begin{pmatrix} l' \\ \varkappa' \end{pmatrix} \tag{6.111}$$

[we confine ourselves to terms quadratic in $u\begin{pmatrix} l \\ \varkappa \end{pmatrix}$]. Here U_0 is the lattice static energy:

$$\Phi_\alpha \begin{pmatrix} l \\ \varkappa \end{pmatrix} = \left. \frac{\partial U}{\partial u_\alpha \begin{pmatrix} l \\ \varkappa \end{pmatrix}} \right|_0 , \tag{6.112}$$

$$\Phi_{\alpha\beta} \begin{pmatrix} l & l' \\ \varkappa & \varkappa' \end{pmatrix} = \left. \frac{\partial^2 U}{\partial u_\alpha \begin{pmatrix} l \\ \varkappa \end{pmatrix} \partial u_\beta \begin{pmatrix} l \\ \varkappa \end{pmatrix}} \right|_0 . \tag{6.113}$$

The subscript 0 of the derivative means that it is calculated for the equilibrium configuration. Just as in the particular case considered above, it is easy to show that

$$\Phi_\alpha \begin{pmatrix} l \\ \varkappa \end{pmatrix} = 0 , \tag{6.114}$$

since $\Phi_\alpha \begin{pmatrix} l \\ \varkappa \end{pmatrix}$ is the force that acts in the direction of α on an atom located at the point $x\begin{pmatrix} l \\ \varkappa \end{pmatrix}$ and should be equal to zero at equilibrium. Relation (6.114), however, is used only for convenience in the final equation, and the calculations are carried out using the general relation (6.111).

It follows from Eq. (6.113) that $\Phi_{\alpha\beta} \begin{pmatrix} l & l' \\ \varkappa & \varkappa' \end{pmatrix}$ is invariant to permutation of the indices:

$$\Phi_{\alpha\beta} \begin{pmatrix} l & l' \\ \varkappa & \varkappa' \end{pmatrix} = \Phi_{\beta\alpha} \begin{pmatrix} l' & l \\ \varkappa' & \varkappa \end{pmatrix} . \tag{6.115}$$

From the translational invariance of the lattice it follows that if a triad of integers (l_1, l_2, l_3) is added to the index l, the coefficient $\Phi_\alpha \begin{pmatrix} l \\ \varkappa \end{pmatrix}$ is not changed, i.e., $\Phi_\alpha \begin{pmatrix} l \\ \varkappa \end{pmatrix}$ is independent of l. By virtue of the same properties, $\Phi_{\alpha\beta} \begin{pmatrix} l & l' \\ \varkappa & \varkappa' \end{pmatrix}$

does not depend on l and l' separately, but on the difference $l - l'$. We express these properties in the form

$$\Phi_\alpha\binom{l}{\varkappa} = \Phi_\alpha\binom{0}{\varkappa}, \quad \Phi_{\alpha\beta}\begin{pmatrix} l & l' \\ \varkappa & \varkappa' \end{pmatrix} = \Phi_{\alpha\beta}\begin{pmatrix} l - l' & \\ \varkappa & \varkappa' \end{pmatrix}. \tag{6.116}$$

Now let $u\binom{l}{\varkappa} = v$, where v is a vector independent of both l and $\varkappa$. Expression (6.111) then takes the form

$$U = U_0 + \sum_{l,\varkappa,\alpha} \Phi_\alpha\binom{l}{\varkappa} v_\alpha + \frac{1}{2} \sum_{\substack{l,\varkappa,\alpha \\ l',\varkappa,\beta}} \Phi_{\alpha\beta}\begin{pmatrix} l & l' \\ \varkappa & \varkappa' \end{pmatrix} v_\alpha v_\beta \,. \tag{6.117}$$

This choice of the vector and of $\binom{l}{\varkappa}$ correspond to displacing the crystal at distance v, meaning that in this case

$$U = U_0,$$

which requires that the coefficient of each power of U_α be zero separately (since the vector U_α is arbitrary):

$$\sum_{l,\varkappa} \Phi_\alpha\binom{l}{\varkappa} = 0\,, \tag{6.118}$$

$$\sum_{\substack{l,\varkappa \\ l',\varkappa'}} \Phi_{\alpha\beta}\begin{pmatrix} l & l' \\ \varkappa & \varkappa' \end{pmatrix} = 0\,. \tag{6.119}$$

Taking into account in Eq. (6.117) the first term of Eq. (6.116), we have

$$\sum_{\varkappa} \Phi_\alpha\binom{0}{\varkappa} = 0\,. \tag{6.120}$$

It is seen from this relation that even if the forces acting on each atom separately are not zero, the resultant force acting on the unit cell is zero.

We now consider the expression

$$F_\alpha\binom{l}{\varkappa} = -\Phi_\alpha\binom{l}{\varkappa} + \sum_{l',\varkappa',\beta} \Phi_{\alpha\beta}\begin{pmatrix} l & l' \\ \varkappa & \varkappa' \end{pmatrix} u_\beta\binom{l'}{\varkappa'}. \tag{6.121}$$

We choose $u_\beta\binom{l'}{\varkappa'} = v_\alpha$, which corresponds to displacement of the crystal as a whole by a vector v. The force acting on an individual atom should remain unchanged in this case. Therefore,

$$\sum_{l',\varkappa'} \Phi_{\alpha\beta}\begin{pmatrix} l & l' \\ \varkappa & \varkappa' \end{pmatrix} = 0 = \sum_{l',\varkappa'} \Phi_{\alpha\beta}\begin{pmatrix} 0 & l' \\ \varkappa & \varkappa' \end{pmatrix}. \tag{6.122}$$

Let us discuss the properties that follow from the symmetry of the potential-energy function and of its derivatives in an infinitely small rotation of the crystal. To this end, we choose $u_\alpha\binom{l}{\varkappa}$ in the form

$$u_\alpha\binom{l}{\varkappa} = \sum_\beta \omega_{\alpha\beta} x_\beta\binom{l}{\varkappa}, \tag{6.123}$$

where $\omega_{\alpha\beta}$ are certain parameters that are elements of an infinitesimal antisymmetric matrix

$$\omega_{\alpha\beta} = -\omega_{\beta\alpha}.$$

We substitute Eq. (6.123) in Eq. (6.111) and retain only the linear terms

$$U = U_0 + \sum_{l,\varkappa,\alpha,\beta} \Phi_\alpha\binom{l}{\varkappa}\omega_{\alpha\beta} x_\beta\binom{l}{\varkappa}.$$

From the invariance of the crystal potential energy to rotations, it follows that

$$\sum_{l,\varkappa} \Phi_\alpha\binom{l}{\varkappa} x_\beta\binom{l}{\varkappa} = \sum_{l,\varkappa} \Phi_\beta\binom{l}{\varkappa} x_\alpha\binom{l}{\varkappa}. \tag{6.124}$$

We substitute now Eq. (6.123) in Eq. (6.121):

$$F_\alpha\binom{l}{\varkappa} = -\Phi_\alpha\binom{l}{\varkappa} - \sum_{l',\varkappa',\beta,\gamma} \Phi_{\alpha\beta}\begin{pmatrix} l & l' \\ \varkappa & \varkappa' \end{pmatrix}\omega_{\beta\gamma} x_\gamma\binom{l'}{\varkappa'}.$$

On the other hand, $F_\alpha\binom{l}{\varkappa}$ should be transformed in a rotation like the α component of the vector

$$F_\alpha\binom{l}{\varkappa} = F_\alpha\binom{l}{\varkappa} + \sum_\beta \omega_{\alpha\beta} F_\beta\binom{l}{\varkappa} = \sum_\beta (\delta_{\alpha\beta} + \omega_{\alpha\beta}) F_\beta\binom{l}{\varkappa}.$$

Using the last two equations, we have

$$\sum_{l',\varkappa',\beta,\gamma} \left\{\Phi_{\alpha\beta}\begin{pmatrix} l & l' \\ \varkappa & \varkappa' \end{pmatrix}\omega_{\beta\gamma} x_\gamma\binom{l'}{\varkappa'} = \sum_\beta \omega_{\alpha\beta}\Phi_\beta\binom{l}{\varkappa}\right\}.$$

Since $\omega_{\alpha\beta}$ is arbitrary, this equation should be valid for the coefficients at all $\omega_{\alpha\beta}$, and we obtain therefore, taking Eq. (6.122) into account,[50]

$$\sum_{l',\varkappa'} \left\{\Phi_{\alpha\beta}\begin{pmatrix} l & l' \\ \varkappa & \varkappa' \end{pmatrix} x_\gamma\binom{l'}{\varkappa'} - \Phi_{\alpha\gamma}\begin{pmatrix} l & l' \\ \varkappa & \varkappa' \end{pmatrix} x_\beta\binom{l'}{\varkappa'}\right\}$$
$$= \delta_{\alpha\beta}\Phi_\gamma\binom{l}{\varkappa} - \delta_{\alpha\gamma}\Phi_\beta\binom{l}{\varkappa}. \tag{6.125}$$

Constraints are imposed on the force constants by the symmetry and the structure of the crystal. We write down the most general symmetry operation that brings the crystal in coincidence with itself[50]:

$$\{S|\,V(S) + x(m)\}x\binom{l}{\varkappa} = Sx\binom{l}{\varkappa} + V(S) + x(m) = x\binom{L}{K}, \tag{6.126}$$

where S is a real orthogonal matrix representation of the rotation operation (this operation is one of the operations of the crystal's point group), $V(S)$ is the displacement vector and is not equal to zero if the symmetry elements of the crystal include screw axes and slip planes, and $x\binom{L}{K}$ is the position occupied by the atoms after the transformation. It is easy to show that the atomic force constants are subject to the following transformation rules:

$$\Phi_\alpha\binom{L}{K} = \sum_\mu S_{\alpha\mu}\Phi_\mu\binom{l}{\varkappa}, \tag{6.127}$$

$$\Phi_{\alpha\beta}\begin{pmatrix} L & L' \\ K & K' \end{pmatrix} = \sum_{\mu,\nu} S_{\alpha\mu} S_{\beta\gamma}\Phi_{\mu\nu}\begin{pmatrix} l & l' \\ \varkappa & \varkappa' \end{pmatrix}. \tag{6.128}$$

From these expressions we can obtain for each type of crystal structure independent nonzero elements of the first- and second-rank tensors $\Phi_\alpha\binom{l}{\varkappa}$ and $\Phi_{\alpha\beta}\begin{pmatrix} l & l' \\ \varkappa & \varkappa' \end{pmatrix}$. For a system of particles interacting via central forces, we have

$$\Phi_{\alpha\beta}\begin{pmatrix} l & l' \\ \varkappa & \varkappa' \end{pmatrix} = -\left.\frac{\partial\Phi(r)}{\partial x_\alpha \partial x_\beta}\right|_{r=r_{\varkappa\varkappa'}} = \Phi_{\alpha\beta}\begin{pmatrix} l' & l \\ \varkappa' & \varkappa \end{pmatrix}. \tag{6.129}$$

We proceed now to calculate the thermodynamic functions. The thermodynamics of a crystal can be formulated by calculating the statistical integral

$$Z = \int \exp[-\beta H] dp_1 \cdots dp_N \, dq_1 \cdots dq_N$$

$$= \int \exp[-\beta H] dp_1 \cdots dp_N \prod_{l,\varkappa} du\binom{l}{\varkappa}. \tag{6.130}$$

In this approximation it is easy to set apart the dependence of the partition function on the temperature, and also the static part of the interaction potential. We introduce for this purpose the new variables

$$u\binom{l}{\varkappa} = \beta^{-1/2}\bar{u}\binom{l}{\varkappa}, \quad du\binom{l}{\varkappa} = \beta^{-3/2} d\bar{u}\binom{l}{\varkappa},$$

$$p_i = \beta^{-1/2}\bar{p}_i, \quad dp_i = \beta^{-3/2} d\bar{p}_i . \tag{6.131}$$

As a result, expression (6.130) can be written in the form

$$Z = \exp[-\beta U_0]\beta^{-3N}\Gamma_0, \tag{6.131'}$$

where Γ_0 does not depend on the temperature. Hence we see that it is not difficult to find an exact expression for the internal energy. Indeed, according to Eq. (1.110),

$$F = -\theta \ln\left\{\frac{1}{N!\,h^{3N}} \exp[-\beta U_0]\beta^{-3N}\Gamma_0\right\},$$

meaning that

$$U = -\theta^2 \frac{\partial(F/\theta)}{\partial\theta} = U_0 + 3N\theta. \tag{6.132}$$

The heat capacity at constant volume is

$$C_V = \left(\frac{\partial U}{\partial T}\right)_V = 3Nk\,. \tag{6.132'}$$

Next,

$$C_p - C_V = \left[\left(\frac{\partial U_0}{\partial U}\right)_\theta + p\right]\left(\frac{\partial V}{\partial T}\right)_p = 0\,,$$

since

$$\left(\frac{\partial U_0}{\partial V}\right)_\theta + p = 0$$

if $\Phi_{\alpha\beta}\begin{pmatrix} l & l' \\ \varkappa & \varkappa' \end{pmatrix}$ is independent of the volume. Indeed, in this case F_0 in Eq. (6.131) is also independent of the volume, meaning that

$$p = -\frac{\partial F}{\partial V} = -\frac{\partial U_0}{\partial V}\,. \tag{6.133}$$

Thus,

$$C_p = C_V = 3Nk\,. \tag{6.134}$$

In addition, the isothermal elastic modulus ε_T is equal to

$$\varepsilon_T = V\frac{\partial^2 U_0}{\partial V^2}\,, \tag{6.135}$$

and the entropy is equal to

$$S = -k\frac{\partial F}{\partial \theta} = kN\theta \ln\left\{\left(\frac{l\theta}{h}\right)^3 F_0\right\}\,. \tag{6.136}$$

Hence we see that S is a function of only the temperature. This shows directly that the adiabatic and isothermal elastic moduli are equal:

$$\varepsilon_T = \varepsilon_S\,. \tag{6.137}$$

Thus, in the approximation considered, the crystal heat capacity (both at constant pressure, which is usually measured in experiment, and at constant volume) obeys the Petit–Dulong law, which can be written on the basis of Eq. (6.134) in the following form (per mole of the substance):

$$C = C_p = C_V = 5.96 \text{ cal mol K} \tag{6.137'}$$

($k = 1.38\times 10^{-14}$ erg/K, $N_A = 6.022\times 10^{22}$ per mole, 1 cal $= 4.184\times 10^7$ erg)

or

$$C = 3R = 24.94 \text{ J} \,/\, \text{K}^{-1}\,\text{mol}^{-1} \tag{6.138}$$

(R is the gas constant). Table XIII lists the heat capacities at constant pressure for a number of elements under normal conditions. We see that for almost all substances the experimental data agree with the calculated heat capacities. A discrepancy exists for diamond, in which the quantum effects are appreciable even at the temperature employed (298.15 K).

Table XIII. Heat capacities at constant pressure for elements under normal conditions.

Substance	Cp (J K^{-1} mol^{-1})	Substance	Cp (J K^{-1} mol^{-1})
Aluminum	24.35	Copper	24.52
Bismuth	25.52	Sodium	28.12
Tungsten	24.8	Tin (white)	26.36
Germanium	23.4	Platinum	25.69
Gold	25.23	Lead	26.44
Cadmium	26.32	Silver	25.49
		Carbon (diamond)	6.12

Considerably above the Einstein temperature we have

$$T_E = \frac{\hbar\omega_E}{k}, \tag{6.138'}$$

where ω_E is the harmonic-oscillator frequency in the Einstein approximation for a crystal (see the preceding section), the heat capacity of many crystals obeys the Dulong and Petit law. But at very high temperatures the anharmonic contributions to the heat capacity can no longer be neglected, so that the heat capacity is not a constant and the equality $C_p = C_V$ is likewise not valid. At low temperatures the heat capacity begins to decrease and tends to zero, but not exponentially as in the Einstein model, but cubically:

$$C = \frac{12N\nu\pi^4}{5}\left(\frac{\theta}{\theta_D}\right)^3, \tag{6.139}$$

where $\theta_{D/k}$ is the Debye temperature, connected with the Einstein temperature by the relation

$$T_E = \sqrt{\frac{3}{5}}\frac{\theta_D}{k} = 0.775\,\frac{\theta_D}{k}, \tag{6.140}$$

and ν is the number of atoms per unit cell.

For a number of substances, however, for example, for solid inert-gas crystals, the heat capacity cannot reach the Dulong and Petit value, and contributions from anharmonic terms are important. A quantum-mechanical crystal-lattice treatment in the harmonic approximation is needed, as is allowance for anharmonic terms.

34. The reciprocal lattice

The reciprocal-lattice concept plays an important role in the analysis of periodic structures, including consideration of diffraction in a crystal, investigation of functions with Bravais-lattice periodicity, and others.

Let R denote the set of all points making up the Bravais lattice. If R starts out from one of the lattice sites, then

$$R = R(l) = \sum_{\alpha=1}^{3} l_\alpha a_\alpha, \quad l = (l_1, l_2, l_3)\,, \tag{6.141}$$

where l_α are integers. We set in correspondence with the Bravais lattice a certain periodic structure, which we call the reciprocal lattice and which is based on three vectors b_β ($\beta = 1, 2, 3$):

$$\begin{aligned} b_1 &= 2\pi \frac{a_2 \times a_3}{a_1(a_2 \times a_3)}\,, \\ b_2 &= 2\pi \frac{a_3 \times a_1}{a_1(a_3 \times a_3)}\,, \\ b_3 &= 2\pi \frac{a_1 \times a_2}{a_1(a_2 \times a_3)}\,. \end{aligned} \tag{6.142}$$

The vectors b_β satisfy the relation

$$b_\beta a_\alpha = 2\pi\delta_{\alpha\beta}\,. \tag{6.143}$$

Any reciprocal-lattice vector k can be expressed in the form

$$k = k_1 b_1 + k_2 b_2 + k_3 b_3\,. \tag{6.144}$$

Let $U(r)$ be a function characterizing some properties of an ideal crystal and having the same periodicity, i.e.,

$$U(r + R) = U(r)\,, \tag{6.145}$$

where R is defined by Eq. (6.141). Fourier expansion of $U(r)$ yields

$$U(r) = \sum_q U_q \exp[i2\pi qr] = \sum_k \bar{U}_k \exp[ikr]\,. \tag{6.146}$$

By virtue of Eq. (6.145) we should have

$$\sum_k \bar{U}_k \exp[jkr]\exp[ikR] = \sum_k \bar{U}_k \exp[ikr]\,, \tag{6.147}$$

therefore,

$$\exp[ikR] = 1.$$

Hence,

$$kR = 2\pi n\,,$$

where n is an integer. If k is defined by Eq. (6.144), then

$$k_1 l_1 + k_2 l_2 + k_3 l_3 = n\,.$$

Thus, k_1, k_2, and k_3 should be integers; this means in fact that k is the reciprocal-lattice vector.

Since the reciprocal lattice is a Bravais lattice, it is possible to construct for it a reciprocal lattice which can be easily shown to be the direct Bravais lattice. Let us find the reciprocal lattice of a simple cubic Bravais lattice, in which the side of the cubic unit cell is equal to a. Inasmuch as in this case,

$$a_1 = ai, \quad a_2 = aj, \quad a_3 = ak\,, \tag{6.148}$$

we have according to Eq. (6.142),

$$b_1 = \frac{2\pi}{a} i, \quad b_2 = \frac{2\pi}{a} j, \quad b_3 = \frac{2\pi}{a} k, \tag{6.149}$$

i.e., the reciprocal lattice is a simple cubic lattice with side $2\pi/a$.

We proceed now to determine the reciprocal lattice of an fcc Bravais lattice whose arbitrary cubic cell has a side a. Substituting Eq. (6.5) in Eq. (6.142) we have

$$b_1 = \frac{2\pi}{a}(j + k - i), \quad b_2 = \frac{2\pi}{a}(k + i - j), \quad b_3 = \frac{2\pi}{a}(i + j - k). \tag{6.150}$$

The reciprocal lattice in this case is thus a body-centered Bravais lattice whose arbitrary cubic cell has a side $4\pi/a$.

Since the reciprocal of a reciprocal lattice is the direct one, it follows that the reciprocal of a bcc lattice with arbitrary cubic-cell side length a is an fcc crystal lattice with arbitrary cubic cell side $4\pi/a$. Let us prove now that the vectors b_1, b_2, and b_3 satisfy the relation

$$b_1(b_2 \times b_3) = \frac{(2\pi)^3}{a_1(a_2 \times a_3)}. \tag{6.151}$$

Consider the vector product

$$b_2 \times b_3 = \frac{(2\pi)^2[(a_3 \times a_1) \times (a_1 \times a_2)]}{[a_1(a_2 \times a_3)]^2}$$

$$= \frac{(2\pi)^2\{[a_3(a_1 \times a_2)]a_1 - [a_1(a_1 \times a_2)]a_3\}}{[a_1(a_2 \times a_3)]^2} = (2\pi)^2 a_1/[a_1(a_2 \times a_3)]. \tag{6.152}$$

We have used here the equation for a double vector product and also the orthogonality of a_α. Next,

$$b_1(b_2 \times b_3) = \frac{(2\pi)^3(a_2 \times a_3)a_1}{[a_1(a_2 \times a_3)]^2} = \frac{(2\pi)^3}{a_1(a_2 \times a_3)},$$

which is identical with Eq. (6.151). Thus, if the initial triad of vectors a_α is noncoplanar and right-handed, the vector triad b_α is also noncoplanar and right-handed.

For a reciprocal lattice, as for the direct, we can introduce a Wigner–Seitz cell, which is called in this case the first Brillouin zone. The second and higher Brillouin zones are unit cells of a different kind, and appear in the theory of electronic levels in a periodic potential.

We have seen above that any vector of the direct lattice is defined by relation (6.141). Each site is characterized by a set of three numbers: l_1, l_2, and l_3. A special symbol is used for the site—the three numbers l_1, l_2, and l_3 in a pair of square brackets: $[[l_1 l_2 l_3]]$.* A crystallographic direction is defined as the direction of a straight line that passes at least through two lattice sites. By

*It is customary to use in place of the indices l_1, l_2, and l_3 the numbers m, n, and l, which we shall use from now on.

virtue of symmetry, an infinite number of lattice sites should lie on this line. If it is assumed that the straight line passes through the origin [[000]], the crystallographic direction is determined by the site closest to the origin on this line. If this site is defined by the integers m, n, and l, the direction is designated $[mnl]$.

The numbers m, n, and l are called the Miller indices of a given crystallographic direction and of all the directions parallel to it. If one of the numbers is negative, it is marked by a superior bar. Thus, for example, if $m = 2$, $n = -1$, and $l = 0$, the symbol for the crystallographic direction is $[2\bar{1}0]$. The crystallographic coordinate axes have the Miller indices [100], [010], and [001].

An arbitrary plane that contains not less than three lattice points (sites) on one straight line is called an atomic plane. By virtue of the translational invariance of the Bravais lattice, any plane contains an infinite number of lattice points. They form on this plane a two-dimensional Bravais lattice. We define the family of atomic planes of a lattice as the aggregate of parallel equidistant planes which, taken together, contain all the points of the Bravais lattice. It is easily seen that it is possible to construct for any lattice many families of atomic planes, i.e., this subdivision is not single valued.

Let us take a certain family of lattice planes whose position is uniquely defined by a unit normal vector n (which is the same for parallel planes), and also by the distance to the origin. Let the distance between planes be d. We introduce a vector $k = 2\pi n/d$, which can be easily seen to be the reciprocal lattice vector: $\exp[ikr]$ is constant in all the atomic planes of the considered family, and it is equal to its value on the plane passing through the origin, i.e., $\exp[ikr] = 1$. Since this equality holds for all the lattice points and the latter are all contained in the family of the planes, this means that k is a reciprocal-lattice vector. Let us show now that k is the smallest reciprocal-lattice vector perpendicular to the chosen family of planes. In fact, assume that there is a reciprocal-lattice vector k_1 such that $k_1 \| k$ and $|k_1| < |k|$. This vector generates a plane wave of wavelength $d_1 = 2\pi/|k_1| > d$. This plane wave cannot therefore have constant values on all the planes of the given family, and consequently the equality

$$\exp[ikr] = 1 \tag{6.153}$$

will not hold if the vector k is replaced by the vector k_1. We have proved that if there exists a family of atomic planes spaced d apart, then there exists reciprocal-lattice vectors perpendicular to these planes, and the length of the smallest of them is $2\pi/d$.

We now consider a certain family of parallel reciprocal-lattice vectors. Let the smallest of them have a length $|k|$ (we have chosen for the sake of argument one smallest vector). We consider in real space a set of planes for which relation (6.153) is satisfied. The planes of this family are perpendicular to the vector k, and the distance between them is

$$d = 2\pi/|k| \,. \tag{6.154}$$

The Bravais-lattice vectors should satisfy Eq. (6.153); therefore, its points should lie on the considered planes. Thus, the family of atomic planes should be contained in the considered family of planes, i.e., it should be its subset. Since the distance between planes is d, if only the nth plane were to contain Bravais-lattice points, by virtue of the foregoing proof there would correspond to this plane a reciprocal-lattice vector of length $2\pi/nd$, i.e., a vector k/n, thus contradicting our conditions. Consequently, for any reciprocal-lattice vector k there exists a family of atomic planes that are perpendicular to it and the distance between them is d. These planes are perpendicular to the vector k, and the length of the smallest vector parallel to k is equal to $2\pi/d$.

As noted above, the orientation of the plane is described by specifying the vector normal to it. We have shown that for any family of atomic planes there exist reciprocal-lattice vectors normal to this family. To be specific, the normal to a given plane is chosen to be the smallest reciprocal-lattice vector perpendicular to it, and the coordinates of this vector are called the Miller indices of the given plane.

In other words, if

$$k = hb_1 + kb_2 + lb_3 , \tag{6.155}$$

the Miller indices are the numbers h, k, and l. By virtue of the definition of the reciprocal lattice, h, k, and l are integers, and since k is the shortest of the vectors parallel to it, h, k, and l are mutually prime—they have no integer common divisor other than unity.

We consider by way of example a simple cubic Bravais lattice. Its reciprocal lattice is also simple cubic. Thus, the Miller indices are chosen to be the coordinates of the vector normal to the plane (we assume that the coordinate frame was chosen in the usual manner). In practice, fc and bc Bravais lattices are described with the aid of an arbitrary cubic cell, i.e., the problem of designating the atomic planes reduces to the preceding one. If noncubic crystals are considered, however, it becomes important to define the Miller indices as the coordinates of the normal in the system defined by the reciprocal lattice (and not by a direct one).

Another interpretation of the Miller indices of a direct lattice can also be used. Since the lattice plane with Miller indices h, k, and l is perpendicular to the reciprocal-lattice vector defined by Eq. (6.155), we have for this plane

$$kr = C = \text{const.} \tag{6.156}$$

This plane intersects the coordinate axes (i.e., the axes crossing the principal vectors a_1, a_2, a_3) at the points l_1a_1, l_2a_2, l_3a_3. The values of l_1, l_2, and l_3 are determined from the equation of the plane (6.156):

$$k(l_i a_i) = C, \quad i = 1, 2, 3, \tag{6.157}$$

whence

$$l_1 = \frac{C}{2\pi h}, \quad l_2 = \frac{C}{2\pi k}, \quad l_3 = \frac{C}{2\pi l}, \tag{6.158}$$

since

$$ka_1 = 2\pi h, \quad ka_2 = 2\pi k, \quad ka_3 = 2\pi l\,.$$

It is seen from Eq. (6.158) that

$$h{:}k{:}l = \frac{1}{l_1}{:}\frac{1}{l_2}{:}\frac{1}{l_3}\,.$$

The Miller indices are therefore defined in crystallography in the following manner: Numbers inversely proportional to the intercepts of the atomic plane on the crystallographic axes are determined, and the Miller indices are taken to be integers proportional to these numbers and having no common factor. The Miller indices of the atomic plane are designated as

$$(hkl)\,.$$

Equivalent families of planes [i.e., of all the planes equivalent to the (hkl) plane by virtue of the crystal symmetry], are designated by

$$\{hkl\}\,.$$

The equivalent planes in a cubic crystal are thus (100), (010), and (001) and are designated $\{100\}$. Directions are analogously designated $\langle hkl\rangle$. Examples of equivalent directions in a cubic crystal are likewise [100], [010], [001], $[\bar{1}00]$, [010], $[00\bar{1}]$. They are designated $\langle 100\rangle$.

35. Harmonic lattice vibrations

Although formulation of the thermodynamics of a crystal in the harmonic approximation does not require consideration of the lattice dynamics, the results that follow will be useful for a quantum-dynamic treatment of a harmonic crystal.

35.1 One-dimensional system

Assume N particles of mass m located at points $R = na$ along a straight line (n is an integer and a is the distance between the nearest neighbors). We denote the displacement of the ith particles from the equilibrium position by $u(a_i)$. We assume for simplicity that only nearest neighbors interact. Let the interaction potential be expandable in a Taylor series. Then

$$U = U_0 + U_2 = U_0 + \tfrac{1}{2}\, k \sum_i \,[u(a_i) - u(a_{i+1})]^2\,, \tag{6.159}$$

where U_0 is the static part of the potential energy, $k = \Phi''(a)$ and $\Phi(r)$ is the interaction potential.

The equation of motion of the particles is

$$m\frac{d^2u(a_i)}{dt^2} = -\frac{\partial U}{\partial u(a_i)} = -\frac{\partial U_2}{\partial u(a_i)}$$

or

$$m\frac{d^2u(a_i)}{dt^2} = -k[2u(a_i) - u(a_{i-1}) - u(a_{i+1})] . \qquad (6.159')$$

This set of equations is identical with the set of equations of motion of N particles interconnected by springs of stiffness k. Since we are considering a system with a large number of particles, the boundary conditions are unimportant and can be chosen from convenience considerations. The most frequently used are the Born–Karman periodic boundary conditions, the analytic expression for which is in our case

$$u(a_{N+1}) = u(a), \quad u(a_0) = u(a_N) . \qquad (6.160)$$

We seek the solution of the system (6.150) in the form

$$u(a_i, t) \sim e^{i(ka_i - \omega t)} . \qquad (6.161)$$

The following condition must be met by virtue of Eq. (6.160):

$$\exp[ika_N] = 1. \qquad (6.162)$$

It is easily seen that

$$a_i = ia,$$

and it follows from Eq. (6.162) that

$$k = \frac{2\pi}{a}\frac{i}{N}, \qquad (6.163)$$

where i is an integer. There are only N physically different solutions satisfying the condition (6.143). We assume that

$$-\pi/2 \leqslant k < \pi/2, \qquad (6.164)$$

which is equivalent to the requirement that the vector k be located in the first Brillouin zone.

Substituting Eq. (6.161) in Eq. (6.159) we get

$$\begin{aligned} -m\omega^2 \exp[i(ka_i - \omega t)] \\ = -k[2 - \exp[-ika] - \exp[ika]]\exp[i(ka_i - \omega t)] \\ = -2k(1 - \cos ka)\exp[i(ka_i - \omega t)] . \end{aligned} \qquad (6.165)$$

Hence it follows directly that

$$\omega = \omega(k) = \sqrt{\frac{2k(1 - \cos ka)}{m}} = 2\sqrt{\frac{k}{m}}\left|\sin\frac{ka}{2}\right| . \qquad (6.166)$$

From Eq. (6.161) we find that the actual displacements of the particles are described by the real and imaginary parts of the sought solution

$$u(a_i, t) \sim \{\cos(ka_i - \omega t), \sin(ka_i - \omega t)\} . \qquad (6.167)$$

Since $\omega(k)$ is an even function, we choose only the positive root in Eq. (6.166). We have thus N values of k, each with its own ω according to Eq. (6.166), and it is clear from Eq. (6.167) that we have obtained altogether $2N$ linearly inde-

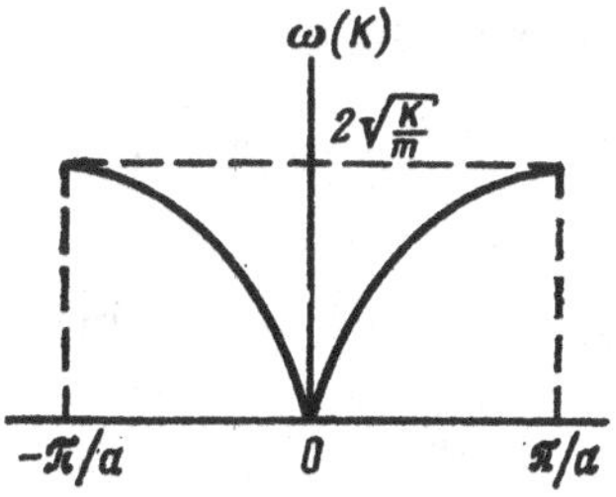

Figure 35.

pendent solutions, i.e., the problem is solved. To determine the explicit form of $u(a_i, t)$ we must specify the initial values of the coordinates and velocities of all the particles.

The plot of $\omega(k)$, i.e., of the frequency ω vs the wave vector k, is called the dispersion curve. It is shown for our case in Fig. 35. If $k \ll \pi/2$, we have

$$\omega = a\sqrt{\frac{k}{m}}|k| .$$

Therefore, the phase and group velocities C and $v = \partial\omega/\partial k$ of the wave propagating along the chain are equal in this approximation to

$$C = v = a\sqrt{\frac{k}{m}} .$$

With increase of k, however, the group velocity decreases to zero (as $|k| \to \pi/2$).

We now consider a linear system of particles whose mutual interaction is not restricted to nearest neighbors. In this case we have in lieu of Eq. (6.158),

$$u = u_0 + \tfrac{1}{2} \sum_{i<j} \Phi''(a_{i-j})[u(a_i) - u(a_j)]^2 . \tag{6.168}$$

The equation of motion takes the form

$$m\frac{d^2u(a_i)}{dt^2} = - \sum_{i \neq j} \Phi''(a_{i-j})[u(a_i) - u(a_j)] . \tag{6.169}$$

We put

$$\Phi''(a_{i-j}) = A_{i-j} = A_{j-i} , \tag{6.170}$$

which is legitimate since $\Phi(r)$ is even, and also

$$A_0 = - \sum_{j \neq 0} A_j , \tag{6.171}$$

from which we get

$$\sum_j A_j = 0 . \tag{6.172}$$

In this notation, Eq. (6.169) takes the form

$$m\frac{d^2u(a_i)}{dt^2} = \sum_j A_{i-j}u(a_j) \tag{6.173}$$

As before, we seek the solution of this equation in the form (6.161), and obtain

$$-m\omega^2 = \sum_j A_{i-j}\exp[-ik(a_j - a_i)] = \sum_l A_l \exp[-ikal]\,. \tag{6.174}$$

Since A_i is real and $A_{-i} = A_i$, the expression on the right-hand side is also real. In addition, assuming that the right-hand side of Eq. (6.174) is negative (this ensures stability of the oscillations), we obtain for the frequency the relation

$$\begin{aligned}\omega(k) &= \frac{1}{\sqrt{m}}\sqrt{-\sum_l A_l \exp[-ikal]}\\ &= \frac{2}{\sqrt{m}}\sqrt{\sum_{l\geqslant 1} A_l \sin^2(kal/2)}\,.\end{aligned} \tag{6.175}$$

As the interaction potential decreases rapidly with distance, the behavior of $\omega(k)$ is determined mainly by A_1, which must be positive. The equilibrium distance between the particles must therefore not exceed the distance to the inflection point of the potential $\Phi(r)$. If the chain is stretched by an external force, its oscillations become unstable at a certain chain length. Shrinkage of the chain, on the other hand, increases the oscillation stability.

Consider a one-dimensional Bravais lattice whose unit cell contains two atoms, assumed for simplicity to have the same mass m. The atoms locations are

$$a_i = ia$$

and

$$a_i + d, \quad d \leqslant a\,. \tag{6.176}$$

We assume that only nearest neighbors interact, and that $\Phi(d) > \Phi(a)$ by virtue of Eq. (6.176). In the harmonic approximation we have

$$\begin{aligned}U = U_0 + U_2 = U_0 &+ \tfrac{1}{2}\,k\sum_i [u_1(a_i) - u_2(a_i)]^2\\ &+ \tfrac{1}{2}\,G\sum_i [u_2(a_i) - u_1(a_{i+1})]^2\,,\end{aligned} \tag{6.177}$$

where $K = \Phi''(d) \geqslant \Phi''(a) = G$. Here $u_1(a_i)$ and $u_2(a_i)$ are the displacements of the particles from the sites a_i and $a_i + d$, respectively.

The set of equations of motion constitutes in this case

$$m\frac{d^2u_1(a_i)}{dt^2} = -\frac{\partial U_2}{\partial u_1(a_i)} = -K[u_1(a_i) - u_2(a_i)] - G[u_1(a_i) - u_2(a_{i-1})] ,$$
$$m\frac{d^2u_2(a_i)}{dt^2} = -\frac{\partial U_2}{\partial u_2(a_i)} = -K[u_2(a_i) - u_1(a_i)] - G[u_2(a_i) - u_1(a_{i+1})] . \quad (6.178)$$

We seek a solution in the form

$$u_1(a_i) = \alpha_1 \exp[i(ka_i - \omega t)] ,$$
$$u_2(a_i) = \alpha_2 \exp[i(ka_i - \omega t)] , \quad (6.179)$$

which also contains besides ω and k the constants α_1 and α_2 which must be determined. As in the preceding case, periodic Born–Karman boundary conditions lead to N nonequivalent values of k. Substituting Eq. (6.179) in Eq. (6.28) we obtain after simple transformations the set of equations

$$[m\omega^2 - (K+G)]\alpha_1 + [K + G\exp[-ika]]\alpha_2 = 0,$$
$$[K + G\exp[ika]]\alpha_1 + [m\omega^2 - (K+G)]\alpha_2 = 0, \quad (6.180)$$

which is linear in α_1 and α_2 and has a nontrivial solution only if the determinant made up of the coefficients of α_1 and α_2 vanishes. This yields the equation

$$[m\omega^2 - (K+G)]^2 = |K + G\exp[-ika]|^2$$
$$= K^2 + G^2 + 2KG\cos ka. \quad (6.181)$$

We hence obtain two values of ω^2 at which the equation in question is satisfied:

$$\omega^2 = \frac{K+G}{m} \pm \frac{1}{m}\sqrt{K^2 + G^2 + 2KG\cos ka} . \quad (6.182)$$

It is easily seen that

$$\frac{\alpha_2}{\alpha_1} = \mp\frac{K + G\exp[ika]}{|K + G\exp[ika]|} . \quad (6.183)$$

Thus, in contrast to the case of one particle per unit cell, we have here two values of ω for each k. The two plots of ω vs k are shown in Fig. 36. The lower $\omega = \omega(k)$ curve is similar in form to the dispersion curve for a monatomic Bravais lattice. It is called the acoustic mode, since $\omega = Ck$ as $ka \to 0$, as is typical of sound waves. The upper curve is called the optical mode since electromagnetic radiation can interact in the crystal with the short-wave lattice vibrations.

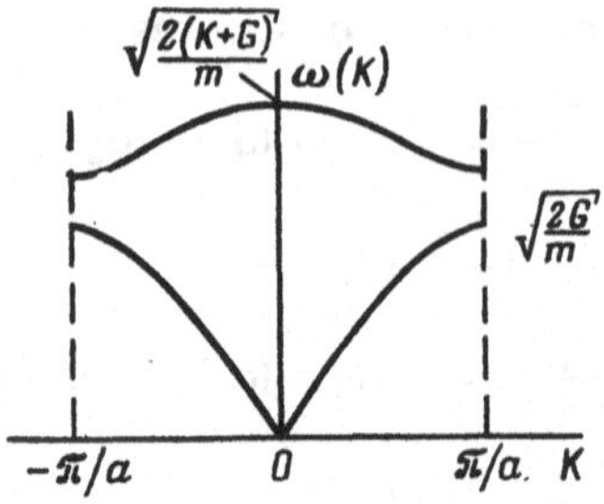

Figure 36.

Let $k \ll \pi/a$; we have then from Eq. (6.175),

$$\omega_0 = \sqrt{\frac{2(K+G)}{m}} - O(ka)^2, \tag{6.184}$$

$$\omega_A = \sqrt{\frac{KG}{2m(K+G)}}(ka), \tag{6.185}$$

where ω_0 corresponds to the optical mode and ω_A to the acoustic one, and according to Eq. (6.183) we have

$$\frac{\varepsilon_2}{\varepsilon_1} = \mp 1. \tag{6.186}$$

The " + " sign corresponds here to acoustic vibrations, when both atoms move in phase, and the " − " sign to optical ones, when the atoms move in counterphase. For $k = \pi/a$ we have

$$\omega_0 = \sqrt{\frac{2k}{m}}, \quad \omega_A = \sqrt{\frac{2G}{m}}.$$

If $K \gg G$, then

$$\omega_0 = \sqrt{\frac{2k}{m}}\left[1 + O\left(\frac{G}{K}\right)\right], \tag{6.187}$$

$$\omega_A = \sqrt{\frac{2G}{m}}\left|\sin\frac{ka}{2}\right|\left[1 = O\left(\frac{G}{K}\right)\right]. \tag{6.188}$$

The frequency of the optical mode is thus independent of K (to first order in G/K), i.e., independent molecular vibrations take place in each cell. The frequency of the acoustic mode, on the other hand, corresponds to a system of particles of mass $2m$ spaced apart. If $K = G$, we again have the case of a single-particle Bravais lattice with a lattice constant $a/2$, where a is the length of the unit cell. In place of the Brillouin zone $-\pi/a \leqslant k < \pi/a$ for the two-particle Bravais lattice we have the Brillouin zone $-2\pi/a \leqslant k < 2\pi/a$ for a single-particle lattice.

35.2 Three-dimensional system

We begin with the simplest case of a one-particle three-dimensional lattice. To simplify the algebra, we transform in Eq. (6.106) to matrix notation:

$$U_2 = \frac{1}{2} \sum_{i,j} u(a_i) D(a_i - a_j) u(a_j) , \tag{6.189}$$

where the matrices $u(a_i)$, $D(a_i - a_j)$ are chosen such that the resultant expression is equivalent to Eq. (6.106).

By virtue of the definition of the coefficients $D_{\mu\nu}(a_i - a_j)$,

$$D_{\mu\nu}(a_i - a_j) = \frac{\partial^2 U}{\partial u_\mu(a_i) \partial u_\nu(a_i)} . \tag{6.190}$$

They have the property

$$D_{\mu\nu}(a_i - a_j) = D_{\nu\mu}(a_j - a_i) . \tag{6.191}$$

Since the Bravais lattice is invariant to inversion, one more symmetry property is present:

$$D_{\mu\nu}(a_{i-j}) = D_{\mu\nu}(a_i - a_j), \quad D_{\mu\nu}(a_j - a_i) = D_{\mu\nu}(a_{j-i}) . \tag{6.192}$$

Using the property (6.191), we can replace the second equation above by

$$D_{\mu\nu}(a_i - a_j) = D_{\nu\mu}(a_i - a_j) . \tag{6.193}$$

The invariance of the Bravais lattice to translations gives rise to one more symmetry property:

$$\sum_i D_{\mu\nu}(a_i) = 0, \tag{6.194}$$

or, in matrix form,

$$\sum_i D(a_i) = 0. \tag{6.195}$$

The symmetry types considered here are particular cases of the lattice symmetry types obtained in general form in Sec. 34.

We now rewrite the particle equation of motion in matrix notation:

$$m \frac{d^2 u(a_i)}{dt^2} = - \sum_j D(a_i - a_j) u(a_j) . \tag{6.196}$$

We seek the solution of this equation in plane-wave form

$$u(a_i, t) = \varepsilon \exp[i(ka_i - \omega t)] , \tag{6.197}$$

where ε is called the polarization vector of the normal mode and must be determined. The Born–Karman periodic boundary solutions take in this case the form

$$u(R + N_l a_l) = u(R) , \tag{6.198}$$

where N_l $(l = 1, 2, 3)$ are numbers satisfying the condition $N = N_1N_2N_3$ (obviously, every one of them must be large). In contrast to the one-dimensional case we now have for k the relation

$$\exp[ikN_l a_l] = 1, \quad l = 1, 2, 3. \tag{6.199}$$

Let the vector k be given by

$$k = \sum_{i=1}^{3} x_i b_i , \tag{6.200}$$

where b_i are reciprocal-lattice vectors that satisfy the relation

$$b_i a_i = 2\pi\delta_{ij} . \tag{6.201}$$

Equation (6.199) is then reduced to

$$\exp(2\pi i N_l x_l) = 1, \tag{6.202}$$

whence

$$x_l = \frac{n_l}{N_l} , \tag{6.203}$$

where n_l are integers. Thus, the allowed values of the vector are

$$k = \sum_{l=1}^{3} \frac{n_l}{N_l} b_l . \tag{6.204}$$

If the vectors k are to correspond to physically different solutions, all must be in the same Brillouin zone, usually taken to be the first. There are thus only N nonequivalent values of the vector k having the form (6.204).

We now substitute Eq. (6.197) in Eq. (6.196):

$$m\omega^2\varepsilon = D(k)\varepsilon , \tag{6.205}$$

where

$$D(k) = \sum_i D(a_i)\exp[-ika_i] . \tag{6.206}$$

We transform this expression, using Eqs. (6.192) and (6.195), into

$$\begin{aligned} D(k) &= \frac{1}{2}\sum_i D(a_i)[\exp[-ika_i] + \exp[ika_i] - 2] \\ &= \sum_i D(a_i)[\cos[ka_i] - 1] = -2\sum_i D(a_i)\sin^2\left(\frac{ka_i}{2}\right). \end{aligned} \tag{6.207}$$

We see thus that $D(k)$ is an even function of k [the matrix $D(k)$ is real]. It follows also from Eq. (6.93) that $D(k)$ is symmetric. Therefore, Eq. (6.205) has as a solution three real eigenvectors ε_1, ε_2, ε_3 that satisfy the normalization condition

$$\varepsilon_s(k)\varepsilon_l(k) = \delta_{sl}, \quad s, l = 1, 2, 3 , \tag{6.208}$$

and are solutions of the equations

$$D(k)\varepsilon_s(k) = \lambda_s(k)\varepsilon_s(k) . \tag{6.209}$$

We can thus find for each vector k three vectors $\varepsilon_s(k)$ corresponding to three normal modes, and also the frequencies

$$\omega_s = \sqrt{\frac{\lambda_s(k)}{m}}\,. \tag{6.210}$$

Let us investigate the behavior of ω_s for small ka_i. In this case

$$D(k) \approx -\frac{|k|^2}{2}\sum_i (\hat{k}a_i)D(a_i), \quad \hat{k} = \frac{k}{|k|}\,. \tag{6.211}$$

Therefore,

$$\omega_s(k) = C_s(\hat{k})|k|\,, \tag{6.212}$$

$C_s(k)$ are quantities determined from Eq. (6.210) under the condition that the matrix whose eigenvalues are $\lambda_s(k)$ is

$$-\frac{1}{2m}\sum_i (\hat{k}a_i)^2 D(a_i)\,. \tag{6.213}$$

Thus, at small $|k|$, the frequency is a linear function of the length of the vector for all values of s. In the general case $C_s(k)$ depends both on the number s of the mode and on the direction of k.

If an isotropic medium is considered, the solution for a given direction k can be chosen such that one mode is polarized along the propagation direction and the other two perpendicular to this direction. For an anisotropic crystal, this choice of solutions is possible only for certain directions of the vector k.

Consider now an arbitrary lattice. Starting from Eq. (6.111), we write the equations of motion in the form

$$m_\varkappa \frac{d^2 u_\alpha\begin{pmatrix} l \\ \varkappa \end{pmatrix}}{dt^2} = -\frac{\partial U}{\partial u_\alpha\begin{pmatrix} l \\ \varkappa \end{pmatrix}} \tag{6.214}$$

or

$$m_\varkappa \frac{d^2 u_\alpha\begin{pmatrix} l \\ \varkappa \end{pmatrix}}{dt^2} = -\sum_{l',\varkappa',\beta} \Phi_{\alpha\beta}\begin{pmatrix} l & l' \\ \varkappa & \varkappa' \end{pmatrix} u_\beta\begin{pmatrix} l' \\ \varkappa' \end{pmatrix}. \tag{6.215}$$

Here $\Phi_{\alpha\beta}\begin{pmatrix} l & l' \\ \varkappa & \varkappa' \end{pmatrix}$ is a force acting in the direction of the α axis on a particle located at a point $x\begin{pmatrix} l \\ \varkappa \end{pmatrix}$ defined by Eq. (6.109), when a particle located at the point $x\begin{pmatrix} l' \\ \varkappa' \end{pmatrix}$ is displaced by a unit segment in the direction of the β axis. We seek the solution of Eq. (6.215) in the form

$$u_\alpha\begin{pmatrix} l \\ \varkappa \end{pmatrix} = \frac{1}{\sqrt{m_\varkappa}} u_\alpha(\varkappa)\exp[-i\omega t + 2\pi i k x(l)] \tag{6.216}$$

[$u_\alpha(\varkappa)$ is independent of l]. We substitute Eq. (6.216) in Eq. (6.215):

$$\omega^2 u_\alpha(\varkappa) = \sum_{\varkappa',\beta} D_{\alpha\beta}\begin{pmatrix} k \\ \varkappa\varkappa' \end{pmatrix} u_\beta(\varkappa') . \tag{6.217}$$

Here $D_{\alpha\beta}\begin{pmatrix} k \\ \varkappa\varkappa' \end{pmatrix}$ is given by

$$D_{\alpha\beta}\begin{pmatrix} k \\ \varkappa\varkappa' \end{pmatrix} = \frac{1}{\sqrt{m_\varkappa m'_\varkappa}} \sum_l \Phi_{\alpha\beta}\begin{pmatrix} l \\ \varkappa\varkappa' \end{pmatrix} \exp[-2\pi i k x(l)] , \tag{6.218}$$

where we have used the property that $\Phi_{\alpha\beta}\begin{pmatrix} l & l' \\ \varkappa & \lambda' \end{pmatrix}$ depends only on the difference $l - l'$. We have thus a set of $3r$ linear homogeneous equations relative to $u_\alpha(\varkappa)$ (r is the number of particles per unit cell). For this set to have nonvanishing solutions, the determinant of the solution coefficients must be equal to zero, i.e.,

$$\left| D_{\alpha\beta}\begin{pmatrix} k \\ \varkappa\varkappa' \end{pmatrix} - \omega^2 \delta_{\alpha\beta}\delta_{\varkappa\varkappa'} \right| = 0. \tag{6.219}$$

From this equation we determine $3r$ quantities $\omega_j^2(k)$ ($j = \overline{1, 3r}$) for each k. In addition, according to Eq. (6.128),

$$D^*_{\alpha\beta}\begin{pmatrix} k \\ \varkappa\varkappa' \end{pmatrix} = D_{\beta\alpha}\begin{pmatrix} k \\ \varkappa'\varkappa' \end{pmatrix} \tag{6.220}$$

(the asterisk here denotes the complex conjugate). The coefficients $D_{\alpha\beta}\begin{pmatrix} k \\ \varkappa\varkappa' \end{pmatrix}$ can be gathered into a $3r\times 3r$ matrix [using the pairs $(\alpha, \varkappa)$ and $(\beta, \varkappa')$ as the indices] that is Hermitian by virtue of Eq. (6.220). Therefore, $\omega_j^2(k)$ are real and $\omega_j(k)$ are either pure real or pure imaginary, but the latter possibility must be rejected, since it leads to instability of the lattice. To exclude this possibility from the very outset, the chosen force constants $\Phi_{\alpha\beta}\begin{pmatrix} l & l' \\ \varkappa & \varkappa' \end{pmatrix}$ must be such that the matrix $D_{\alpha\beta}\begin{pmatrix} k \\ \varkappa\varkappa' \end{pmatrix}$ considered above has a positive principal minor.

In the general case we thus have $3r$ modes:

$$\omega = \omega_j(k), \quad j = 1,2,\ldots,3r . \tag{6.221}$$

An expression in closed form can be obtained for $\omega_j(k)$ only for a few simple models of the crystal.

We choose any $\omega_j^2(k)$ from its $3r$ values. We have a vector $\left(\varkappa \middle| \begin{matrix} k \\ j \end{matrix}\right)$ whose components satisfy Eq. (6.217), and $\omega_j^k(k)$ can be represented in this case in the form[50]

$$\omega_i^2(k) l_\alpha\left(\varkappa \middle| \begin{matrix} k \\ j \end{matrix}\right) = \sum_{\varkappa',\beta} D_{\alpha\beta}\begin{pmatrix} k \\ \varkappa, \varkappa' \end{pmatrix} l_\beta\left(\varkappa' \middle| \begin{matrix} k \\ j \end{matrix}\right) . \tag{6.222}$$

Since this set of equations defines the vector $l\left(\varkappa\middle|\begin{matrix}k\\j\end{matrix}\right)$ accurate to an arbitrary constant factor, the latter can be chosen such as to orthonormalize the components of the vector $l\left(\varkappa\middle|\begin{matrix}k\\j\end{matrix}\right)$:

$$\sum_{\varkappa,\alpha} l_\alpha\left(x_\varkappa\middle|\begin{matrix}k\\j\end{matrix}\right) l_\alpha\left(\varkappa\middle|\begin{matrix}k\\j'\end{matrix}\right) = \delta_{jj'} ,$$

$$\sum_j l_\beta^*\left(\varkappa'\middle|\begin{matrix}k\\j\end{matrix}\right) l_\alpha\left(\varkappa\middle|\begin{matrix}k\\j\end{matrix}\right) = \delta_{\alpha\beta}\delta_{\varkappa\varkappa'} . \tag{6.223}$$

From Eq. (6.218) we see that

$$D_{\alpha\beta}^*\begin{pmatrix}k\\\varkappa\varkappa'\end{pmatrix} = D_{\alpha\beta}\begin{pmatrix}-k\\\varkappa\varkappa'\end{pmatrix} . \tag{6.224}$$

From this relation, as well as from the complex conjugate of Eq. (2.222), follow two possibilities:

$$l_\alpha\left(\varkappa\middle|\begin{matrix}k\\j\end{matrix}\right) = l_\alpha^*\left(\varkappa\middle|\begin{matrix}-k\\j\end{matrix}\right) \tag{6.225}$$

or

$$l_\alpha\left(\varkappa\middle|\begin{matrix}k\\j\end{matrix}\right) = -l_\alpha^*\left(\varkappa\middle|\begin{matrix}-k\\j\end{matrix}\right) . \tag{6.226}$$

We use hereafter the relation (6.225).[50,184] In addition, we can put

$$\omega_j^2(k) = \omega_j^*(-k) . \tag{6.227}$$

The eigenvectors $l\left(\varkappa\middle|\begin{matrix}k\\j\end{matrix}\right)$ for Bravais lattices are real. They can be complex only if the unit cell contains more than one particle.

We show now that three out of the $3r$ solutions $\omega_j(k)$ tend to zero as $k \to 0$. According to Eqs. (6.218) and (6.222) we have at $k = 0$ the equality

$$\sum_{l,\beta,\varkappa'} \frac{\Phi_{\alpha\beta}\begin{pmatrix}l\\\varkappa\varkappa'\end{pmatrix}}{m_\varkappa} \frac{l_\beta\left(\varkappa'\middle|\begin{matrix}0\\j\end{matrix}\right)}{\sqrt{m_{\varkappa'}}} = \omega_j^2(0)\frac{l_\alpha\left(\varkappa\middle|\begin{matrix}0\\j\end{matrix}\right)}{\sqrt{m_\varkappa}} . \tag{6.228}$$

Let $\frac{1}{\sqrt{m_{\varkappa'}}} l_\beta\left(\varkappa'\middle|\begin{matrix}0\\j\end{matrix}\right)$ be independent of $\varkappa'$ ($\beta = 1, 2, 3$). The left-hand side of this equality is then zero by virtue of Eq. (6.122), and since we are considering the nontrivial solutions $l\left(\varkappa\middle|\begin{matrix}k\\j\end{matrix}\right)$, we get

$$\omega_j^2(0) = 0 . \tag{6.229}$$

We therefore have for each α one solution for which Eq. (6.229) is satisfied, i.e., there are three such solutions. These oscillations satisfy the relation[50]

$$\frac{l\left(\varkappa\middle|\begin{matrix}0\\ j\end{matrix}\right)}{\sqrt{m_\varkappa}} = \frac{l\left(\varkappa'\middle|\begin{matrix}0\\ j\end{matrix}\right)}{\sqrt{m_{\varkappa'}}} = u\left(\begin{matrix}l\\ \varkappa\end{matrix}\middle|\begin{matrix}0\\ j\end{matrix}\right) = u\left(\begin{matrix}l\\ \varkappa'\end{matrix}\middle|\begin{matrix}0\\ j\end{matrix}\right), \tag{6.230}$$

where $u\begin{pmatrix}l & k\\ \lambda & j\end{pmatrix}$ is the displacement of the $\varkappa$th particle of the jth unit cell from the equilibrium position, if the particle oscillates at a frequency $\omega_j(k)$. These oscillations are called acoustic. In our case all the particles in each unit cell move in phase, and the oscillations have the same amplitude. The $(3r-3)$ oscillations whose frequencies do not vanish at $k=0$ are called optical.

As shown above, as $k\to 0$ the unit-cell particles move in phase at zero frequency. For arbitrary $\omega_j^2(k)$ they are determined by the force constants $\Phi_{\alpha\beta}\begin{pmatrix}l & l'\\ \varkappa & \varkappa'\end{pmatrix}$. Thus, considering the solutions obtained in the limit of small k, we can find the connection between $\Phi_{\alpha\beta}\begin{pmatrix}l & l'\\ \varkappa & \varkappa'\end{pmatrix}$ and the macroscopic elastic constants of a solid, which are defined by a tensor $E_{\alpha\mu\tau\nu}$. To investigate this aspect of the problem, we consider the case of a simple Bravais lattice. It is easily seen that the harmonic part of the potential energy can be written in the form

$$U_2 = -\tfrac{1}{4}\sum_{i,j}\{u(a_i)-u(a_j)\}D(a_i-a_j)\{u(a_i)-u(a_j)\}\,. \tag{6.231}$$

In elasticity theory, a solid is regarded as a continuous medium. Any deformation of the solid is then described by a continuous field of displacements $u(q)$. A connection between the theory of lattice vibrations and the continual elasticity theory can be found only for lattice deformations whose scale is large compared with the characterisic effective radius of the intermolecular forces. Since we are considering displacements $u(a_i)$ that change little from cell to cell, we can introduce a smooth function $u(q)$ that coincides with $u(a_i)$ if $q=a_i$. We can therefore write (with good accuracy) for large wavelengths

$$u(a_i) = u(a_j) + (a_i - a_j)\nabla u(q)|_{q=a_i}\,. \tag{6.232}$$

With this expression taken into account, Eq. (6.231) takes the form

$$U_2 = \frac{1}{2}\sum_{i,\mu,\nu,\sigma,\tau}\left[\frac{\partial}{\partial x_\sigma}u_\mu(a_i)\right]\left[\frac{\partial}{\partial x_\varepsilon}u_\nu(a_i)\right]E_{\sigma\mu\tau\nu}\,, \tag{6.233}$$

where

$$E_{\sigma\mu\tau\nu} = -\tfrac{1}{2}\sum_i a_{i\sigma}D_{\mu\nu}(a_i)a_{i\tau}\,. \tag{6.234}$$

Since $U(a_i)$ varies slowly, we can transform Eq. (6.233) into

$$U_2 = \frac{1}{2}\sum_{\sigma\tau\mu\nu}\int dq\left[\frac{\partial}{\partial x_\sigma}u_\mu(q)\right]\left[\frac{\partial}{\partial x_\varepsilon}u_\nu(q)\right]\overline{E}_{\sigma\mu\tau\nu}\,, \tag{6.235}$$

$$\overline{E}_{\sigma\mu\tau\nu} = \frac{1}{v}E_{\sigma\mu\tau\nu}\,, \tag{6.236}$$

where v is the unit-cell volume.

By virtue of the definition of $E_{\sigma\mu\tau\nu}$ we have

$$E_{\sigma\mu\tau\nu} = E_{\sigma\nu\tau\mu} = E_{\tau\mu\nu\sigma} . \qquad (6.237)$$

This decreases the number of independent parameters that determine $E_{\sigma\mu\tau\nu}$ to 36. In addition, we recognize that rigid rotation of the crystal does not alter its energy. Let

$$u(a_i) = \delta\omega \times a_i, \quad \delta\omega = \delta\omega\hat{u} . \qquad (6.238)$$

Substituting this expression in Eq. (6.233), we see that for the multiplier of $\delta\omega$ to vanish it is necessary that U_2 depend on the derivatives via their symmetrized combination

$$E_{\sigma\mu} = \frac{1}{2}\left(\frac{\partial u_\mu}{\partial x_\sigma} + \frac{\partial u_\sigma}{\partial x_\mu}\right). \qquad (6.239)$$

Thus, Eq. (6.235) should take the form

$$U_2 = \tfrac{1}{2} \int dq \sum_{\sigma\mu\tau\nu} [E_{\sigma\mu} C_{\sigma\mu\tau\nu} E_{\tau\nu}] , \qquad (6.240)$$

where

$$C_{\sigma\mu\tau\nu} = -\frac{1}{8v} \sum_i [a_{i\sigma} D_{\mu\nu} a_{i\tau} + a_{i\mu} D_{\sigma\nu} a_{i\tau} + a_{i\sigma} D_{\mu\tau} a_{i\nu} + a_{i\mu} D_{\sigma\tau} a_{i\nu}] . \qquad (6.241)$$

It follows directly from the definition of $C_{\sigma\mu\tau\nu}$ that

$$C_{\sigma\mu\tau\nu} = C_{\mu\sigma\tau\nu} = C_{\sigma\mu\nu\tau} , \qquad (6.241')$$

and from Eq. (6.193) that

$$C_{\sigma\mu\tau\nu} = C_{\tau\nu\sigma\mu} . \qquad (6.242)$$

The number of independent components of the tensor $C_{\sigma\mu\tau\nu}$ is finally equal to 21. Further decrease of the number of the independent components $C_{\sigma\mu\tau\nu}$ is possible if account is taken of the specific form of the crystal structure. Thus the maximum number of components necessary to determine $C_{\sigma\mu\tau\nu}$ for each of the following crystalline systems is triclinic, 21; monoclinic, 13; rhombic, 9; hexagonal, 5; cubic, 3 for all point groups; tetragonal (C_4, C_4h, S_4), 7; tetragonal (C_{4v}, D_4, D_{2d}), 6; rhombohedral (C_3, S_8), 7; and rhombohedral (C_{3v}, D_3, D_{3d}), 6.

Table XIV lists the elastic constants of a number of cubic crystals[169] (the elastic constants are given in units of 10^{12} dyn cm^{-2} at $T = 300$ K).

Here

$$C_{11} = C_{xxxx} = C_{yyyy} = C_{zzzz} ,$$
$$C_{12} = C_{xxyy} = C_{yyzz} = C_{zzxx} ,$$
$$C_{44} = C_{xyxy} = C_{yzyz} = C_{zxzx} .$$

Table XIV. Elastic constants of cubic crystals.

Material	C_{11}	C_{12}	C_{44}
Li (78 K)	0.148	0.125	0.108
Na	0.070	0.061	0.045
Cu	1.68	1.21	0.75
Ag	1.24	0.93	0.46
Au	1.86	1.57	0.42
Al	1.07	0.61	0.28
Pb	0.46	0.39	0.144
Ge	1.29	0.48	0.67
Si	1.66	0.64	0.80
V	2.29	1.19	0.43
Ta	2.67	1.61	0.82
Nb	2.47	1.35	0.285
Fe	2.34	1.36	1.18
Ni	2.45	1.40	1.25
LiCl	0.494	0.228	0.246
NaCl	0.487	0.124	0.126
KF	0.656	0.146	0.125
PbCl	0.361	0.062	0.047

36. Normal coordinates of crystal vibrations*

The Hamiltonian of the system in question is

$$H = T + U , \tag{6.243}$$

where T is given by Eq. (6.110) and U by Eq. (6.111). According to the assumptions above, Eq. (6.243) is a positive-definite quadratic form that can be diagonalized. We take $u_\alpha \begin{pmatrix} l \\ \varkappa \end{pmatrix}$ in the form

$$u_\alpha \begin{pmatrix} l \\ \varkappa \end{pmatrix} = \frac{1}{\sqrt{Nm_\varkappa}} \sum_{k,j} l_\alpha \left(\varkappa \middle| \begin{matrix} k \\ j \end{matrix} \right) Q \begin{pmatrix} k \\ j \end{pmatrix} \exp[2\pi i k x(l)] \, . \tag{6.244}$$

For $u_\alpha \begin{pmatrix} l \\ \varkappa \end{pmatrix}$ to be real, the following condition must be met for $Q \begin{pmatrix} k \\ j \end{pmatrix}$:

$$Q \begin{pmatrix} -k \\ j \end{pmatrix} = Q^* \begin{pmatrix} k \\ j \end{pmatrix} . \tag{6.245}$$

Therefore,

$$T = \frac{1}{2N} \sum_{\substack{l, \varkappa, \alpha \\ k, k', j, j'}} l_\alpha \left(\varkappa \middle| \begin{matrix} k \\ j \end{matrix} \right) l_\alpha \left(\varkappa \middle| \begin{matrix} k' \\ j' \end{matrix} \right) \frac{dQ \begin{pmatrix} k \\ j \end{pmatrix}}{dt} \cdot \frac{dQ \begin{pmatrix} k' \\ j' \end{pmatrix}}{dt} \exp[2\pi i (k + k') x(l)] \, . \tag{6.246}$$

*In the exposition of this material we follow Ref. 50.

But

$$\sum_{l} \exp[2\pi i k x(l)] = N\Delta(k), \tag{6.247}$$

where

$$\Delta k = 0$$

if $k = 0$ or is equal to the reciprocal-lattice vector, and $\Delta(k) = 1$ in all other cases. Consequently, in view of Eq. (6.247) the relation (6.246) takes the form

$$T = \frac{1}{2} \sum_{k,j} \frac{dQ^*\begin{pmatrix} k \\ j \end{pmatrix}}{dt} \frac{dQ\begin{pmatrix} k \\ j \end{pmatrix}}{dt}. \tag{6.248}$$

We transform now the expression for U_2:

$$U_2 = \frac{1}{2} \sum_{\substack{l,\varkappa,\alpha \\ l',\varkappa',\beta}} \Phi_{\alpha\beta}\begin{pmatrix} l & l' \\ \varkappa & \varkappa' \end{pmatrix} \frac{1}{N\sqrt{m_\alpha m_\beta}}$$

$$\times \sum_{\substack{k,k' \\ j,j'}} l_\alpha\left(\varkappa \middle| \begin{matrix} k \\ j \end{matrix}\right) l_\beta\left(\varkappa' \begin{matrix} k' \\ j' \end{matrix}\right) Q\begin{pmatrix} k \\ j \end{pmatrix} Q\begin{pmatrix} k' \\ j' \end{pmatrix}. \tag{6.249}$$

Recognizing that

$$\exp\{2\pi i k'[x(l) - x(l')]\} = 1,$$

and multiplying by this expression the right-hand side of Eq. (6.249), we obtain with allowance for Eqs. (6.247) and (6.218):

$$U_2 = \frac{1}{2} \sum_{\substack{\varkappa,\alpha,k,k' \\ \varkappa',\beta,\, j,j'}} D_{\alpha\beta}\begin{pmatrix} k' \\ \varkappa\varkappa' \end{pmatrix} l_\beta\left(\varkappa' \middle| \begin{matrix} k' \\ j' \end{matrix}\right) l_\alpha\left(\varkappa \middle| \begin{matrix} -k' \\ j \end{matrix}\right) Q\begin{pmatrix} -k' \\ j \end{pmatrix} Q\begin{pmatrix} k' \\ j' \end{pmatrix} \tag{6.250}$$

and get furthermore, in view of Eqs. (6.222), (6.223), and (6.224),

$$U = \frac{1}{2} \sum_{j,k} \omega_j^2(k) Q^*\begin{pmatrix} k \\ j \end{pmatrix} Q\begin{pmatrix} k \\ j \end{pmatrix}. \tag{6.251}$$

In the new coordinates, the Hamiltonian (6.243) thus takes the form (we omit the static part of the energy, which can be very easily taken into account)

$$H = \frac{1}{2} \sum_{k,j} \left\{ \frac{dQ^*\begin{pmatrix} k \\ j \end{pmatrix}}{dt} \frac{dQ\begin{pmatrix} k \\ j \end{pmatrix}}{dt} + \omega_j^2(k) Q^*\begin{pmatrix} k \\ j \end{pmatrix} Q\begin{pmatrix} k \\ j \end{pmatrix} \right\}. \tag{6.252}$$

It is also easy to rewrite in the new coordinates the system Lagrangian L on the basis of which we determine the momentum that is the canonical conjugate of the coordinate $Q^*\begin{pmatrix} k \\ j \end{pmatrix}$:

$$P\binom{k}{j} = \frac{\partial L}{\partial \left[dQ^*\binom{k}{j} \quad dt \right]} = \frac{dQ^*\binom{k}{j}}{dt}. \tag{6.253}$$

For the Hamilton equations of motion

$$\frac{dQ\binom{k}{j}}{dt} = \frac{\partial H}{\partial P^*\binom{k}{j}},$$

$$\frac{dP\binom{k}{j}}{dt} = \frac{\partial H}{\partial Q^*\binom{k}{j}} = \omega_j^2(k)\frac{dQ\binom{k}{j}}{dt} \tag{6.254}$$

we obtain

$$\frac{d^2Q\binom{k}{j}}{dt^2} + \omega_j^2(k)Q\binom{k}{j} = 0. \tag{6.255}$$

It is clear therefore that $Q\binom{k}{j}$ is a periodic function of the time with frequency $\omega_j(k)$. Such coordinates are called normal.

In the general case we have for $Q\binom{k}{j}$ the equation

$$Q\binom{k}{j} = \frac{1}{\sqrt{2}}\left[q_1\binom{k}{j} + iq_2\binom{k}{j}\right], \tag{6.256}$$

where $q_\lambda\binom{k}{j}$ $(\lambda = 1, 2)$ are real variables. By virtue of Eq. (2.45) we have

$$q_1\binom{k}{j} = q_1\binom{-k}{j},$$
$$q_2\binom{k}{j} = -q_2\binom{-k}{j}. \tag{6.257}$$

Thus, only half of all the coordinates $\left\{q_\lambda\binom{k}{j}\right\}$ are linearly independent. The independent coordinates are obtained by taking into account only those vectors k which lie in the first Brillouin zone on one side of an arbitrary plane. This operation will be designated below by $k > 0$ (this is only an arbitrary designation since it is of course meaningless to speak of a positive or negative vector). The Hamiltonian (6.243) reduces in the new coordinates $q_\lambda\binom{k}{j}$ to the form

$$
\begin{aligned}
H = \frac{1}{2} \sum_{\substack{K>0 \\ j}} \Bigg\{ & \frac{d\left(\begin{smallmatrix} -k \\ j \end{smallmatrix}\right)}{dt} \frac{dQ\left(\begin{smallmatrix} k \\ j \end{smallmatrix}\right)}{dt} + \omega_j^2(k) Q\left(\begin{smallmatrix} -k \\ j \end{smallmatrix}\right) Q\left(\begin{smallmatrix} k \\ j \end{smallmatrix}\right) \\
& + \frac{dQ\left(\begin{smallmatrix} k \\ j \end{smallmatrix}\right)}{dt} \frac{dQ\left(\begin{smallmatrix} -k \\ j \end{smallmatrix}\right)}{dt} + \omega_j^2(k) Q\left(\begin{smallmatrix} k \\ j \end{smallmatrix}\right) Q\left(\begin{smallmatrix} -k \\ j \end{smallmatrix}\right) \Bigg\}
\end{aligned}
$$

$$
= \frac{1}{2} \sum_{k>0} \sum_j \sum_{\lambda=1,2} \left\{ \left(\frac{dq_\lambda\left(\begin{smallmatrix} k \\ j \end{smallmatrix}\right)}{dt} \right)^2 + \omega_j^2(k) q_\lambda^2\left(\begin{smallmatrix} k \\ j \end{smallmatrix}\right) \right\}. \tag{6.258}
$$

Solving the system (6.244) we obtain

$$
Q\left(\begin{smallmatrix} k \\ j \end{smallmatrix}\right) = \frac{1}{\sqrt{N}} \sum_{l,\varkappa,\alpha} l_\alpha^*\left(\varkappa \Big| \begin{smallmatrix} k \\ j \end{smallmatrix}\right) \sqrt{m_\varkappa}\, u_\alpha\left(\begin{smallmatrix} l \\ \varkappa \end{smallmatrix}\right) \exp[-2\pi i k x(l)], \tag{6.259}
$$

whence

$$
P\left(\begin{smallmatrix} k \\ j \end{smallmatrix}\right) = \frac{1}{\sqrt{N}} \sum_{l,\varkappa,\alpha} l_\alpha^*\left(\varkappa \Big| \begin{smallmatrix} k \\ j \end{smallmatrix}\right) \frac{P_\alpha\left(\begin{smallmatrix} l \\ \varkappa \end{smallmatrix}\right)}{\sqrt{m_\varkappa}} \exp[-2\pi i k x(l)]. \tag{6.260}
$$

Here

$$
P_\alpha\left(\begin{smallmatrix} l \\ \varkappa \end{smallmatrix}\right) = m_\varkappa \frac{du_\alpha\left(\begin{smallmatrix} l \\ \varkappa \end{smallmatrix}\right)}{dt}.
$$

The entire foregoing analysis was in the context of classical mechanics. For a quantum-mechanical approach, the coordinate and momentum of each particle must be assigned a certain self-adjoint operator. Then

$$
\left[u_\alpha\left(\begin{smallmatrix} l \\ \varkappa \end{smallmatrix}\right), P_\beta\left(\begin{smallmatrix} l' \\ \varkappa' \end{smallmatrix}\right) \right] = i\hbar \delta_{ll'} \delta_{\varkappa\varkappa'} \delta_{\alpha\beta}. \tag{6.261}
$$

From Eqs. (6.259) and (6.260) we get

$$
\left[Q^*\left(\begin{smallmatrix} k \\ j \end{smallmatrix}\right), P\left(\begin{smallmatrix} k' \\ j' \end{smallmatrix}\right) \right] = i\hbar \Delta(k - k')\delta_{jj'},
$$

$$
\left[Q\left(\begin{smallmatrix} k \\ j \end{smallmatrix}\right), P^*\left(\begin{smallmatrix} k' \\ j' \end{smallmatrix}\right) \right] = i\hbar \Delta(k - k')\delta_{jj'}. \tag{6.262}
$$

The commutators for other Q and P combinations are zero.

Transforming to the operator form in Eq. (6.258), we obtain the Hamiltonian operator that leads to the Schrödinger equation

$$
\frac{1}{2} \sum_{k,j} \left\{ -\hbar^2 \frac{\partial^2}{\partial q^2\left(\begin{smallmatrix} k \\ j \end{smallmatrix}\right)} + \omega_j^2(k) q^2\left(\begin{smallmatrix} k \\ j \end{smallmatrix}\right) \right\} \psi = E\psi. \tag{6.263}
$$

The variables here are separable, and the wave function ψ takes the form

$$\psi = \prod_{k,j} \psi_{n_j(k)}\left[q\binom{k}{j}\right], \tag{6.264}$$

$n_j(k)$ is the quantum number of an oscillator with index $\binom{k}{j}$, and the energy can be represented in the form

$$E_{\{n\}} = \sum_{k,j} E_{n_j(k)}, \tag{6.265}$$

where

$$E_{n_j(k)} = [n_j(k) + \tfrac{1}{2}]\hbar\omega_j(k),$$
$$n_j(k) = 0, 1, 2, \ldots . \tag{6.266}$$

This problem is frequently solved by transforming to the second-quantization representation. This is done by changing from the operators $Q\binom{k}{j}$ and $P\binom{k}{j}$ to the second-quantization operators. This is most conveniently done by going to the second-quantization representation in Eq. (6.258). We note first of all that Eq. (6.258) can be also rewritten in the form

$$H = \frac{1}{2}\sum_{k,j}\left[Y^2\binom{k}{j} + \omega_j^2(k)X^2\binom{k}{j}\right], \tag{6.267}$$

where $Y\binom{k}{j}$ and $X\binom{k}{j}$ are readily related to the $q_\lambda\binom{k}{j}$ and their derivatives. We replace now the dynamic variables in Eq. (6.267) by operators that satisfy the following commutation relations:

$$\left[X\binom{k}{j}, Y\binom{k'}{j'}\right] = i\hbar\delta_{kk'}\delta_{jj'},$$
$$\left[X\binom{k}{j}, X\binom{k'}{j'}\right] = \left[Y\binom{k}{j}, Y\binom{k'}{j'}\right] = 0. \tag{6.268}$$

We choose $X\binom{k}{j}$ and $Y\binom{k}{j}$ in the form

$$X\binom{k}{j} = \sqrt{\frac{\hbar}{2\omega_j(k)}}\,(a_{kj}^+ + a_{kj}),$$

$$Y\binom{k}{j} = i\sqrt{\frac{\hbar\omega_j(k)}{2}}\,(a_{kj}^+ - a_{kj}), \tag{6.269}$$

where a_{kj} and a_{kj}^+ are Hermitian-adjoint operators. In the general case it would be necessary to choose $X\binom{k}{j}$ and $Y\binom{k}{j}$ to be linear functions of a_{kj}^+ and a_{kj} with arbitrary coefficients chosen to satisfy the condition that relation (6.268) be valid and that the Hamiltonian be reducible to standard form in the second-quantization representation. According to Eqs. (6.269) and (6.268) the operators a_{kj}^+ and a_{kj} satisfy the following commutation relations:

$$[a_{kj}, a_{k'j'}^+] = \delta_{kk'}\delta_{jj'} ,$$
$$[a_{kj}, a_{k'j'}] = [a_{kj}^+, a_{k'j'}^+] = 0. \tag{6.270}$$

The Hamiltonian of the system can be represented in the form

$$H = \tfrac{1}{2} \sum_{k,j} \hbar\omega_j(k)[a_{kj}^+ a_{kj} + a_{kj}a_{kj}^+] , \tag{6.271}$$

or, taking Eq. (6.270) into account,

$$H = \tfrac{1}{2} \sum_{k,j} \hbar\omega_j(k)[a_{kj}^+ a_{kj} + \tfrac{1}{2}] . \tag{6.272}$$

The equation of motion for the operator a_{kj} can be easily determined:

$$\frac{da_{kj}}{dt} = \frac{i}{\hbar}[H, a_{kj}] = \frac{i}{\hbar}\hbar\omega_j(k)[a_{kj}^+ a_{kj}, a_{kj}]$$
$$= i\omega_j(k)[a_{kj}^+, a_{kj}]a_{kj} = -i\omega_j(k)a_{kj} . \tag{6.273}$$

We have similarly for a_{kj}^+,

$$\frac{da_{kj}^+}{dt} = \frac{i}{\hbar}[H, a_{kj}^+] = i\omega_j(k)a_{kj}^+ . \tag{6.274}$$

It is shown in quantum mechanics that the eigenvalues of the operator a_{kj}^+ (a_{kj}) are non-negative natural numbers N_{kj}. Thus, as seen from Eq. (6.272), we obtain the same system energy as given by the solution of the Schrödinger equation.

We determine now the thermodynamic properties of the system in question. The partition function is in this case

$$Z = \sum_{n_j(k)} \exp[-\beta E_{\{n_j(k)\}}] = \prod_{k,j} \{\exp[-\beta\hbar\omega_j(k)/2] + \exp[-3\beta\hbar\omega_j(k)/2]$$
$$+ \exp[-5\beta\hbar\omega_j(k)/2] + \cdots\} = \prod_{k,j} \frac{\exp[-\beta\hbar\omega_j(k)/2]}{\{1-\exp[-\beta\hbar\omega_j(k)]\}} , \tag{6.275}$$

and the free energy

$$F = -\theta \ln Z = \theta \sum_{k,j} \ln\left\{2 \sinh\left[\frac{\beta\hbar\omega_j(k)}{2}\right]\right\} \tag{6.276}$$

or

$$F = N\varepsilon_0 + \theta \sum_{k,j} \ln\{1 - \exp[-\beta\hbar\omega_j(k)]\} , \tag{6.277}$$

where

$$\varepsilon_0 = \sum_{k,j} \frac{\hbar\omega_j(k)}{2N} . \tag{6.278}$$

Here $N\varepsilon_0$ is the energy of the zero-point oscillations, depends on the specific volume (density) of the system, and is independent of temperature. As already noted it is easy to take into account the static energy of the crystal lattice. To this end we define ε_0 not by Eq. (6.278), but by the expression

$$\varepsilon_0 = \frac{1}{N} U_0 + \sum_{k,j} \frac{\hbar\omega_j(k)}{2N}. \tag{6.279}$$

In this case, too, ε_0 depends only on the system density and equals the energy per unit cell at $T = 0$ K.

From Eq. (6.276) it is easy to determine the basic thermodynamic functions, viz.,

the internal energy

$$E = F - \theta\left(\frac{\partial F}{\partial \theta}\right)_V = -\frac{\partial F}{\partial \beta} = \sum_{k,j}\left\{\frac{\hbar\omega_j(k)}{2} + \frac{\hbar\omega_j(k)}{\exp[\beta\hbar\omega_j(k)] - 1}\right\}, \tag{6.280}$$

the heat capacity at constant volume

$$C_V = k\left(\frac{\partial E}{\partial \theta}\right)_V = k\sum_{k,j}\left(\frac{\beta\hbar\omega_j(k)}{2}\right)^2 \Big/ \sinh^2\left(\frac{\beta\hbar\omega_j(k)}{2}\right), \tag{6.281}$$

the entropy

$$S = -k\left(\frac{\partial F}{\partial \theta}\right)_V = k\sum_{k,j}\left\{\frac{\hbar\omega_j(k)}{2}\coth\left(\frac{\beta\hbar\omega_j(k)}{2}\right) - \ln\left[2\sinh\left(\frac{\beta\hbar\omega_j(k)}{2}\right)\right]\right\}, \tag{6.282}$$

etc.

Let us investigate the behavior of the thermodynamic functions of a crystal lattice at low temperatures. In this case the largest contribution to the partition function is made by the low-frequency terms, for which $\hbar\omega_j(k) \sim \theta$. This takes place for the three acoustic modes $\omega_j(k)$ as $ka \to 0$. Here we have acoustic oscillations, since $\omega = Uk$, where U is the speed of sound. In this case the sound wavelength is $\lambda \sim U/\omega$, with $\lambda \gg a$, meaning $\omega \ll U/a$. At temperatures

$$\theta \ll \hbar U/a \tag{6.283}$$

the main contribution to the partition function is made by the terms corresponding to the acoustic oscillations. Note that in the limit of a large crystal the summation over k can be replaced by integration, for in this case the values of k are closely spaced. The interval k, $k + dk$ contains $V[4\pi k^2 dk/(2\pi)^3]$ natural oscillations with a given polarization. Sound waves propagating in an isotropic solid can be longitudinal (velocity U_l) or transverse (velocity U_t) and can have two independent polarization directions. We express k in terms of the frequency ω and determine the total number of possible sound oscillations [in the interval $(\omega, \omega + d\omega)$]

$$V\frac{\omega^2 d\omega}{(2\pi)^2}\left(\frac{1}{U_l^3} + \frac{1}{U_t^3}\right). \tag{6.284}$$

The average sound velocity is given by

$$\bar{U}=\left[\frac{1}{3}\left(\frac{1}{U_t^3}+\frac{1}{U_l^3}\right)\right]^{-1/3}. \tag{6.285}$$

Expression (6.284) can be represented in this case in the form

$$V\frac{3\omega^2 d\omega}{2\pi^2\bar{U}^3}. \tag{6.286}$$

For a crystal we can also introduce U by the known procedure. Thus, relation (2.286) is valid for a crystal. Three acoustic oscillation modes exist in a medium, the velocity of each being dependent on the direction.

In accordance with the foregoing, we change in Eq. (6.277) from summation to integration, taking into account the rapid convergence of the integrals at small T, and choosing therefore the integration limits to be 0 and ∞:

$$F=N\varepsilon_0+\theta\frac{3V}{2\pi^2\bar{U}^3}\int_0^\infty \ln(1-\exp[-\hbar\omega/\theta])\omega^2\,d\omega. \tag{6.287}$$

We introduce in the integrand the dimensionless integration variable

$$x=\hbar\omega/\theta.$$

Then

$$F=N\varepsilon_0+\frac{3V\theta^4}{2\pi^2(\hbar\bar{U})^3}\int_0^\infty \ln(1-e^{-x})x^2\,dx. \tag{6.288}$$

Integration by parts yields

$$\begin{aligned}F&=N\varepsilon_0+\frac{3V\theta^4}{2\pi^2(\hbar\bar{U})^3}\left\{\ln(1-e^{-x})\frac{x^3}{3}\bigg|_0^\infty-\frac{1}{3}\int_0^\infty\frac{x^3dx}{e^x-1}\right\}\\&=N\varepsilon_0-\frac{V\theta^4}{2\pi^2(\hbar\bar{U})^3}I,\end{aligned} \tag{6.289}$$

where

$$I=\int_0^\infty\frac{x^3\,dx}{e^x-1}=\int_0^\infty\frac{x^3e^{-x}}{1-e^{-x}}dx=\int_0^\infty x^3\sum_{n=1}^\infty e^{-nx}\,dx.$$

We now calculate the integral

$$\int_0^\infty x^3e^{-nx}\,dx=\frac{3}{n}\int_0^\infty x^2e^{-nx}\,dx=\cdots=\frac{6}{n^4}.$$

We substitute this result in Eq. (6.289):

$$I=6\sum_{n=1}^\infty\frac{1}{n^4}=6\frac{\pi^4}{90}=\frac{\pi^4}{15}$$

(Ref. 185). We thus obtain for the free energy

$$F=N\varepsilon_0-\frac{V\theta^4\pi^2}{30(\hbar\bar{U})^3}. \tag{6.290}$$

With the aid of this expression, we get the entropy

$$S = -k\left(\frac{\partial F}{\partial \theta}\right) = Vk\frac{2\pi^3\theta^3}{15(\hbar\bar{U})^3}, \tag{6.291}$$

the energy

$$E = N\varepsilon_0 + V\frac{\pi^2\theta^4}{(\hbar\bar{U})^3}, \tag{6.292}$$

and the heat capacity at constant volume:

$$C_V = \frac{2\pi^2 k}{5(\hbar\bar{U})^3}\theta^3 V. \tag{6.293}$$

Recognizing that

$$C_p - C_V = -\theta\frac{(\partial p/\partial\theta)_V^2}{(\partial p/\partial V)_\theta} = -\theta\frac{(\partial S/\partial V)_\theta^2}{(\partial p/\partial V)_\theta}$$

and taking Eq. (6.291) into account, we get

$$C_p - C_V \sim \theta^7. \tag{6.294}$$

Thus, as $\theta \to 0$, we can write, with high accuracy,

$$C_p = C_V = C = \frac{2\pi^2 k}{5(\hbar\bar{U})^3}\theta^3 V. \tag{6.295}$$

As θ tends to zero, the heat capacity of the crystal decreases in accord with a cubic law, in good agreement with the experimental data. The range of validity of Eq. (6.295) depends on the type of lattice, amounting to several times ten degrees for crystals with simple cubic lattice and smaller as a rule for more complicated structures. It is easy to show that the heat capacity as $\theta \to 0$ is proportional to θ^2 for a two-dimensional system and to θ for a one-dimensional one. We have considered the crystal properties governed by its lattice vibrations. In a number of cases it is necessary to take into account the contributions made to the thermodynamic properties and to the motion of the free electrons. The electronic contribution, proportional to θ, to the specific heat of the crystal predominates therefore over the contribution of the lattice vibrations at very low temperatures, on the order of a few degrees Kelvin.

We now consider the thermodynamics of the crystal in the high-temperature limit. It is convenient to use for this purpose the relation (6.276), inasmuch as at

$$\beta\hbar\omega_j(k) \ll 1$$

we have the series expansion

$$\ln\left\{2\sinh\left[\frac{\beta\hbar\omega_j(k)}{2}\right]\right\} = \ln\left\{2\left[\frac{\beta\hbar\omega_j(k)}{2}\right] + \frac{1}{3!}\left[\frac{\beta\hbar\omega_j(k)}{2}\right]^3 + \cdots\right\}$$
$$= \ln[\beta\hbar\omega_j(k)] + [\beta\hbar\omega_j(k)]^2/24 + \cdots.$$

Substituting this expression in Eq. (6.276), we get

$$F = \sum_{k,j} \theta \ln[\beta\hbar\omega_j(k)] + \frac{\theta}{24}\sum_{k,j}[\beta\hbar\omega_j(k)]^2 + \cdots . \quad (6.296)$$

We determine the "geometric mean" frequency $\overline{\omega}$ with the aid of the relation

$$\ln \overline{\omega} = \frac{1}{3Nr}\sum_{k,j} \ln \omega_j(k), \quad (6.297)$$

and also introduce the notation

$$\gamma = \frac{1}{72Nr}\sum_{k,j}[\hbar\omega_j(k)]^2 . \quad (6.298)$$

It is necessary to add to expression (6.296) for the free energy the static energy of the lattice, and obtain by virtue of Eqs. (6.297) and (6.298),

$$F = N\varepsilon_0 + 3Nr\theta \ln\frac{\hbar\overline{\omega}}{\theta} + 3Nr\gamma/\theta + \cdots , \quad (6.299)$$

where

$$\varepsilon_0 = U_0/N . \quad (6.300)$$

This yields the system entropy

$$S = -k\left(\frac{\partial F}{\partial \theta}\right)_V = -3Nrk \ln\frac{\hbar\overline{\omega}}{\theta} + 3Nrk + 3Nr\gamma k/\theta^2 + \cdots , \quad (6.301)$$

the energy

$$\begin{aligned} E = F + \theta S/k = N\varepsilon_0 &+ 3Nr\theta ln\frac{\hbar\overline{\omega}}{\theta} + 3Nr\gamma/\theta + 3Nr\theta \\ &- 3Nrk \ln\frac{\hbar\overline{\omega}}{\theta} - 3Nr\gamma/\theta + \cdots \\ &= N\varepsilon_0 + 3Nr\theta + 6Nr\gamma/\theta + \cdots , \end{aligned} \quad (6.302)$$

and the heat capacity

$$C_V = 3Nrk - 6Nr\gamma k/\theta^2 - \cdots . \quad (6.303)$$

Solids usually have $C_V \simeq C_p$, therefore Eq. (6.303) determines the heat capacity of a crystal at both constant volume and constant pressure:

$$C = C_p = C_V = 3Nrk - 6Nrk\gamma/\theta^2 - \cdots . \quad (6.304)$$

At high temperatures we can neglect the terms inversely proportional to θ. We therefore obtain from Eq. (6.304),

$$C = 3Nrk , \quad (6.305)$$

i.e., the Dulong and Petit law holds in this case, as expected.

A sufficiently complete calculation of the thermodynamic properties of the system is thus possible in the limits of low and high temperatures. For intermediate temperatures, however, the calculations become quite complicated. It is therefore natural to replace the general phonon spectrum by some simplified one that leads to correct asymptotic values at low and high temperatures. The solution of this interpolation problem is of course not unique, but it permits an approximately correct description of the heat capacity in the intermediate region.

The temperature distribution at low temperatures is determined by Eq. (2.286). At high temperatures, all $3Nr$ oscillations are excited. We extrapolate relation (2.286), which is valid at low temperatures, to the high-temperature region. We assume that the spectrum begins with $\overline{\omega} = 0$ and ends with $\omega = \omega_m$ determined from the equation for the total number of oscillations:

$$\frac{3V}{2\pi^2 \overline{U}^3} \int_0^{\omega_m} \omega^2 \, d\omega = 3Nr . \tag{6.306}$$

Hence

$$\omega_m = \overline{U} \left(\frac{6\pi^2 Nr}{V} \right)^{1/3} . \tag{6.307}$$

We introduce the Debye temperature connected with ω_m by the relation

$$\theta_D = \hbar \omega_m . \tag{6.308}$$

In this case the crystal free energy takes the form

$$F = N\varepsilon_0 + 9Nr\theta \left(\frac{\theta}{\theta_D} \right)^3 \int_0^{\theta_D/\theta} x^2 \ln(1 - e^{-x}) dx. \tag{6.309}$$

We integrate by parts:

$$\int_0^{\theta_D/\theta} x^2 \ln(1 - e^{-x}) dx = \frac{x^3}{3} \ln(1 - e^{-x}) \Big|_0^{\theta_D/\theta} - \frac{1}{3} \int_0^{\theta_D/\theta} \frac{x^3 \, dx}{e^x - 1}$$

$$= \frac{1}{3} \left(\frac{\theta_D}{\theta} \right)^3 \ln(1 - e^{-\theta_D/\theta}) - \frac{1}{9} \left(\frac{\theta_D}{\theta} \right) D \left(\frac{\theta_D}{\theta} \right), \tag{6.310}$$

where

$$D\left(\frac{\theta_D}{\theta} \right) = 3 \left(\frac{\theta}{\theta_D} \right)^3 \int_0^{\theta_D/\theta} \frac{x^3 \, dx}{e^x - 1} \tag{6.311}$$

is the Debye function. The free energy of the crystal is thus

$$F = N\varepsilon_0 + Nr\theta \left[3 \ln(1 - e^{-\theta_D/\theta}) - D\left(\frac{\theta_D}{\theta} \right) \right] . \tag{6.312}$$

This yields directly the entropy

$$S = -k \left(\frac{\partial F}{\partial \theta} \right)_V = -Nrk [3 \ln(1 - e^{-\theta_D/\theta}) - D(\theta_D/\theta)]$$

$$+ \frac{Nrk\theta_D}{\theta} \left[\frac{3e^{-\theta_D/\theta}}{1 - e^{-\theta_D/\theta}} + \frac{\theta_D^2}{\theta} \frac{dD(\theta_D/\theta)}{d\theta} \right] . \tag{6.313}$$

Table XV. Temperature dependences of specific heat according to Debye theory.

z	$\bar{C}_v$	z	$\bar{C}_v$	z	$\bar{C}_v$
0.11	0	0.35	0.687	0.70	0.905
0.05	0.009 74	0.40	0.746	0.75	0.917
0.10	0.075 8	0.45	0.791	0.80	0.926
0.15	0.213	0.50	0.825	0.85	0.934
0.20	0.369	0.55	0.852	0.90	0.941
0.25	0.503	0.60	0.874	0.05	0.947
0.30	0.608	0.65	0.891	1.00	0.952

Taking the derivative

$$\frac{dD(\theta_D/\theta)}{d\theta} = -\frac{3}{e^{\theta_D/\theta}-1}\frac{\theta_D}{\theta^2} + \frac{3}{\theta}D\left(\frac{\theta_D}{\theta}\right) \tag{6.314}$$

and substituting it in Eq. (6.313), we have

$$\begin{aligned} S &= -Nrk\left[3\ln(1-e^{-\theta_D/\theta}) - D\left(\frac{\theta_D}{\theta}\right)\right] + 3NrkD\left(\frac{\theta_D}{\theta}\right) \\ &= -Nrk\left[3\ln(1-e^{-\theta_D/\theta}) - 4D(\theta_D/\theta)\right]. \end{aligned} \tag{6.315}$$

We determine now the system energy:

$$\begin{aligned} E = F + TS &= N\varepsilon_0 + Nr\theta\left[3\ln(1-e^{-\theta_D/\theta}) - D(\theta_D/\theta)\right] \\ &- Nr\theta\left[3\ln(1-e^{-\theta_D/\theta}) - 4D(\theta_D/\theta)\right] = N\varepsilon_0 + 3Nr\theta D(\theta_D/\theta), \end{aligned}$$

and also the heat capacity

$$C_V = \left(\frac{\partial E}{\partial T}\right)_V = 3Nk\left\{D\left(\frac{\theta_D}{\theta}\right) + \theta\frac{dD(\theta_D/\theta)}{d\theta}\right\}. \tag{6.316}$$

Introducing the variable

$$Z = \theta_D/\theta, \tag{6.317}$$

we can rewrite Eq. (6.326) in the form

$$C_V = 3Nk\left\{D(\theta_D/\theta) - \frac{\theta_D}{\theta}D'\left(\frac{\theta_D}{\theta}\right)\right\}, \tag{6.318}$$

where

$$D'\left(\frac{\theta_D}{\theta}\right) = \frac{dD(z)}{dz}. \tag{6.319}$$

Table XV lists the temperature dependences of the specific heat according to the Debye theory

$$\overline{C}_V = \frac{C_V}{3Nk} = D(z) - \frac{\theta_D}{\theta}D'(z). \tag{6.320}$$

Equation (6.320) yields correct asymptotic values of the specific heat of the crystal at low and high temperature. Thus, as $\theta \to 0$ and $z \to \infty$ we have, taking Eq. (6.289) into account,

$$D(z) \to \frac{\pi^4}{5z^3}.$$

Therefore,

$$C_V = \frac{12Nr_k\pi^4}{5}\left(\frac{\theta}{\theta_D}\right)^3. \tag{6.321}$$

At high temperatures $\theta \to 0$, therefore,

$$D(z) = 1 - \tfrac{3}{8}z + \tfrac{1}{20}z^2 - \cdots.$$

Substituting this expression in Eq. (6.318) we get

$$C_V = 3Nrk\left[1 - \frac{1}{20}\left(\frac{\theta_D}{\theta}\right)^2 - \cdots\right]. \tag{6.322}$$

We see that the temperature dependence of the crystal heat capacity is determined only by a single parameter θ_D. As a result, in this approximation the dependence of $C_V/3Nrk$ is one and the same for all substances. Such an approximation is justified for bodies with simple crystal lattices; it does not hold for more complicated structures in view of the complexity of the oscillation spectrum.

Expression (6.318) is sometimes regarded not as an interpolation formula, but as a general relation. Therefore, to reconcile in the entire range of the calculated and experimental values of the heat capacity, θ_D is regarded as a function of temperature. The experimentally determined heat capacity is sometimes cited as a function $\theta_D(\theta)$ that can be recalculated into heat capacity by using Table XV.

37. Density of the number of vibrations

It has already been mentioned that since the number of atoms in a crystal is large, the entire range of natural frequencies is densely populated. It is therefore more effective to consider the frequency distribution function rather than individual frequencies.

We introduce the quantity $G(\omega^2)d\omega^2$ and define it as the number of squares of the frequencies in the interval $(\omega^2, \omega^2 + d\omega^2)$ as $d\omega \to 0$. We also introduce the quantity $g(\omega)d\omega$, equal to the number of frequencies in the interval $\omega, \omega + d\omega$ as $d\omega \to 0$. Obviously, the relation between $g(\omega)$ and $G(\omega^2)$ is

$$g(\omega) = 2\omega G(\omega^2). \tag{6.323}$$

In accord with the foregoing, we introduce the distribution function for the jth mode, which we normalize to $1/3r$:

$$\int_0^{\omega_L(j)} g_j(\omega)d\omega\,\frac{1}{3r}\,. \tag{6.324}$$

In this case the general distribution function is

$$g(\omega) = \sum_{j=1}^{3r} g_j(\omega) \tag{6.325}$$

and

$$\int_0^{\omega_L} g(\omega)d\omega = \sum_{j=1}^{3r}\int_0^{\omega_L} g_j(\omega)d\omega = 1\,, \tag{6.326}$$

where $\omega_L(j)$ is the highest frequency of the jth mode

$$\omega_L = \max_j \omega_L(j)\,. \tag{6.327}$$

As shown in the preceding section, in the harmonic approximation the thermodynamic functions of a crystal are additive functions of the normal vibration frequencies. It is therefore possible to replace the integration in the expressions for them by averaging over the frequency spectrum. In the upshot we have expressions for the free energy

$$F = N\varepsilon_0 + 3Nr\theta\int_0^{\omega_L}\ln\left\{2\sinh\frac{\hbar\omega}{2\theta}\right\}g(\omega)d\omega\,, \tag{6.328}$$

for the entropy

$$S = 3Nrk\int_0^{\omega_L}\left[\frac{\hbar\omega}{2\theta}\coth\frac{\hbar\omega}{2\theta} - \ln\left\{2\sinh\frac{\hbar\omega}{2\theta}\right\}\right]g(\omega)d\omega\,, \tag{6.329}$$

for the energy

$$E = \frac{3rN}{2}\int_0^{\omega_L}\hbar\omega\coth\frac{\hbar\omega}{2\theta}\,g(\omega)d\omega\,, \tag{6.330}$$

and for the heat capacity

$$C_V = 3Nrk\int_0^{\omega_L}\left(\frac{\hbar\omega}{2\theta}\right)^2\operatorname{csch}^2\frac{\hbar\omega}{2\theta}\,g(\omega)d\omega\,. \tag{6.331}$$

To determine the thermodynamic functions of a crystal we must therefore know the function $g(\omega)$. It has been the subject of many calculations for various models. Starting from the definition of $g(\omega)$—the density of the number of vibrations—we have

$$g(\omega) = \sum_j\int\frac{dk}{(2\pi)^3}\,\delta[\omega - \omega_j(k)]\,. \tag{6.332}$$

This expression can be written in a somewhat different form. By definition, $g(\omega)d\omega$ is the number of normal vibrations in the region $(\omega, \omega + d\omega)$, and in the limit of a large crystal it is the volume in k space contained between the two equal-frequency surfaces ω and $\omega + d\omega$, divided by $(2\pi)^3$. To determine the explicit form of $d\omega$ we note that since ω is a constant-energy surface,

$\omega = \text{const}$, the gradient $\nabla_k \omega(k)$ of $\omega(k)$ with respect to k is a vector normal to this surface and having an absolute value equal to the rate of change of $\omega(k)$ in the normal direction. Hence

$$\omega + d\omega = \omega + dk\ \nabla_k \omega(k)$$

or

$$d\omega = qk\nabla_k \omega(k)\,.$$

The distance between two infinitely close surfaces, measured along the normal to them at a given point, is therefore

$$\frac{d\omega}{|\nabla_k \omega(k)|}\,. \tag{6.333}$$

We denote an element of an equal-frequency surface by df_{kj} [we recognize that there are $3r$ modes of the $\omega_j(k)$ vibrations]. Integrating over the surface within the limits of one Brillouin zone (the first one is usually chosen), we have

$$g(\omega) = \sum_j \int \frac{df_{kj}}{(2\pi)^3} \cdot \frac{1}{|\nabla_k \omega_j(k)|}\,. \tag{6.334}$$

Since the function $\omega_j(k)$ is periodic, $g(\omega)$ contains singularities, because the denominator of the right-hand side of Eq. (6.334) in the integrand vanishes at certain points. The singularities of the function $g(\omega)$ are called van Hove singularities.

We express now the principal approximations used by us earlier in the formalism of the vibration-number density function $g(\omega)$. Thus, in the Debye approximation the dispersion law was assumed linear for the three acoustic modes (we are considering now a crystal with a single-atom Bravais lattice), and the wave vectors of the normal vibrations were assumed to be located in a sphere of radius k_D. From Eq. (6.332) we have then

$$g_D(\omega) = 3\int_{k<k_D} \frac{dk}{(2\pi)^3}\,\delta(\omega - \overline{U}k) = \frac{3}{2\pi^2}\int_0^{k_D} dk\cdot k^2 \delta(\omega - \overline{U}k)$$

$$= \begin{cases} \dfrac{3}{2\pi^2}\dfrac{\omega^2}{\bar{u}^2}\,, & \omega < \omega_D = k_D \overline{U} \\ 0\,, & \omega > \omega_D \end{cases}\,. \tag{6.335}$$

The choice of the quantity k_D ensures equality of the areas below the $g_D(\omega)$ curve and below the exactly calculated $g(\omega)$ curve [the area considered is between the $g(\omega)$ curve and the abscissa axis]. On the other hand, the choice of U ensures equality of $g_D(\omega)$ and $g(\omega)$ as $\omega \to 0$. The last property ensures correct description of heat-capacity behavior at low temperatures in the Debye approximation, and the first property ensures the correct value at high temperatures (the Dulong and Petit law).

In the Einstein approximation we have

$$g_E(\omega) = \int \frac{dk}{(2\pi)^3}\,\delta(\omega - \omega_E) = n\delta(\omega - \omega_E)\,. \tag{6.335'}$$

The form of $g(\omega)$ for crystals is quite complicated. Even if the unit cell contains one atom, the function consists of three acoustic modes. In general, however, the expression for $g(\omega)$ has three acoustic and $3r-3$ optical modes.

38. Crystal defects

A defect in a crystal is usually defined as any region in which the ideal crystal structure is disturbed. Depending on the size of the disturbed region, crystal defects are classified as two-dimensional, one-dimensional, and pointlike. Disturbances in the form of absence of an atom from a lattice site, substitution of an impurity atom for a regular atom, or intrusion of an excess atom in an interstice are point defects. These defects are the causes of the observed electric conductivity of ionic crystals, of the change of their optical properties (including their color), and others.

Dislocations are linear defects comprising disturbance of the regular alternation of the planes. Thus, in an edge dislocation an extra partial plane of atoms is inserted between two neighboring atomic planes.

In the case of a screw dislocation, the crystal is, so to speak, cut along a half-plane and the parts of the lattice on the two sides of the cut are displaced a distance equal to the period. The dislocations strongly influence the strength of the crystal.

We turn to the question of the thermodynamic characteristics of defects. We consider to this end the simplest case—a Schottky defect in a single-atom Bravais lattice. Let the number of vacancies in the crystal be n, and assume that $n \sim N$. In this case n can be estimated by minimizing the corresponding thermodynamic potential, say the Gibbs energy if the crystal is under constant pressure p:

$$G = U - TS + pV .$$

We regard the crystal as a system with $(N+n)$ sites, of which n are free. In first approximation, the crystal volume is $(N+n)v_0$, where v_0 is the volume per atom in the ideal crystal. For each set n we can calculate $F_0(n)$, which we assume in the case when $n/N \ll 1$ to depend only on the number of the vacancies. The number of distributions of a total of n vacancies over $(N+n)$ sites is

$$C_{N+n}^{n} = \frac{(N+n)!}{N!\,n!} .$$

It is therefore necessary to add to the entropy S of a fixed distribution of vacancies a term S^c due to the variability of the vacancy distribution in the crystal:

$$S^c = k \ln \frac{(N+n)!}{N!\,n!} . \tag{6.336}$$

Using Stirling's formula

$$\ln I \simeq I[\ln(I/e)] ,$$

we transform Eq. (6.336) into

$$S^c = k\left[(N+n)\ln\left(\frac{N+n}{e}\right) - N\ln\frac{N}{e} - n\ \ln\frac{n}{e}\right] + \cdots$$
$$= k\left[n - n\ln\frac{n}{N}\right] + \cdots. \tag{6.337}$$

The general expression for the Gibbs energy takes in our approximation the form

$$G(n) = F_0(n) - kT\ln\left[n - n\ln\frac{n}{N}\right] + p(N+n)v_0\,. \tag{6.338}$$

The derivative of this expression with respect to n is

$$\frac{\partial G}{\partial n} = \frac{\partial F_0}{\partial n} + pv_0 - \theta\ln\frac{N}{n}\,. \tag{6.339}$$

For $n/N \ll 1$ we have

$$\frac{\partial F_0}{\partial n} \approx \left.\frac{\partial F_0}{\partial n}\right|_{n=0} = \bar{E}\,,$$

where $\bar{E}$ is independent of n. Equating (6.339) to zero, we determine the value of n at which G is a minimum:

$$n = Ne^{-(\bar{E}+pv_0)/\theta}\,. \tag{6.340}$$

In fact,

$$\frac{\partial^2 G}{\partial h^2} = \theta > 0\,,$$

i.e., the sufficient condition for the existence of a minimum is met.

Let us estimate $\bar{E}$. To this end we write down the expression for the partition function

$$e^{-\beta F_0} = \sum_i e^{-\beta E_i}\,,$$

where the E_i are determined for the Hamiltonian in the harmonic approximation. Then

$$F_0 = U_0 + F_2\,,$$

where U_0 is the static energy of the lattice with the vacancies, and F_2 is the free energy of the lattice and is governed by the quadratic part of the Hamiltonian. Since F_2 is as a rule small compared with U_0, we can put

$$\bar{E} \approx E_0 = \left.\frac{\partial U_0}{\partial n}\right|_{n-0}\,, \tag{6.341}$$

where E_0 is the energy needed to remove one atom from the lattice. Under ordinary pressures the term pv_0 can be neglected compared with E_0 and we ultimately have

$$n = N e^{-\beta E_0}. \tag{6.342}$$

Thus, n/N is small but not zero (E_0 is of the order of several electron volts for most crystals).

Point defects can also be of types different from the ones considered above. If the assumption made for Schottky defects are also valid here, Eq. (6.342) has an obvious generalization:

$$n_j = N_j e^{-\beta \bar{E}_j}, \quad \bar{E}_j = \left.\frac{\partial F_0}{\partial n_j}\right|_{n_j} = 0, \tag{6.343}$$

where N_j is the number of sites that can be occupied by a defect of type j.

Relations (6.343) are valid also when the number of defects of one type is independent of the presence of other defects. In the opposite case the problem must be solved anew. Thus, in the case of an ionic crystal the presence of only one vacancy leads to an excess negative charge that must be offset by a point defect of another type, so as to leave the crystal electrically neutral in the general case:

$$\sum_i q_i n_i = 0, \tag{6.344}$$

where q_i is the charge of a defect of type i. In this case the problem of the conditional extremum can be reduced to the preceding problem by means of a Lagrange multiplier λ, by determining the minimum of the quantity

$$G + \lambda \left(\sum_j q_j n_i \right). \tag{6.345}$$

As a result we have

$$n_j = N_j e^{-\beta(\bar{E}_j + \lambda q_j)}, \tag{6.346}$$

where λ is determined by using Eq. (6.344).

As a rule, the difference between the minimum value of the energy $\bar{E}_j$ and the value $\bar{E}_k$ closest to it is much larger than θ, i.e.,

$$|\bar{E}_k - \bar{E}_j|/\theta \gg 1.$$

To each sign of the charge there corresponds, therefore, in this case, defects of one type, with

$$n_+ = N_+ e^{-\beta(\bar{E}_+ + \lambda e)}, \quad \bar{E}_+ = \min_{q_j = +e} (\bar{E}_j),$$

$$n_- = N_- e^{-\beta(\bar{E}_- - \lambda e)}, \quad \bar{E}_- = \min_{q_j = -e} (\bar{E}_j). \tag{6.347}$$

Since

$$\begin{aligned} n_i &\ll n_+, \quad q_i = +e, \\ n_i &\ll n_+, \quad q_i = -e, \end{aligned} \tag{6.348}$$

it follows from Eq. (6.344) that

$$n_+ = n_- \tag{6.349}$$

with high degree of accuracy. On the other hand, from Eq. (6.347) we have

$$n_+ n_- = N_+ N_- e^{-\beta(E_+ + E_-)} \tag{6.350}$$

and from Eq. (6.349),

$$n_+ = n_- = \sqrt{N_+ N_-}\, e^{-(\beta/2)(E_+ + E_-)}. \tag{6.351}$$

In ionic crystals, when the densities of the negative and positive ion vacancies are equal, the defects produced are called Schottky defects (for example, in alkali-halide crystals). On the other hand, when the densities of the interstitial ions and of their vacancies are equal, we have Frenkel defects (for example, in silver halides).

In thermodynamic equilibrium, the probability of formation of linear and two-dimensional defects is practically zero, since the probability of formation of a linear defect is proportional to $N^{1/3}$ and that of a two-dimensional defect is proportional to $N^{2/3}$. Linear and two-dimensional defects in crystals are apparently metastable configurations,[169] but since the time to establish equilibrium is long, they are, so to speak, quenched. Only point defects are usually taken into account in the determination of the thermodynamic properties of crystals. Other defect types require special treatment.

Chapter 7 Correlation theory of crystals

It was noted in the preceding chapter that to develop a statistical theory of crystals anharmonic terms must be taken into account in the system Hamiltonian. Although the Einstein approximation does in fact make it possible to take these terms into account, the neglect of the particle-motion correlation leads in a number of cases to an appreciable discrepancy between the theoretical and experimental data.

To develop a statistical theory of a crystal with account taken of the correlation of its particle motion, we derive a chain of equations for the correlation functions (density matrices) and devise a method of solving the equations of this chain. The resultant equations for the unary and binary distribution functions (density matrices) are used next to construct the thermodynamics of crystalline systems. We investigate also anharmonic effects in crystals.

39. Equation chain for correlation functions

Consider a system of N identical particles in a volume V at a temperature T with a pair-interaction potential $\Phi(r)$, where r is the distance between the particles. Let the position of each particle be determined by a vector q with Cartesian coordinates q^α $(\alpha = 1,2,3)$.

To take the crystal-particle motion correlations into account, we transform from the partial distribution functions ρ_s to the correlation functions $g_2,\ldots,g_s$ with the aid of the relations

$$\rho_2(q_1,q_2) = \rho_1(q_1)\rho_1(q_2) + g_2(q_1,q_2), \tag{7.1}$$

$$\rho_3(q_1,q_2,q_3) = \rho_1(q_1)\rho_1(q_2)\rho_1(q_3) + \rho_1(q_1)g_2(q_2,q_3) + \rho_1(q_2)g_2(q_1,q_3) + \rho_1(q_3)g_2(q_1,q_2) + g_3(q_1,q_2,q_3), \tag{7.2}$$

$$\begin{aligned}\rho_4(q_1,q_2,q_3,q_4) = {}& \rho_1(q_1)\rho_1(q_2)\rho_1(q_3)\rho_1(q_4)\\ &+ \rho_1(q_1)\rho_1(q_2)g_2(q_3,q_4) + \rho_1(q_1)\rho_1(q_3)g_2(q_2,q_4)\\ &+ \rho_1(q_1)\rho_1(q_4)g_2(q_2,q_3) + \rho_1(q_2)\rho_1(q_3)g_2(q_1,q_4)\\ &+ \rho_1(q_2)\rho_1(q_4)g_2(q_1,q_3) + \rho_1(q_3)\rho_1(q_4)g_2(q_1,q_2)\\ &+ \rho_1(q_1)g_3(q_2,q_3,q_4) + \rho_1(q_2)g_3(q_1,q_3,q_4)\end{aligned}$$

$$+\rho_1(q_3)g_3(q_1,q_2,q_4)+\rho_1(q_4)g_3(q_1,q_2,q_3)$$
$$+g_2(q_1,q_2)g_2(q_3,q_4)+g_2(q_1,q_3)g_2(q_2,q_4)$$
$$+g_2(q_1,q_4)g_2(q_2,q_3)+g_4(q_1,q_2,q_3,q_4),\ \text{etc.} \tag{7.3}$$

Substituting Eqs. (7.1)–(7.3) in Eq. (1.144), we obtain a chain of equations for the correlation functions. The first three equations of this chain are

$$\theta\frac{\partial\rho_1(q_1)}{\partial q_1^\alpha}+\rho_1(q_1)\int\frac{\partial\Phi(1,2)}{\partial q_1^\alpha}\rho_1(q_2)dq_2+\int\frac{\partial\Phi(1,2)}{\partial q_1^\alpha}g_2(q_1,q_2)dq_2=0, \tag{7.4}$$

$$\theta\frac{\partial g_2(q_1,q_2)}{\partial q_1^\alpha}+\frac{\partial\Phi(1,2)}{\partial q_1^\alpha}\rho_1(q_1)\rho_1(q_2)+\frac{\partial\Phi(1,2)}{\partial q_1^\alpha}g_2(q_1,q_2)$$
$$+g_2(q_1,q_2)\int\frac{\partial\Phi(1,3)}{\partial q_1^\alpha}\rho_1(q_3)dq_3+\rho_1(q_1)\int\frac{\partial\Phi(1,3)}{\partial q_1^\alpha}g_2(q_2,q_3)dq_3$$
$$+\int\frac{\partial\Phi(1,3)}{\partial q_1^\alpha}g_3(q_1,q_2,q_3)dq_3=0, \tag{7.5}$$

$$\theta\frac{\partial g_3(q_1,q_2,q_3)}{\partial q_1^\alpha}+\frac{\partial\Phi(1,2)}{\partial q_1^\alpha}[g_2(q_2,q_3)\rho_1(q_1)+g_2(q_3,q_1)\rho_1(q_2)]$$
$$+\frac{\partial\Phi(1,3)}{\partial q_1^\alpha}[g_2(q_1,q_2)\rho_1(q_3)+g_2(q_2,q_3)\rho_1(q_1)]$$
$$+\frac{\partial\,[\Phi(1,2)+\Phi(1,3)]}{\partial q_1^\alpha}g_3(q_1,q_2,q_3)+g_2(q_1,q_2)\frac{\partial}{\partial q_1^\alpha}\int\Phi(1,4)g_2(q_3,q_4)dq_4$$
$$+g_2(q_1,q_3)\frac{\partial}{\partial q_1^\alpha}\int\Phi(1,4)g_2(q_2,q_4)dq_4+g_3(q_1,q_2,q_3)\frac{\partial}{\partial q_1^\alpha}\int\Phi(1,4)\rho_1(q_4)dq_4$$
$$+\rho_1(q_1)\frac{\partial}{\partial q_1^\alpha}\int\Phi(1,4)g_3(q_2,q_3,q_4)dq_4+\int\frac{\partial\Phi(1,4)}{\partial q_1^\alpha}g_4(q_1,q_2,q_3,q_4)dq_4=0, \tag{7.6}$$

where

$$\Phi(i,j)=\Phi(|q_i-q_j|).$$

We shall also need from the chain (1.144) three equations in which the distribution function ρ is replaced by the correlation functions g only in the integral terms. These equations contain ρ_1, ρ_2, ρ_3, g_2, g_3, g_4 and are of the form

$$\theta\frac{\partial\rho_1(q_1)}{\partial q_1^\alpha}+\frac{\partial u(q_1)}{\partial q_1^\alpha}\rho_1(q_1)+\int\frac{\partial\Phi(1,2)}{\partial q_1^\alpha}g_2(q_1,q_2)dq_2=0, \tag{7.7}$$

$$\theta\frac{\partial\rho_2(q_1,q_2)}{\partial q_1^\alpha}+\frac{\partial\,[\Phi(1,2)+u(q_1)+u(q_2)]}{\partial q_1^\alpha}\rho_2(q_1,q_2)$$
$$+\rho_1(q_1)\int\frac{\partial\Phi(1,3)}{\partial q_1^\alpha}g_3(q_2,q_3)dq_3+\rho_1(q_2)\int\frac{\partial\Phi(1,3)}{\partial q_1^\alpha}$$
$$\times g_2(q_1,q_3)dq_3+\int\frac{\partial\Phi(1,3)}{\partial q_1^\alpha}g_3(q_1,q_2,q_3)dq_3=0, \tag{7.8}$$

$$\theta\frac{\partial\rho_3(q_1,q_2,q_3)}{\partial q_1^\alpha}+\frac{\partial\,[U_3+u(q_1)+u(q_2)+u(q_3)]}{\partial q_1^\alpha}\rho_3(q_1,q_2,q_3)$$
$$+\rho_2(q_1,q_2)\int\frac{\partial\Phi(1,4)}{\partial q_1^\alpha}g_2(q_3,q_4)dq_4+\rho_2(q_2,q_3)\int\frac{\partial\Phi(1,4)}{\partial q_1^\alpha}g_2(q_1,q_4)dq_4$$
$$+\rho_2(q_1,q_3)\int\frac{\partial\Phi(1,4)}{\partial q_1^\alpha}g_2(q_2,q_4)dq_4+\rho_1(q_1)\int\frac{\partial\Phi(1,4)}{\partial q_1^\alpha}g_3(q_2,q_3,q_4)dq_4$$
$$+\rho_1(q_2)\int\frac{\partial\Phi(1,4)}{\partial q_1^\alpha}g_3(q_1,q_2,q_3,q_4)dq_4+\rho_1(q_3)\int\frac{\partial\Phi(1,4)}{\partial q_1^\alpha}g_3(q_1,q_2,q_4)dq_4$$
$$+\int\frac{\partial\Phi(1,4)}{\partial q_1^\alpha}g_4(q_1,q_2,q_3,q_4)dq_4=0, \tag{7.9}$$

where

$$u(q)=\int\Phi(|q-q'|)\rho_1(q')dq'. \tag{7.10}$$

From Eq. (1.145) we follow the normalization conditions for the correlation functions:

$$\int g_s(q_1,\ldots,q_s)dq_1\cdots dq_s=0. \tag{7.11}$$

The equation chains (7.4)–(7.6) and (7.7)–(7.10) for the correlation functions are the starting point for the considered correlation theory of the crystalline state.

The approach employed has much in common with the analysis given in Sec. 14. At the same time, however, it leads to a better understanding of the character of the assumptions made to obtain the solution, and permits the functions to be normalized in such a way that the particle localization in a definite cell is taken into account in each order of the approximation. This condition is met with a high degree of accuracy for an overwhelming majority of crystals within the framework of the equilibrium theory of the crystalline state.

40. Solution of initial equations

We proceed now to solve the initial equations (7.4)–(7.9) for the correlation functions.

Equations (7.4)–(7.6) were solved for a crystal in Ref. 52 by the following approach. Since the crystal is well described in the approximation in which the binary distribution function is multiplicative, the correlation terms are regarded as corrections to the self-consistent-field approximation. Accordingly, the first two terms of Eq. (7.5) play the principal role, and the remainder are corrections. As a result we get the closed set of equations (7.4) and (7.5) which is solved in Ref. 52 by successive approximations.

A second approach to the solution of Eqs. (7.4)–(7.9) was proposed in Ref. 56. It is based on the assumption that the integral terms with the correlation functions $g_s(q_1,...,q_s)$ are small compared with other terms of the equations.* Substituting the solutions obtained in the initial equations, we can directly verify that such an approximation is correct. This solution method can also be used in the quantum case. In addition, the distribution functions obtained in this manner for a homogeneous phase go over, as we shall show, into the corresponding equations obtained in Ref. 2. We describe here the second method of solving the initial equations. We obtain the zeroth approximation of the solutions of Eqs. (7.4)–(7.9) for a gas by dropping their integral terms. In the case of a gas $\rho_1(q) = C = \text{const}$, and in the zeroth approximation we have from Eqs. (7.1)–(7.9):

$$\theta \frac{\partial \rho_1^0}{\partial q_1^\alpha} = 0,$$

$$\theta \frac{\partial g_2^0(q_1,q_2)}{\partial q_1^\alpha} + C^2 \frac{\partial \Phi(1,2)}{\partial q_1^\alpha} + \frac{\partial \Phi(1,2)}{\partial q_1^\alpha} g_2^0(q_1,q_2) = 0, \tag{7.12}$$

$$\theta \frac{\partial g_3^0(q_1,q_2,q_3)}{\partial q_1^\alpha} + C \left\{ \frac{\partial \Phi(1,2)}{\partial q_1^\alpha} [g_2^0(q_2,q_3) + g_2^0(q_1,q_3)] \right.$$

$$\left. + \frac{\partial \Phi(1,3)}{\partial q_1^\alpha} [g_2^0(q_1,q_2) + g_0^2(q_2,q_3)] \right\}$$

$$+ \frac{\partial [\Phi(1,2) + \Phi(1,3)]}{\partial q_1^\alpha} g_3^0(q_1,q_2,q_3) = 0.$$

This yields, when account is taken of Eq. (7.11),

$$\rho_1^0 = C = \frac{N}{V}, \quad g_2^0 = C^2 \left\{ \exp\left[-\frac{1}{\theta} \Phi(1,2) \right] - 1 \right\},$$

$$g_3^0 = C^3 \left\{ \exp\left[-\frac{1}{\theta} U_3 \right] - \frac{1}{C^2} [g_2^0(q_1,q_2) \right.$$

$$\left. + g_2^0(q_1,q_3) + g_2^0(q_2,q_3)] - 1 \right\}. \tag{7.13}$$

For the determination of the correlation function in first-order approximation, we have from Eqs. (7.5) and (7.13) the equation

$$\theta \frac{\partial q_2^1}{\partial q_1^\alpha} + C^2 \frac{\partial \Phi(1,2)}{\partial q_1^\alpha} + \frac{\partial \Phi(1,2)}{\partial q_1^\alpha} g_2^1(q_1,q_2)$$

$$+ C^3 \int \frac{\partial \Phi(1,3)}{\partial q_1^\alpha} \exp\left[-\frac{1}{\theta} U_3 \right] dq_3 = 0. \tag{7.14}$$

*These integral terms, which contain the correlation functions $g_s(q_1,...,q_s)$, will hereafter be called correlation integrals.

Since

$$\frac{1}{\theta}\cdot\frac{\partial\Phi(1,3)}{\partial q_1^\alpha}\exp\left[-\frac{1}{\theta}U_3\right]$$
$$=-\exp\left[-\frac{1}{\theta}\Phi(1,2)\right]\frac{\partial}{\partial q_1^\alpha}\exp\left[-\sum_{1\le i\le 2}\frac{\Phi(i,3)}{\theta}\right],$$

we have from Eq. (7.14),

$$\theta\frac{\partial g_2^1}{\partial q_1^\alpha}+C^2\frac{\partial\Phi(1,2)}{\partial q_1^\alpha}+\frac{\partial\Phi(1,2)}{\partial q_1^\alpha}g_2^1(q_1,q_2)$$
$$=C^3\exp\left[-\frac{1}{\theta}\Phi(1,2)\right]\frac{\partial}{\partial q_1^\alpha}\int\exp\left[-\frac{1}{\theta}\sum_{1\le i\le 2}\Phi(i,3)\right]dq_3.$$

We put

$$g_2^1(q_1,q_2)=C_2^1(q_1,q_2)\exp\left[-\frac{1}{\theta}\Phi(1,2)\right]-C^2,$$

and obtain then

$$\frac{\partial C_2^1}{\partial q_1^\alpha}=C^3\frac{\partial}{\partial q_1^\alpha}\int\exp\left[-\frac{1}{\theta}\sum_{1\le i\le 2}\Phi(i,3)\right]dq_3.$$

This equation can be integrated directly. Account must be taken here of the symmetry of $C_2^1(q_1,q_2)$ with respect to the arguments. The integration constant is determined from the normalization condition (7.11). We have consequently

$$g_2^1(q_1,q_2)=C^2\exp\left[-\frac{1}{\theta}\Phi(|q|)\right]$$
$$\times\left[1+C\int f(|q-q'|)f(|q'|)dq'\right]-C^2,$$

where

$$f(|q|)=\exp\left[-\frac{1}{\theta}\Phi(|q|)\right]-1$$

and

$$q=q_1-q_2.$$

According to Eq. (1.143), the Bogolyubov distribution function $F_2(q_1,q_2)$ is connected with the parent function $\rho_2(q_1,q_2)$ by the relation

$$F_2(q_1,q_2)=v^2\rho_2(q_1,q_2).$$

Therefore, we obtain for the binary function $F_2^0(q_1,q_2)$ in the zeroth and first approximations, respectively,

$$F_2^0(q_1,q_2)=v^2[\rho_1(q_1)\rho_1(q_2)+g_2^0]=\exp\left[-\frac{1}{\theta}\Phi(|q_1-q_2|)\right], \tag{7.15}$$

$$F_2^1(q_1,q_2)=v^2[\rho_1(q_1)\rho_1(q_2)+g_2^1]$$

$$=\exp\left[-\frac{1}{\theta}\Phi(|q_1-q_2|)\right]\left[1+\frac{1}{v}\int f(|q_1-q_2-q'|)f(|q'|)dq'\right],$$

which coincides with the approximations of $F_2(q_1,q_2)$ obtained in Ref. 2 for a gas.

To determine the unary $\rho_1(q_1)$ and binary $\rho_2(q_1,q_2)$ distribution functions of a crystal, with allowance for the collective vibrations of its particles (i.e., with allowance for the correlations between the crystal particles), we begin simultaneously with Eqs. (7.4)–(7.6) and (7.7)–(7.9).

It is known that the distribution functions $\rho_1(q_1)$ and $\rho_2(q_1,q_2)$ are sufficient to find the thermal and caloric equations of state of a system. We shall solve Eqs. (7.4)–(7.9) by using the aforementioned singularity of the correlation integrals contained in these equations. We transform in Eqs. (7.4)–(7.9) from the functions $\rho_1, \rho_2, g_2, g_3, g_4$ to the functions p_1, p_2, t_2, t_3, t_4 by using the relations

$$\rho_1(q_1)=p_1(q_1)\exp[-\beta u(q_1)],$$

$$\rho_2(q_1,q_2)=p_2(q_1,q_2)\exp\left[-\beta\sum_{1\le i\le 2}u(q_i)\right],$$

$$g_2(q_1,q_2)=t_2(q_1,q_2)\exp\left[-\beta\sum_{1\le i\le 2}u(q_i)\right], \tag{7.16}$$

$$g_3(q_1,q_2,q_3)=t_3(q_1,q_2,q_3)\exp\left[-\beta\sum_{1\le i\le 3}u(q_j)\right],$$

$$g_4(q_1,q_2,q_3,q_4)=t_4(q_1,q_2,q_3,q_4)\exp\left[-\beta\sum_{1\le i\le 4}u(q_i)\right],$$

where

$$\beta=1/\theta.$$

Substituting Eq. (7.16) in Eqs. (7.5)–(7.9), we get

$$\theta\frac{\partial p_1(q_1)}{\partial q_1^\alpha}+\int\frac{\partial\Phi(1,2)}{\partial q_1^\alpha}t_2(q_1,q_2)\exp[-\beta u(q_2)]dq_2=0, \tag{7.17}$$

$$\theta\frac{\partial p_2(q_1,q_2)}{\partial q_1^\alpha}+\frac{\partial\Phi(1,2)}{\partial q_1^\alpha}p_2(q_1,q_2)$$

$$+p_1(q_1)\int\frac{\partial\Phi(1,3)}{\partial q_1^\alpha}t_2(q_2,q_3)\exp[-\beta u(q_3)]dq_3$$

$$+p_1(q_2)\int\frac{\partial\Phi(1,3)}{\partial q_1^\alpha}t_2(q_1,q_3)\exp[-\beta u(q_3)]dq_3$$

$$+\int\frac{\partial\Phi(1,3)}{\partial q_1^\alpha}t_3(q_1,q_2,q_3)\exp[-\beta u(q_3)]dq_3=0, \tag{7.18}$$

$$\theta\frac{\partial t_2(q_1,q_2)}{\partial q_1^\alpha}+\frac{\partial\Phi(1,2)}{\partial q_1^\alpha}p_1(q)p_1(q_2)+\frac{\partial\Phi(1,2)}{\partial q_1^\alpha}t_2(q_1,q_2)$$
$$+p_1(q_1)\int\frac{\partial\Phi(1,3)}{\partial q_1^\alpha}t_2(q_2,q_3)\exp[-\beta u(q_3)]dq_3$$
$$+\int\frac{\partial\Phi(1,3)}{\partial q_1^\alpha}t_3(q_1,q_2,q_3)\exp[-\beta u(q_3)]dq_3=0, \tag{7.19}$$

$$\theta\frac{\partial t_3(q_1,q_2,q_3)}{\partial q_1^\alpha}+\frac{\partial\Phi(1,2)}{\partial q_1^\alpha}[t_2(q_2,q_3)p_1(q_1)+p_1(q_2)t_2(q_1,q_3)]$$
$$+\frac{\partial\Phi(1,3)}{\partial q_1^\alpha}[t_2(q_1,q_2)p_1(q_3)+t_2(q_2,q_3)p_1(q_1)]+\frac{\partial U_3}{\partial q_1^\alpha}t_3(q_1,q_2,q_3)$$
$$+t_2(q_1,q_2)\frac{\partial}{\partial q_1^\alpha}\int\Phi(1,4)t_2(q_3,q_4)\exp[-\beta U(q_4)]dq_4$$
$$+t_2(q_1,q_4)\frac{\partial}{\partial q_1^\alpha}\int\Phi(1,4)t_2(q_2,q_4)\exp[-\beta u(q_4)]dq_4$$
$$+p_1(q_1)\frac{\partial}{\partial q_1^\alpha}\int\Phi(1,4)t_3(q_2,q_3,q_4)\exp[-\beta u(q_4)]dq_4$$
$$+\int\frac{\partial\Phi(1,4)}{\partial q_1^\alpha}t_4(q_1,q_2,q_3,q_4)\exp[-\beta u(q_4)]dq_4=0, \tag{7.20}$$

under the normalization conditions

$$\int p_s(q_1,\dots,q_s)\exp\Big[-\beta\sum_{1\le i\le s}u(q_i)\Big]dq_1\cdots dq_s=N^s,$$
$$\int t_s(q_1,\dots,q_s)\exp\Big[-\beta\sum_{1\le i\le s}u(q_i)\Big]dq_1\cdots dq_s=0 \tag{7.21}$$
$$(s=1,2,3,4,\dots).$$

We seek the solution of Eqs. (7.17)–(7.20) in the form

$$\begin{aligned}p_s(q_1,\dots,q_s)&=p_s^0(q_1,\dots,q_s)+p_s^1(q_1,\dots,q_s)+\cdots,\\ t_s(q_1,\dots,q_s)&=t_s^0(q_1,\dots,q_s)+t_s^1(q_1,\dots,q_s)+\cdots,\end{aligned} \tag{7.22}$$

where p_s^0 and t_s^0 are determined without allowance for the correlation integrals, which we assume to be proportional to some small parameter; p_s^1 and t_s^1 are proportional to the first power of this parameter, and so on.

For the zeroth approximations we have from Eqs. (7.17)–(7.20) the set of equations

$$\frac{\partial p_1^0(q_1)}{\partial q_1^\alpha}=0,$$

$$\theta\frac{\partial p_2^0(q_1,q_2)}{\partial q_1^\alpha}+\frac{\partial\Phi(1,2)}{\partial q_1^\alpha}p_2^0(q_1,q_2)=0,$$

$$\theta\frac{\partial t_2^0(q_1,q_2)}{\partial q_1^\alpha}+\frac{\partial\Phi(1,2)}{\partial q_1^\alpha}p_1^0(q_1)p_1^0(q_2)+\frac{\partial\Phi(1,2)}{\partial q_1^\alpha}t_2^0(q_1,q_2)=0,$$

$$\theta\frac{\partial t_3^0(q_1,q_2,q_3)}{\partial q_1^\alpha}+\frac{\partial\Phi(1,2)}{\partial q_1^\alpha}[t_2^0(q_2,q_3)p_1^0(q_1)+t_2^0(q_1,q_3)p_1^0(q_2)]$$
$$+\frac{\partial\Phi(1,3)}{\partial q_1^\alpha}[t_2^0(q_1,q_2)p_1^0(q_3)+t_2^0(q_2,q_3)p_1^0(q_1)]+\frac{\partial U_3}{\partial q_1^\alpha}t_3^0(q_1,q_2,q_3)=0. \tag{7.23}$$

Taking the normalization conditions (7.21) into account, we obtain the following solution for the set of equations (7.23):

$$p_1^0(q)=A_0=\text{const}, \tag{7.24}$$

$$t_2^0(q_1,q_2)=B_0\exp[-\beta\Phi(1,2)]-(p_1^0)^2, \tag{7.25}$$

$$t_3^0(q_1,q_2,q_3)=C_0\exp[-\beta U_3]-B_0p_1^0\exp[-\beta\Phi(1,2)]$$
$$+\exp[-\beta\Phi(1,3)]+\exp[-\beta\Phi(2,3)]+2(p_1^0)^3], \tag{7.26}$$

where

$$A_0=\frac{N}{\int\exp[-\beta u(q)]dq}, \tag{7.27}$$

$$B_0=\frac{N^2}{\int\exp\{-\beta[\Phi(1,2)+u(q_1)+u(q_2)]\}dq_1\,dq_2}, \tag{7.28}$$

$$C_0=\frac{N^3}{\int\exp\{-\beta[U_3+\Sigma_{1\leqslant i\leqslant s}u(q_i)]\}dq_1\,dq_2\,dq_3}. \tag{7.29}$$

A few remarks concerning the solution are in order. It is known[33] that in the analysis of a crystal it is taken into account that two particles cannot occupy the same crystal-lattice site. This fact is allowed for in the determination of the self-consistent potential (7.10). A similar circumstance must be taken into account when a self-consistent potential is obtained from the second equation of the Bogolyubov chain.The integration here is with respect to the positions of the third particle, with the positions of the first and second particles excluded. That is to say, the self-consistent potential (7.10) is calculated for a particle not only without integrating over the cell containing this particle, but also disregarding one more cell, so that we obtain not $u(q)$ but $\tilde{u}(q)$. We can, however, use $u(q)$ in place of $\tilde{u}(q)$, as is done in the introduced definitions (7.16), provided that the function $\Phi(|q_1-q_2|)$ in the corresponding solutions is replaced by the function $\widetilde{\Phi}(|q_1-q_2|)$ obtained from the relation

$$\phi(|q_1-q_2|)+\tilde{u}(q_1)+\tilde{u}(q_2)=\widetilde{\Phi}(|q_1-q_2|)+u(q_1)+u(q_2). \tag{7.30}$$

Similar reasoning can be used also for the case of three particles. For simplicity, we shall distinguish in the calculation between u and $\tilde{u}$, $\tilde{\tilde{u}}$,...; to allow for the exclusion of self-action, we must change in the final result from Φ to $\tilde{\Phi}$.

We note one more feature of our problem. For a crystal, by virtue of the localization of its particles, we have for any of their configurations $q_1,...,q_2$ (here q_i varies only in the region of motion of the ith atom)

$$\int \cdots \int \rho_s(q_1,...,q_s)dq_1 \cdots dq_s = 1. \tag{7.31}$$

This condition is valid only for the crystal phase, whereas the normalization condition (1.145) is valid for an arbitrary phase. For the first approximations of the functions $p_1(q_1)$ and $p_2(q_1,q_2)$, which are needed for the subsequent calculation of $\rho_1(q_1)$ and $\rho_2(q_1,q_2)$, we obtain from Eqs. (7.17) and (7.18) the equations

$$\theta \frac{\partial p_1^1(q_1)}{\partial q_1^\alpha} + \int \frac{\partial \Phi(1,2)}{\partial q_1^\alpha} t_2^0(q_1,q_2)\exp[-\beta u(q_2)]dq_2 = 0,$$

$$\theta \frac{\partial p_2^1(q_1,q_2)}{\partial q_1^\alpha} + \frac{\partial \Phi(1,2)}{\partial q_1^\alpha} p_2^1(q_1,q_2)$$
$$+ \int \frac{\partial \Phi(1,3)}{\partial q_1^\alpha} t_3^0(q_1,q_2,q_3)\exp[-\beta u(q_3)]dq_3 = 0,$$

which take, upon substitution of the zeroth approximations $t_2^0(q_1,q_2)$ and $t_3^0(q_1,q_2,q_3)$ from Eqs. (7.25) and (7.26), the form

$$\theta \frac{\partial p_1^1(q_1)}{\partial q_1^\alpha} + \int \frac{\partial \Phi(1,2)}{\partial q_1^\alpha} \{B_0 \exp[-\beta\Phi(1,2)]$$
$$-(p_1^0)^2\}\exp[-\beta u(q_2)]dq_2 = 0,$$

$$\theta \frac{\partial p_2^1(q_1,q_2)}{\partial q_1^\alpha} + \frac{\partial \Phi(1,2)}{\partial q_1^\alpha} p_2^1(q_1,q_2)$$
$$+ \int \frac{\partial \Phi(1,3)}{\partial q_1^\alpha} \{C_0 \exp[-\beta U_3] - B_0 p_1^0 \exp[-\beta\Phi(1,2)]\}$$
$$\times \exp[-\beta u(q_3)]dq_3 = 0.$$

After integrating these equations with allowance for Eqs. (7.21) and (7.22), we obtain

$$p_1^1(q_1) = \int \{B_0 \exp[-\beta\Phi(1,2)]$$
$$+ (p_1^0)^2\beta\Phi(1,2)\} \exp[-\beta u(q_2)]dq_2 - A_1, \tag{7.32}$$

where

$$A_1 = \frac{\int \{B_0 \exp[-\beta\Phi(1,2)] + (p_1^0)^2\beta\Phi(1,2)\}}{\int \exp[-\beta u(q)]dq} \times \frac{\exp\{-\beta\,[u(q_1)+u(q_2)]\}dq_1\,dq_2}{\int \exp[-\beta u(q)]dq};$$

and

$$p_2^1(q_1,q_2) = \int \{C_0 \exp[-\beta U_3] + B_0 p_1^0 \beta\,[\Phi(1,3) + \Phi(2,3)]\exp[-\beta\Phi(1,2)]\}\exp[-\beta u(q_3)]dq_3 - B_1 \exp[-\beta\Phi(1,2)], \tag{7.33}$$

where

$$B_1 = \frac{\int\{C_0 \exp[-\beta U_3] + B_0 p_1^0 \beta\,[\Phi(1,3)+\Phi(2,3)]}{\int \exp\{-\beta\,[\Phi(1,2)+u(q_1)]} \frac{\times \exp - \beta\Phi(1,2)\}\exp[-\beta\,\Sigma_{i=1}^3 u(q_i)]dq_1\,dq_2\,dq_3]}{+u(q_2)]\}dq_1\,dq_2}.$$

In the first approximation, $p_1(q_1)$ and $p_2(q_1,q_2)$ are thus equal to

$$p_1(q_1) = p_1^0 + p_1^1 = p_1^0 + \int \{B_0 \exp[-\beta\Phi(1,2)] + (p_1^0)^2\beta\Phi(1,2)\}\exp[-\beta u(q_2)]dq_2 - A_1, \tag{7.34}$$

$$p_2(q_1,q_2) = p_2^0 + p_2^1 = B_0 \exp[-\beta\Phi(1,2)] + \int \{C_0 \exp[-\beta U_3] + B_0 p_1^0 \beta\,[\Phi(1,3)+\Phi(2,3)] \times \exp[-\beta\Phi(1,2)]\}\exp[-\beta u(q_3)]dq_3 - B_1 \exp[-\beta\Phi(1,2)]. \tag{7.35}$$

Note that p_1^1 and p_2^2 were determined by us without changing from Φ to $\tilde{\Phi}$. We shall use the expressions obtained for $p_1(q_1)$ and $p_2(q_1,q_2)$ to determine the distribution functions $\rho_1(q_1)$ and $\rho_2(q_1,q_2)$ of the crystal.

41. Unary and binary distribution functions of a crystal

We determine now the unary and binary distribution functions of a crystal. On the basis of Eqs. (7.16) and (7.25), these functions have in the zeroth approximation the form

$$\rho_1^0(q_1) = p_1^0 \exp[-\beta u(q_1)],$$
$$\rho_2^0(q_1,q_2) = B_0 \exp\{-\beta\,[\Phi(1,2) + u(q_1) + u(q_2)]\} \tag{7.36}$$

and we obtain on the basis of Eqs. (7.16), (7.25), (7.32), and (7.33) the first-approximation expressions for the unary and binary distribution functions:

$$\rho_1^1(q_1) = (p_1^0 + p_1^1)\exp[-\beta u(q_1)]$$

$$= \Big(p_1^0 + \int \{B_0 \exp[-\beta\Phi(1,2)]$$

$$+ (p_1^0)^2\beta\Phi(1,2)\}\exp[-\beta u(q_2)]dq_2 - A_1\Big)\exp[-\beta u(q_1)], \quad (7.37)$$

$$\rho_2^1(q_1,q_2) = (p_2^0 + p_2^1)\exp\{-\beta\,[u(q_1) + u(q_2)]\}$$

$$= B_0 \exp\{-\beta\,[\Phi(1,2) + u(q_1) + u(q_2)]\}$$

$$+ \int C_0 \exp[-\beta U_3] + B_0\, p_1^0\beta\{\Phi(1,3)$$

$$+ \Phi(2,3)\exp[-\beta\Phi(1,2)]\}\exp\{[-\beta u(q_3)]dq_3$$

$$- B_1 \exp[-\beta\Phi(1,2)]\}\exp\{-\beta\,[u(q_1) + u(q_2)]\}. \quad (7.38)$$

The expressions obtained for $\rho_1(q_1)$ and $\rho_2(q_1,q_2)$ in the zeroth and first approximations are functionals of the unary distribution function $\rho_1(q_1)$. As follows from Eqs. (7.36) and (7.37), the single-particle distribution function is determined by a corresponding nonlinear integral equation. The zeroth approximation for $\rho_1(q_1)$ corresponds in correlation theory to the self-consistent-field theory, in which the first equation of Eq. (7.36) takes the form of the integral equation[33]

$$\theta \ln \lambda\rho_1(q) + \int \Phi(|q - q'|)\rho_1(q')dq' = 0. \quad (7.39)$$

This equation was solved in Ref. 33 by interaction and by a variational method, and the solution was used to formulate the thermodynamics of a crystal in the self-consistent-field approximation.

We now obtain explicit expressions for the single-particle distribution function in the zeroth (7.36) and first (7.37) approximations. Our basic premise is, as in Ref. 33, that in a crystal the parent single-particle distribution function can be represented in the form

$$\rho_1(q) = \sum_i \omega(q - a_i), \quad (7.40)$$

where a_i is the radius vector of the ith lattice site, and the function is finite and given by

$$\omega(q - a_i)\omega(q - a_j) = 0$$

if $i \neq j$.

The function $\omega(q - a_i)$ is the specific distribution function of one definite particle and determines the probability of its location near the ith lattice site. Therefore,

$$\int \omega(q - a_i)dq = 1. \quad (7.41)$$

The range of particle motion is quite limited,[33] so that in the zeroth approximation we can set it equal, with sufficient accuracy, to a Dirac δ function. Substituting next Eq. (7.40) on the right-hand side of the first equation of Eq. (7.36), we obtain the first iteration for $\rho_1^0(q)$. The boundary of the integration region is determined, as in Ref. 33, from the condition that the free energy be a minimum. The self-consistent potential $u(q)$ near the kth site is obviously equal to[33]

$$u^k(q) = \sum_{i\neq k} \Phi(|q - a_i|), \tag{7.42}$$

and the unary distribution function near the kth site (i.e., the specific distribution function) is in the zeroth and first approximations of correlation theory:

$$\rho_1^{0k}(q) = p_1^{0k} \exp[-\beta u^k(q)], \tag{7.43}$$

where

$$p_1^{0k} = 1 \Big/ \int \exp[-\beta u^k(q)]\, dq; \tag{7.44}$$

$$\rho_1^{1k}(q_1) = p_1^{0k} + \sum_{i\neq k} \int \{B_0 \exp[-\beta\Phi(1,2)] + (p_1^{0k})^2 \beta\Phi(1,2)\} \exp[-\beta u^l(q_2)]\, dq_2 - A_1^k \exp[-\beta u^k(q_1)], \tag{7.45}$$

where

$$A_1^k = \frac{\Sigma_{l\neq k} \int \{B_0 \exp[-\beta\Phi(1,2)] + (p_1^{0k})^2 \beta\Phi(1,2)\}}{\int \exp[-\beta u^k(q)]\, dq} \times \frac{\exp\{-\beta\,[u^k(q_1) + u^l(q_2)]\}\, dq_1\, dq_2}{\int \exp[-\beta u^k(q)]\, dq}. \tag{7.46}$$

The unary distribution function of the crystal (i.e., the unary single-particle distribution function) is given by the expression

$$\rho_1(q_1) + \sum_{k=1}^{N} \rho_1^k(q_1). \tag{7.47}$$

We turn now to the binary distribution function. In the zeroth approximation we have for the specific binary distribution function

$$\rho_2^{0kl}(q_1,q_2) = B_0 \exp\{-\beta\,[\Phi(1,2) + u^k(q_1) + u^l(q_2)]\}, \tag{7.48}$$

and in the first approximation

$$\rho_2^{1kl}(q_1,q_2) = B_0 \exp\{ -\beta [\Phi(1,2) + u^k(q_1) + u^l(q_2)] \}$$
$$+ \Big(\sum_{n\neq k,l} \int \{ C_0 \exp[-\beta U_3] + B_0 \rho_1^{0n} [\Phi(1,3)$$
$$+ \Phi(2,3)] \exp[-\beta\Phi(1,2)] \} \exp[-\beta u^n(q_3)] dq_3$$
$$- B_1 \exp[-\beta\Phi(1,2)] \Big) \exp\{ -\beta [u^k(q_1) + u^l(q_2)] \}. \tag{7.49}$$

The binary distribution function of the crystal (i.e., its parent binary distribution function) is defined as

$$\rho_2(q_1,q_2) = \sum_{k<l} \rho_2^{kl}(q_1,q_2). \tag{7.50}$$

We have thus obtained the final expressions for $\rho_1(q_1)$ and $\rho_2(q_1,q_2)$ of a crystal in the zeroth and first approximations of correlation theory.

We now show that the zeroth approximation (7.48) for the binary distribution function is equal (accurate to the first term of the corresponding expansion) to the expression obtained for this distribution in Ref. 52. Indeed, expanding Eq. (7.48) in powers of $\beta\Phi(|q_1 - q_2|)$ (it must be borne in mind that Φ has here the meaning of $\tilde{\Phi}$), we have accurate to the linear term

$$\rho_2^{0kl}(q_1 q_2) = \rho_1^{0k}(q_1)\rho_1^{0l}(q_2)\Big\{1 - \beta \Big[\Phi(|q_1 - q_2|)$$
$$- \frac{1}{2} \int \Phi(|q_1' - q_2'|)\rho_1^0(q_1')\rho_1^0(q_2') dq_1' dq_2' \Big]\Big\} \tag{7.51}$$

in full agreement with the corresponding expression obtained in Ref. 52 for the binary function. The second of second and higher order in $\beta\Phi(|q_1 - q_2|)$ are comparable in magnitude with the correction terms $\rho_2^{1kl}(q_1,q_2)$ in first-order approximation.

It is easily seen that Eq. (7.50) can be written in the form

$$\rho_2^{0kl}(q_1,q_2) = \rho_l^{0k}(q_1)\rho_1^{0l}(q_2)\{1 - \beta [\Phi(|q_1 - q_2|) - \bar{\Phi}]\}, \tag{7.52}$$

where

$$\bar{\Phi} = \tfrac{1}{2} \int \Phi(|q_1 - q_2|)\rho_1^0(q_1)\rho_1^0(q_2) dq_1 \, dq_2. \tag{7.53}$$

It is clear from Eq. (7.51) that the correlation contribution to $\rho_2^{0kl}(q_1,q_2)$, due to allowance for the collective vibrations (CV) is proportional to the deviation of the pair potential from its mean value calculated from the distribution in the self-consistent-field (SCF) approximation.

It is convenient to rewrite expression (7.50) for $\rho_2^{0kl}(q_1,q_2)$ more concisely as a sum of two terms:

$$\rho_2^{0kl}(q_1,q_2) = (\rho_2^{0kl})_{\text{SCF}} + (\rho_2^{0kl})_{\text{CV}}, \tag{7.54}$$

where

$$(\rho_2^{0kl})_{\rm SCF} = \rho_1^{0k}(q_1)\rho_1^{0l}(q_2),$$
$$(\rho_2^{0kl})_{\rm CV} = -\beta\,[\Phi - \bar{\Phi}]\rho_1^{0k}(q_1)\rho_1^{0l}(q_2). \tag{7.55}$$

It is possible to obtain in the same manner a series for the first approximation of the binary function $\rho_2^{0kl}(q_1,q_2)$ and compare it with the corresponding expression obtained in Ref. 52. We obtain here agreement if we retain in the expansion the terms $\sim\beta^2\Phi^2$.

42. Thermal and caloric equations of state of a crystal

We obtain first the equation of state of the crystal phase. It is known that the pressure is expressed in terms of the binary distribution function[2]

$$p = \frac{N\theta}{V} - \frac{1}{6V}\int |q_1 - q_2|\Phi'(|q_1 - q_2|)\rho_2(q_1,q_2)dq_1\,dq_2. \tag{7.56}$$

Substituting Eqs. (7.48) and (7.49) in succession in Eq. (7.55) we obtain the thermal equation of state of a crystal in the zeroth and first approximations of the correlation theory, respectively,

$$p^0 = \frac{N\theta}{V} - \frac{B_0}{6V}\sum_{k\neq l}\int |q_1 - q_2|\Phi'(|q_1 - q_2|)$$
$$\times\exp\{-\beta\,[\Phi(1,2) + u^k(q_1) + u^l(q_2)]\}, \tag{7.57}$$

$$p^1 = p^0 - \frac{1}{6V}\sum_{k\neq l\neq n}\int |q_1 - q_2|\Phi'(|q_1 - q_2|)\Big(\{C_0\exp[-\beta U_3]$$
$$+ B_0\,\rho_1^{0n}\beta\,[\Phi(1,3) + \Phi(2,3)]\exp[-\beta\Phi(1,2)]\}\exp[-\beta u^n(q_3)]dq_3$$
$$- B_1\exp[-\beta\Phi(1,2)]\Big)\exp\{-\beta\,[u^k(q_1) + u^l(q_2)]\}dq_1\,dq_2. \tag{7.58}$$

Averaging the Hamiltonian of the considered system

$$H = \frac{1}{2m}\sum_k p_k^2 + \sum_{k<l}\Phi(|q_k - q_l|)$$

over a canonical ensemble, we obtain an expression for the system's internal energy[2]

$$E = \bar{H} = \tfrac{3}{2}N\theta + \tfrac{1}{2}\int \Phi(1,2)\rho_2(q_1,q_2)dq_1\,dq_2. \tag{7.59}$$

Substituting in Eq. (7.59) the expressions (7.48) and (7.49) for $\rho_2(q_1,q_2)$ we obtain the caloric equation of state of the crystal in the zeroth and first approximations, respectively,

$$E^0 = \tfrac{3}{2} N\theta + \tfrac{1}{2} \sum_{k,l} \int \Phi(1,2) \exp\{ -\beta\, [\Phi(1,2) + u^k(q_1) + u^l(q_2)]\} dq_1\, dq_2, \tag{7.60}$$

$$E^1 = E^0 + \tfrac{1}{2} \sum_{k,l,n} \int \Phi(1,2) \{ c_0 \exp[-\beta U_3] + B_0\, p_1^{0n} \beta\, [\Phi(1,3) + \Phi(2,3)] \exp[-\beta \Phi(1,2)] \exp[-\beta u^0(q_3)] dq_3 - B_1 \exp[-\beta \Phi(1,2)] \} \exp\{ -\beta\, [u^k(q_1) + u^l(q_2)] \} dq_1\, dq_2 \,. \tag{7.61}$$

Knowing the thermal and caloric equations of state, we can use the known thermodynamic relations to find the other thermodynamic functions of the crystal and obtain its complete thermodynamic description.

It is known from thermodynamics that the expressions for the pressure p and for the internal energy E of a system are connected by the differential equation (the thermodynamic-compatibility condition)

$$\theta \left(\frac{\partial p}{\partial \theta} \right)_V = \left(\frac{\partial E}{\partial V} \right)_V + p. \tag{7.62}$$

It is easy to verify that p and E obtained directly from the Gibbs distribution satisfy the relation (7.62). Indeed, the Gibbs configuration distribution of a system of N particles is given by

$$D(q_1,\dots,q_N) = \frac{1}{Q} \exp[-\beta U(q_1,\dots,q_N)],$$

where Q is the configuration integral and $U(q_1,\dots,q_N)$ is the potential energy of the interaction between the particles.

The pressure and the internal energy are, respectively,

$$p = \frac{\theta}{Q} \left(\frac{\partial Q}{\partial V} \right)_\theta,$$

$$E = \frac{3}{2} N\theta + \frac{\theta^2}{Q} \left(\frac{\partial Q}{\partial \theta} \right)_V.$$

Therefore,

$$\theta \left(\frac{\partial p}{\partial \theta} \right)_V = \frac{\theta}{Q} \left(\frac{\partial Q}{\partial \theta} \right)_V - \frac{\theta^2}{Q^2} \left(\frac{\partial Q}{\partial V} \right)_\theta \left(\frac{\partial Q}{\partial \theta} \right)_V + \frac{\theta^2}{Q} \frac{\partial^2 Q}{\partial V \partial \theta},$$

$$\left(\frac{\partial E}{\partial V} \right)_\theta = - \frac{\theta^2}{Q^2} \left(\frac{\partial Q}{\partial \theta} \right)_V \left(\frac{\partial Q}{\partial V} \right)_\theta + \frac{\theta^2}{Q} \frac{\partial^2 Q}{\partial \theta \partial V}.$$

Substituting these expressions in Eq. (7.62), we verify that the latter turns into an identity. The Bogolyubov distribution-functions method yields also expressions for the dynamic functions that meet the thermodynamic-compatibility condition (see Sec. 12).

43. Comparison of the equation of state of a crystal in the self-consistent-field approximation and in correlation theory

To compare the equations of state of a crystal without allowance for the collective vibrations of its particles (the self-consistent-field approximation) and with these vibrations taken into account (correlation theory), we substitute in the exact equations (7.56) and (7.60) the expressions (7.54) for the binary distribution function. The latter expression is more convenient than Eq. (7.48) for the estimate of the contributions of the correlation terms. We obtain then, respectively, the thermal and caloric equations of state of the crystal:

$$p = p_{SCF} + p_{CV}, \tag{7.63}$$

where

$$p_{SCF} = \frac{\theta}{v} - \frac{1}{6V}\int |q_1 - q_2|\Phi'(1,2)\rho_1(q_1)\rho_1(q_2)dq_1\,dq_2, \tag{7.64}$$

$$\begin{aligned} p_{CV} = \frac{\beta}{6V}\Bigg\{\int & |q_1 - q_2|\Phi'(1,2)\Phi(1,2)\rho_1(q_1)\rho_1(q_2)dq_1\,dq_2 \\ & - \int |q_1 - q_2|\Phi'(1,2)\rho_1(q_1)\rho_1(q_2)dq_1\,dq_2 \\ & \times\left[\frac{1}{2N}\int \Phi(1,2)\rho_1(q_1)\rho_1(q_2)dq_1\,dq_2\right]\Bigg\}, \end{aligned} \tag{7.65}$$

and

$$E = E_{SCF} + E_{CV}, \tag{7.66}$$

where

$$E_{SCF} = \frac{3}{2}N\theta + \frac{1}{2}\int \Phi(1,2)\rho_1(q_1)\rho_1(q_2)dq_1\,dq_2, \tag{7.67}$$

$$\begin{aligned} E_{CV} = -\frac{\beta}{2}\int & \left[\Phi(1,2) - \frac{1}{2}\int \Phi(1,2)\rho_1(q_1)\rho_1(q_2)dq_1\,dq_2\right]^2 \\ & \times\rho_1(q_1)\rho_1(q_2)dq_1\,dq_2. \end{aligned} \tag{7.68}$$

It is clear from Eq. (7.68) that the contribution of the correlation terms to the internal energy of the crystal is proportional to the mean-squared deviation of the pair potential $\Phi(|q_1 - q_2|)$, calculated in the self-consistent-field approximation. On the other hand, the contribution of the corresponding correlation terms (7.65) to the expression (7.63) for the pressure is not reducible to a similar form. We shall show, however, that p_{CV} can also be calculated if E_{CV} is known. We introduce the function

$$\Phi_2(|q_1 - q_2|) = \Phi(\lambda|q_1 - q_2|).$$

Obviously,

$$\Phi(|q_1 - q_2|) = \lim_{\lambda \to 1} \Phi_2(|q_1 - q_2|),$$

$$\lim_{\lambda \to 1} \frac{d\Phi_\lambda}{d\lambda} = |q_1 - q_2|\Phi'(|q_1 - q_2|),$$

$$E_{\rm CV}^\lambda = -\frac{\beta}{2}\int\Big[\Phi_\lambda(|q_1 - q_2|) - \frac{1}{2N} \times \int \Phi_\lambda(|q_1 - q_2|)\rho_1(q_1)\rho_1(q_2)dq_1\,dq_2\Big]^2 \rho_1(q_1)\rho_1(q_2)dq_1\,dq_2, \tag{7.69}$$

$$E_{\rm CV} = \lim_{\lambda \to 1} E_{\rm CV}^\lambda,$$

$$\lim_{\lambda \to 1} \frac{dE_{\rm CV}^\lambda}{d\lambda} = -\beta\Big\{\int |q_1 - q_2|\Phi'(1,2)\Phi(1,2) \times \rho_1(q_1)\rho_1(q_2)dq_1\,dq_2 + \int |q_1 - q_2|\Phi'(1,2)\rho_1(q_1)\rho_1(q_2) \times dq_1\,dq_2\Big[\frac{1}{2N}\int \Phi(1,2)\rho_1(q_1)\rho_1(q_2)dq_1\,dq_2\Big]\Big\}. \tag{7.70}$$

Comparing Eq. (7.65) with Eq. (7.70) we get

$$p_{\rm CV} = -\frac{1}{6V}\lim_{\lambda \to 1} \frac{dE_{\rm CV}^\lambda}{d\lambda}.$$

We thus obtain the thermal and caloric equations of state of a crystal in a form suitable for the estimate of the correlation terms:

$$p = p_{\rm SCF} - \frac{1}{6V}\lim_{\lambda \to 1} \frac{dE_{\rm CV}^\lambda}{d\lambda}, \tag{7.71}$$

$$E = E_{\rm SCF} + \lim_{\lambda \to 1} E_{\rm CV}^\lambda. \tag{7.72}$$

44. Free energy of a crystal

We proceed now directly to calculate the free energy of a crystal. We choose $u_1(q)$ in the form

$$u_1(q_1) = u_1^0 + \tfrac{1}{2}m\omega^2[(q_1^1)^2 + (q_1^2)^2 + (q_1^3)^2]. \tag{7.73}$$

This function coincides with the harmonic part of the self-consistent potential $u(q)$. The configuration integral is calculated in this case directly, and according to Sec. 11 the configuration part of the free energy is equal in the self-consistent-field approximation[33] to

$$F_{\rm SCF} = F_0 + \Big\langle U - \sum_i u_1(q_i)\Big\rangle_0, \tag{7.74}$$

where

$$F_0 = -\frac{3}{2} N\theta \ln(2\pi\theta) + \frac{N\theta}{2} \ln(\lambda_1\lambda_2\lambda_3) + U^0, \tag{7.75}$$

U^0 is the static energy of the crystal

$$\langle U_0 \rangle = \frac{N(N-1)}{2} \frac{\int(|q-q'|)\exp\{-\beta[u_1(q_1)+u_1(q_2)]\}dq\,dq'}{\{\int \exp[-\beta u_1(q)]dq\}^2}, \tag{7.76}$$

$$\left\langle \sum_i u_1(q_i) \right\rangle_0 = \frac{N\int u_1(q)\exp[-\beta u_1(q)]dq}{\int \exp[-\beta u_1(q)]dq}. \tag{7.77}$$

From Eq. (2.80) we can obtain an improved expression for the free energy[38]

$$F - F_{\rm SCF} = \tfrac{1}{4} \int \{\Phi(|q_1 - q_2|) - u_1(q_1) - u_1(q_2)\}\{\rho_2(q_1,q_2) - \rho_1^{0h}(q_1)\rho_1^{0h}(q_2)\}dq_1\,dq_2, \tag{7.78}$$

where $\rho_1^{0h}(q)$ is the unary distribution function for the potential $u_1(q)$.

We rewrite Eq. (7.78) in the form

$$F - F_{\rm SCF} = \tfrac{1}{4} \int \Phi(|q_1 - q_2|)\{\rho_2(q_1,q_2) - \rho_1^{0h}(q_1)\rho_1^{0h}(q_2)\}dq_1\,dq_2 - \tfrac{1}{2}\int u_1(q)\{\rho_1(q) - \rho_1^{0h}(q)\}dq. \tag{7.79}$$

In the zeroth approximation we obtain from Eq. (7.79),

$$F - F_{\rm SCF} = \tfrac{1}{4} \int \Phi(|q_1 - q_2|)\{\rho_2^0(q_1,q_2) - \rho_1^{0h}(q_1)\rho_1^{0h}(q_2)\}dq_1\,dq_2 - \tfrac{1}{2}\int u_1(q)\{\rho_1^0(q) - \rho_1^{0h}(q)\}dq. \tag{7.80}$$

Substituting in Eq. (7.78) the expression for $\rho_2^0(q_1,q_2)$ from Eq. (7.51), we have

$$F - F_{\rm SCF} = -\frac{\beta}{4}\left\{\iint \Phi^2(|q_1 - q_2|)\rho_1^0(q_1)\rho_1^0(q_2)dq_1\,dq_2 - \frac{1}{2N}\left[\iint \Phi(|q_1 - q_2|)\rho_1^0(q_1)\rho_1^0(q_2)dq_1\,dq_2\right]^2\right\} + \frac{1}{2}\int [u(q) - u_1(q)]\,[\rho_1^0(q) - \rho_1^{0h}(q)]dq. \tag{7.81}$$

The unary distribution function in Eq. (7.81) is subject to the condition that the self-consistent potential be defined without integration over the positions of the first and second parts. To lift this restriction, we rewrite Eq. (7.81) in the form

$$F - F_{\mathrm{SCF}} = -\frac{\beta}{4}\left\{\int \Phi(|q_1 - q_2|)\tilde{\Phi}(|q_1 - q_2|)\rho_1^0(q_1)\rho_1^0(q_2)dq_1\,dq_2\right.$$
$$-\frac{1}{2N}\left.\left[\int \Phi(|q_1 - q_2|)\rho_1^0(q_1)\rho_1^0(q_2)dq_1\,dq_2\right]\right\}$$
$$\times\left[\int \tilde{\Phi}(|q_1 - q_2|)\rho_1^0(q_1)\rho_1^0(q_2)dq_1\,dq_2\right]$$
$$+\frac{1}{2}\int [u(q) - u_1(q)]\,[\rho_1^0(q) - \rho_1^0(q)]\,dq, \tag{7.82}$$

where $\tilde{\Phi}(|q_1 - q_2|)$ is determined from Eq. (7.30).

We confine ourselves in the allowance for the anharmonicity to terms up to the fourth inclusive. Then

$$\Phi = \Phi_0 + \Phi_1 + \Phi_2 + \Phi_3 + \Phi_4, \tag{7.83}$$
$$\tilde{\Phi} = \tilde{\Phi}_2 + \tilde{\Phi}_3 + \tilde{\Phi}_4, \tag{7.84}$$

where Φ_n and $\tilde{\Phi}_n$ are terms proportional to the nth power of the displacement. From Eqs. (7.82)–(7.84) we have

$$F - F_{\mathrm{SCF}} = -\frac{\beta}{2}(\overline{\Phi_1\tilde{\Phi}_3} + \overline{\Phi_3\tilde{\Phi}_3} + \overline{\tilde{\Phi}_2^2} + \overline{\Phi_2\tilde{\Phi}_4}$$
$$+ \overline{\tilde{\Phi}_2\Phi_4} - \overline{\Phi_2\tilde{\Phi}_4}) + \frac{1}{2}\int [u(q) - u_1(q)]\,[\rho_1^0(q) - \rho_1^{0h}(q)]\,dq, \tag{7.85}$$

where

$$\overline{A} = \tfrac{1}{2}\int A\rho_1^0(q_1)\rho_1^0(q_2)dq_1\,dq_2. \tag{7.86}$$

Note that the terms $\overline{\Phi_3\tilde{\Phi}_3}$ and $\overline{\Phi_2\tilde{\Phi}_4}$ are also of fourth order of smallness,[27] and the term $\Phi_4\tilde{\Phi}_4$ can be neglected.

The calculations yield

$$\overline{\tilde{\Phi}_2\Phi_2} = \Delta_2^2\,\frac{1}{2}\sum_k z_k\left\{(\alpha^4 + 2\alpha^{22})\left(\Phi'' - \frac{1}{a_k}\Phi'\right)^2\right.$$
$$\left.+\frac{2\Phi'}{a_k}\left(\Phi'' - \frac{1}{a_k}\Phi'\right) + \frac{3}{a_k^2}\Phi'^2\right\}, \tag{7.87}$$

$$\Phi'' = \Phi''(a_k),\quad \Phi' = \Phi'(a_k),\quad \alpha_i = \frac{a_i}{a}\quad (i = 1,2,3),\quad a_k = V_k a, \tag{7.88}$$

$$\alpha^{22} = \alpha_1^2\alpha_2^2 + \alpha_2^2\alpha_3^2 + \alpha_3^2\alpha_4^2,\quad \alpha^4 = \alpha_1^4 + \alpha_2^4 + \alpha_3^4\,, \tag{7.89}$$

where

$$\Delta_2 = I^{-1}\iint_{-b}^{b}\int x^2 \exp\Big[-\frac{m\omega^2}{2\theta}(x^2+y^2+z^2)$$

$$-\frac{\gamma_1}{\theta}(x^4+y^4+z^4)-\frac{\gamma^2}{\theta}(x^2y^2+x^2z^2+y^2z^2)\Big]\,dx\,dy\,dz\,, \tag{7.90}$$

$$I = \iint_{-b}^{b}\int \exp\Big[-\frac{m\omega^2}{2\theta}(x^2+y^2+z^2)-\frac{\gamma_1}{\theta}(x^4+y^4+z^4)$$

$$-\frac{\gamma_2}{\theta}(x^2y^2+x^2z^2+y^2z^2)\Big]\,dx\,dy\,dz, \tag{7.91}$$

$$m\omega^2 = \frac{1}{3}\sum_k z_k\left[\Phi''(V_k a)+\frac{2}{(V_k a)}\Phi'(V_k a)\right], \tag{7.92}$$

$$\gamma_1 = \frac{1}{24}\sum_k z_k\left[\alpha_{1k}^4\Phi^{\mathrm{IV}}(V_k a)+6\,\frac{\alpha_{1k}^2(1-\alpha_{1k}^2)}{V_k a}\Phi'''\right]$$

$$-3\left[\frac{1}{V_k a}\Phi'''(V_k a)-\frac{1}{(V_k a)^3}\Phi'(V_k a)\right](3-18\alpha_{1k}^2+15\alpha_{1k}^4), \tag{7.93}$$

$$\alpha_{ik} = \frac{a_{ik}}{V_k a},$$

$$\gamma_2 = \frac{1}{6}\sum_k z_k\Big\{(\alpha_{1k}^2+\alpha_{2k}^2)\Phi^{\mathrm{IV}}(V_k a)$$

$$+\frac{(\alpha_{1k}+\alpha_{2k}+6\alpha_{1k}\alpha_{ik}^2)}{(V_k a)}\Phi'''(V_k a)\Big\}+3\Big[\frac{1}{(V_k a)^2}\Phi''(V_k a)$$

$$-\frac{1}{(V_k a)^3}\Phi'(V_k a)\Big][1-3(\alpha_{1k}^2+\alpha_{2k}^2+15\alpha_{1k}^2\alpha_{2k}^2)]\,. \tag{7.94}$$

In view of the rapid convergence of the integral, we can put

$$I = \iint_{-\infty}^{\infty}\int \exp\Big[-\frac{m\omega^2}{2\theta}(x^2+y^2+z^2)-\frac{\gamma_1}{\theta}(x^4+y^4+z^4)$$

$$+\frac{\gamma_4}{\theta}(x^2y^2+x^2z^2+y^2z^2)\Big]\,dx\,dy\,dz. \tag{7.95}$$

The terms $\overline{\Phi_3\tilde{\Phi}_3}$, $\overline{\Phi_1\tilde{\Phi}_3}$, $\overline{\Phi_2\tilde{\Phi}_4}$, and $\overline{\tilde{\Phi}_2\Phi_4}$ are calculated similarly. These expressions, however, contain in addition to Δ_2 also the quantities

$$\Delta_4 = I^{-1} \iint_{-b}^{b} \int x^4 \exp\Big[-\frac{m\omega^2}{2\theta}(x^2+y^2+z^2) - \frac{\gamma_1}{\theta}(x^4+y^4+z^4) - \frac{\gamma_2}{\theta}(x^2y^2+y^2z^2+z^2x^2) \Big]\, dx\, dy\, dz, \tag{7.96}$$

$$\Delta_{24} = I^{-1} \iint_{-b}^{b} \int x^2y^4 \exp\Big[-\frac{m\omega^2}{2\theta}(x^2+y^2+z^2) - \frac{\gamma_1}{\theta}(x^4+y^4+z^4) - \frac{\gamma_2}{\theta}(x^2y^4+y^2z^2+z^2x^2) \Big] dx\, dy\, dz,$$

$$\Delta_6 = I^{-1} \iint_{-b}^{b} \int x^6 \exp\Big[-\frac{m\omega^2}{2\theta}(x^2+y^2+z^2) - \frac{\gamma_1}{\theta}(x^4+y^4+z^4) - \frac{\gamma_2}{\theta}(x^2y^2+y^2z^2+z^2x_2) \Big]\, dx\, dy\, dz\,. \tag{7.97}$$

We take the expression for b from the self-consistent-field approximation

$$\frac{\partial F_{\text{SCF}}}{\partial b} = 0. \tag{7.98}$$

We conclude with the following remark.

With the free energy F of the crystal known, we can use the expression

$$p = -\frac{\partial F}{\partial V} = -\frac{\partial f}{\partial v} \quad (f = F/N,\ v = V/N)$$

to obtain

$$v = v(p,T)$$

and

$$a = a(p,T), \tag{7.99}$$

where a is the crystal-lattice constant.

Since, however, the determination of the function (7.90) entails differentiation of the approximately obtained expression for the crystal free energy F, which adversely affects the accuracy, it is better to obtain the dependence of the lattice constant on the pressure and on the temperature from the thermal equation of state (7.56).

The method developed in the present section to determine the free energy of a crystal is applicable in the quantum region. In the classical case one can choose for the function $u_1(q)$, besides Eq. (7.73), also the anharmonic part of the self-consistent potential, and even

$$u_1(q) = u(q), \tag{7.100}$$

since the free energy F_0 can also be calculated in this case. This permits anharmonicities of all order to be taken into account, but is fraught with computational difficulties. Therefore, if the anharmonicity needs to be taken into account only up to a certain order, it is necessary to choose $u_1(q)$ in the corresponding approximation.

It was proposed in Ref. 186 to take a self-consistent account of higher-order anharmonicities on the basis of the Green's function method, and the result led to development of a theory of strongly anharmonic crystals.[187–189]

With allowance for anharmonicities up to fourth order inclusive, we have

$$u_1(q) = u_1^0 + \frac{m\omega^2}{2}(x^2 + y^2 + z^2) + \gamma_1(x^4 + y^4 + z^4) \qquad (7.101)$$

$$+ \gamma_2(x^2y^2 + x^2z^2 + y^2z^2),$$

$$F_0^{(4)} = -\frac{3}{2}N\theta \ln(2\pi m\theta) - N\theta \ln \int \exp\left\{-\frac{1}{\theta}\left[\frac{m\omega^2}{\theta}(x^2 + y^2 + z^2)\right.\right.$$

$$\left.\left. + \gamma_1(x^4 + y^4 + z^4) + \gamma_2(x^2y^2 + x^2z^2 + y^2z^2)\right]\right\} dx\, dy\, dz + U^0, \qquad (7.102)$$

and

$$F_{\rm SCF}^{(4)} = F_0^{(4)} + \left\langle U - \sum_i u_1(q_1)\right\rangle_0. \qquad (7.103)$$

We obtain accordingly in the zeroth approximation

$$F - F_{\rm SCF}^{(4)} = \tfrac{1}{4}\int \Phi(|q_1 - q_2|)\{\rho_2^0(q_1,q_2) - \rho_1^0(q_1)\rho_1^0(q_2)\}dq_1\, dq_2. \qquad (7.104)$$

Calculations similar to the foregoing lead ultimately to

$$F - F_{\rm SCF}^{(4)} = \frac{\beta}{2}(\overline{\tilde{\Phi}_3\Phi_1} + \overline{\Phi_3\tilde{\Phi}_3} + \overline{\tilde{\Phi}_2^2} + \overline{\Phi_2\tilde{\Phi}_2} + \overline{\tilde{\Phi}_2\Phi_4} - \overline{\Phi_2}\,\overline{\tilde{\Phi}_4}). \qquad (7.105)$$

Expansion of $F_{\rm SCF}^{(4)}$ in a perturbation-theory series yields for the crystal free energy an expression:

$$F - F_{\rm SCF} = -\frac{\beta}{2}(\overline{\Phi_1\tilde{\Phi}_3} + \overline{\Phi_3\tilde{\Phi}_3} + \overline{\tilde{\Phi}_2^2} + \overline{\Phi_2\tilde{\Phi}_4}$$

$$+ \overline{\tilde{\Phi}_2\Phi_4} - \overline{\Phi_2}\,\overline{\tilde{\Phi}_4}) - \frac{1}{2}\int [u(q) - u_1(q)]\rho_1^{0h}(q)dq, \qquad (7.106)$$

which does not agree with expression (7.85) for the free energy calculated in correlation theory with the aid of the modified Bogolyubov variational principle, viz., the last term of Eq. (7.106) calculated by perturbation theory is higher than the corresponding term in Eq. (7.85).

45. Temperature dependences of the lattice constant, of the expansion coefficient, and of the isothermal elastic modulus

We start with the thermal equation of state (7.63) of the crystal and with expression (7.85) for its free energy.

Equation (7.63) takes the form

$$p = p_{\mathrm{SCF}} - \frac{1}{6V}\int |q_1 - q_2|\Phi'(|q_1 - q_2|)\{\rho_2(q_1,q_2) - \rho_1(q_1)\rho_1(q_2)\}dq_1\, dq_2. \quad (7.107)$$

Substituting here expression (7.51) for $\rho_2(q_1,q_2)$ we get

$$p = p_{\mathrm{SCF}} + \frac{1}{6V\theta}\left\{\int |q_1 - q_2|\Phi'(|q_1 - q_2|)\tilde{\Phi}(|q_1 - q_2|) \times \rho_1^0(q_1)\rho_1^0(q_2)dq_1\, dq_2 - \frac{1}{2N}\left[\int |q_1 - q_2|\Phi'(|q_1 - q_2|) \times \rho_1^0(q_1)\rho_1^0(q_2)dq_1\, dq_2\right]\left[\int \tilde{\Phi}(|q_1 - q_2|)\rho_1^0(q_1)\rho_1^0(q_2)dq_1\, dq_2\right]\right\}. \quad (7.108)$$

We include hereafter anharmonicities up to fourth-order inclusive, in analogy with the expression for the free energy, and obtain

$$p = p_{\mathrm{SCF}} + \frac{\beta}{3V}\{\overline{f_1\tilde{\Phi}_3} + \overline{f_3\tilde{f}_3} + \overline{f_2\tilde{\Phi}_2} + \overline{f_4\tilde{\Phi}_4} + \overline{f_2\tilde{\Phi}_4} - \overline{\tilde{f}_2}\ \overline{\tilde{\Phi}_4}\}, \quad (7.109)$$

where f_1, f_2, f_3, f_4 are the terms in the Taylor expansion of the function

$$f(|q_1 - q_2|) = |q_1 - q_2|\Phi'(|q_1 - q_2|). \quad (7.110)$$

From Eq. (7.109) we can determine a as a function of the temperature and of the pressure: $a = a(\theta,p)$. If p_{SCF} is chosen in an approximation that takes account of anharmonicities up to fourth-order inclusive, we can write

$$p = p_2 = -\frac{1}{3V}\left\{\overline{f_4} - \frac{1}{\theta}[\overline{f_1\tilde{\Phi}_3} + \overline{f_3\tilde{\Phi}_3} + \overline{f_4\tilde{\Phi}_4} + \overline{f_2\tilde{\Phi}_2} + \overline{f_2\tilde{\Phi}_4} - \overline{f_2\tilde{\Phi}_4}]\right\}, \quad (7.111)$$

where p_2 is the pressure calculated for the terms of the expansion of the function $|q_1 - q_2|\Phi'(|q_1 - q_2|)$ up to second-order inclusive.

It can be seen from Eq. (7.111) that allowance for the correlation decreases the anharmonic terms. In addition, in our case anharmonicities of fourth order are also taken into account in p_2, since the averaging is over a distribution function calculated with account taken of anharmonicities up to fourth-order inclusive.

An expression for the pressure in the quasiharmonic approximation was determined in Ref. 33 on the basis of the free energy of the system. We show now that in our case p_2 has the same form in the same approximation. To this end we put $b = \infty$ and neglect fourth-order effects.

The expression for p_2 is

$$p_2 = \frac{\theta}{v} - \frac{1}{6v}\left\{\sum_k z_k f(\nu_k a) + \sum_k \left[f''(\nu_k a) + \frac{2}{\nu_k a} f(\nu_k a)\right]\Delta_2\right\}, \quad (7.112)$$

where

$$\Delta_2 = \frac{\theta}{(1/3)\Sigma_k z_k [\Phi''(\nu_k a) + (2/\nu_k a)\Phi'(\nu_k a)]}. \quad (7.113)$$

We choose $\Phi(r)$ to be a Lennard-Jones potential

$$\Phi(r) = 4\varepsilon\left[\left(\frac{\sigma}{r}\right)^{12} - \left(\frac{\sigma}{r}\right)^{6}\right],$$

where ϵ and σ are constants. Then

$$\sum_k z_k f(\nu_k a) = 24\epsilon\left(\frac{\sigma}{a}\right)^6\left[-2A_{12}\left(\frac{\sigma}{a}\right)^6 + A_6\right], \quad (7.114)$$

$$\Delta_2 = \frac{\theta\sigma^2}{8\varepsilon(\sigma/a)^8[22A_{14}(\sigma/a)^6 - 5A_8]}, \quad (7.115)$$

where $A_6 = 14.454$; $A_8 = 12.802$; $A_{12} = 12.132$; $A_{14} = 12.059$.

Substituting Eqs. (7.114) and (7.115) in Eq. (7.112) we obtain[33]

$$p = \frac{\theta}{v}\frac{20(a/\sigma)^6 - 154(A_{14}/A_8)}{5(a/\sigma)^6 - 22(A_{14}/A_8)} - \frac{1}{v}4\varepsilon A_6\frac{[(a/\sigma)^6 - 2(A_{12}/A_6)]}{(a/\sigma)^{12}} \quad (7.116)$$

or

$$p = \frac{2R}{V}\left(\frac{10CV^2 - 77\alpha}{5CV^2 - 22\alpha}T - \frac{2A_6\varepsilon}{k}\frac{CV^2 - 2\beta}{C^2V^4}\right), \quad (7.117)$$

where V is the molar volume and

$$C = \frac{2}{N^2\sigma^6}, \quad CV^2 = \left(\frac{a}{\sigma}\right)^6,$$

$$\alpha = \frac{A_{12}}{A_8}, \quad \beta = \frac{A_{12}}{A_6}.$$

Note that the succeeding terms for p, calculated from the expression for F and from Eq. (7.107), will, generally speaking, not be equal.

As already noted, it is more natural to determine the pressure by using in the thermal equation of state an approximate expression for the binary distribution function than to start out from the approximate expression for the pressure p, for in the latter case the differentiation lowers the accuracy.

Whereas the determination of p by the latter method can still be justified, the calculation of the isothermal elastic modulus, which is expressed in terms of the second derivative of F, leads to a substantial loss of accuracy. The same applies also to the determination of the heat capacity C_v by using F.

Lattice constant

Let us examine expression (7.109) in greater detail. It yields for a static lattice

$$\frac{pv}{4\varepsilon} = 2\left(\frac{\sigma}{a}\right)^{12} A_{12} - A_6\left(\frac{\sigma}{a}\right)^6, \tag{7.118}$$

whence

$$\left(\frac{\sigma}{a}\right)^6 = \frac{A_6 + \sqrt{A_6^2 + 2(pv/\varepsilon)A_{12}}}{4A_{12}} \tag{7.119}$$

and

$$a = \sigma\sqrt{\frac{4A_{12}}{A_6 + \sqrt{A_6^2 + 2(pv/\varepsilon)A_{12}}}}. \tag{7.120}$$

It is known that a changes very little when the temperature is raised from 0 K to the melting point (T_m). The terms supplementing Eq. (7.20) in the expression for a can therefore be regarded at $0 < T < T_m$ as corrections.

From Eq. (7.109) we have, in the general case,

$$\begin{aligned} a = \sigma\sqrt[6]{4A_{12}/\{A_6 + \sqrt{A_6^2 + 8A_{12}(pv/4\varepsilon - \theta/4\varepsilon}} \\ \overline{\overline{+ 3(\sigma/a)^8[-44(\sigma/a)^6 A_{14} + 5A_8]\Delta_2/\sigma^2)}} \\ \overline{\overline{- 1/12\varepsilon N(\overline{f_1\tilde{\Phi}_3} + \overline{f_3\tilde{\Phi}_3} + \overline{f_2\tilde{\Phi}_2} + \overline{f_2\tilde{\Phi}_4} - \overline{\overline{f}_2\tilde{\Phi}}_4 - \theta f_4)\}}}. \end{aligned} \tag{7.121}$$

Taking the foregoing remark into account, we determine a from Eq. (7.121) by successive approximations, using Eq. (7.120) as the zeroth approximation.

The dependence of $a = a/\sigma$ on the temperature at normal pressure ($p \approx 0$) is determined by curve 1 of Fig. 37. The figure shows also plots of a vs T at $p = 0$, obtained from Eq. (7.121) without allowance for the correlations (by the self-consistent-field method) in the quasiharmonic approximation (curve SCF-2):

$$\begin{aligned} a = \sigma\sqrt[6]{4A_{12}/\left[A_6 + \sqrt{A_6^2 - 8A_{12}(\theta/4\varepsilon}\right.} \\ \overline{\overline{+ 3(\sigma/a)^8[44A_{14}(\sigma/a)^6 - 5A_8]\Delta_2/\sigma^2)]}} \end{aligned} \tag{7.122}$$

and in an approximation that takes the anharmonicities into account up to fourth-order inclusive (curve SCF-4):

$$\begin{aligned} a = \sigma\sqrt[6]{4A_{12}/\left[A_6 + \sqrt{A_6^2 - 8A_{12}(\theta/4\varepsilon}\right.} \\ \overline{\overline{+ 3(a/\sigma)^8[44A_{14}(\sigma/a)^6 - 5A_6]\Delta_2/\sigma^2) + \overline{f}^4/12\varepsilon N]}}, \end{aligned} \tag{7.123}$$

and also with allowance for the correlations in the approximation of the anharmonicities of third- and fourth-order inclusive (plot CV).

Comparing these plots, we see the following: (1) In the self-consistent-field method, at temperatures $\sim T_m$, the quasiharmonic approximation overestimates a, and the fourth-order-anharmonicity approximation underesti-

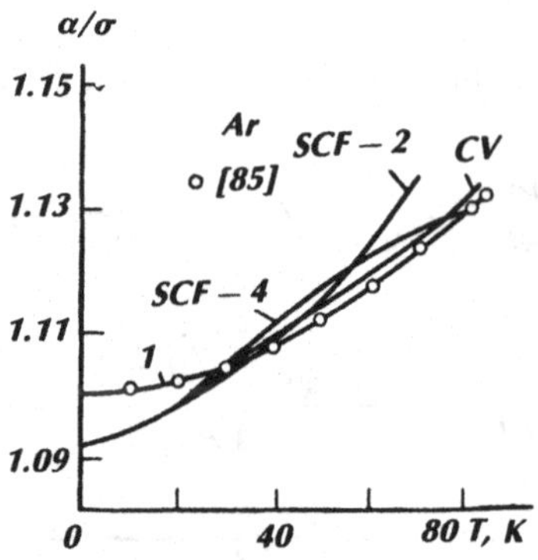

Figure 37.

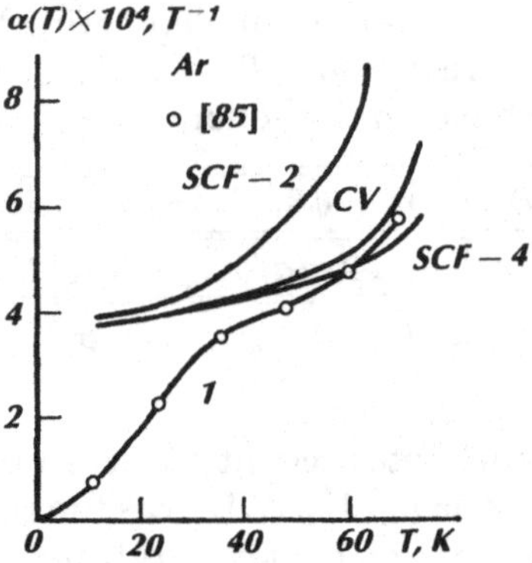

Figure 38.

mates a (the approximation with harmonicities of even higher order increases the deviation from experiment); (2) allowance for correlations improves the agreement between the theoretical curve and the experimental data for the temperature dependence of the lattice constant.

Expansion coefficient

The expression for the linear expansion coefficient is

$$\alpha(T) = \frac{1}{a(0)} \cdot \frac{da(T)}{dT}. \tag{7.124}$$

Differentiating Eqs. (7.121), (7.122), and (7.123) with respect to T, we obtain from Eq. (7.124) expressons for $\alpha(T)$ in the corresponding approximations.

Figure 38 shows the experimental curve (1) and the theoretical curves for the different approximations. In the quasiharmonic approximation we obtain an overestimate of $\alpha(T)$ (curve SCF-2); the fourth-order-anharmonicity approximation improves somewhat the agreement with experiment (curve SCF-4). When the correlations are taken into account the agreement with experiment becomes much better (curve CV).

Isothermal elastic modulus

By definition, the isothermal elastic modulus is

$$\varepsilon_T = -v\left(\frac{\partial p}{\partial v}\right)_T = -\frac{a}{3}\left(\frac{\partial p}{\partial a}\right)_T. \tag{7.125}$$

Using Eq. (7.109), we get

$$\varepsilon_T = p - \frac{a}{3}\left(\frac{\partial(vp_2)}{\partial a}\right)_T - \frac{a}{vN}\frac{\partial}{\partial a}$$
$$\times\Big\{\overline{f}_4 - \frac{1}{\theta}\,[\,\overline{f_1\tilde{\Phi}_3} + \overline{f_3\tilde{\Phi}_3} + \overline{f_4\tilde{\Phi}_2} + \overline{f_4\tilde{\Phi}_2}$$
$$+ + \overline{f_2\tilde{\Phi}_2} + \overline{f_2\tilde{\Phi}_2} + \overline{f_2\tilde{\Phi}_4} - \overline{f}_2\,\overline{\tilde{\Phi}}_4]\Big\}_T, \tag{7.126}$$

whence, in the quasiharmonic approximation,

$$\varepsilon_T = p - \frac{a}{3}\left(\frac{\partial(vp_2)}{\partial a}\right)_T, \tag{7.127}$$

and in the approximation of the fourth-order anharmonicity of the self-consistent-field method

$$\varepsilon_T = p - \frac{a}{3}\left(\frac{\partial(vp_2)}{\partial a}\right)_T - \frac{a}{vN}\left(\frac{\partial}{\partial a}\overline{f}_4\right)_T. \tag{7.128}$$

The experimental data and the numerical results are shown in Fig. 39. We see that the fourth-order-anharmonicity approximation (SCF-4 curve) improves substantially the values of the isothermal elastic modulus compared with the results of the quasiharmonic approximation.

Allowance for the correlation decreases the contribution of the fourth-order anharmonicities, although the curve lies higher than in the case of the quasiharmonic self-consistent-field approximation.

46. Heat capacity of inert-gas crystals

The heat capacity C is a caloric properties of matter:

$$C = \frac{\delta Q}{dT}.$$

Any body has an infinite number of heat capacities ($-\infty < C < \infty$). The most generally used are the heat capacities C_p (at constant pressure) and C_V (at constant volume). C_p is measured directly in experiment, but C_V is easier to

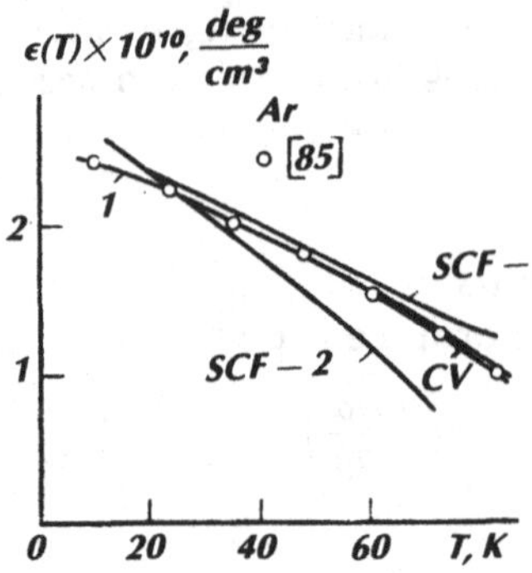

Figure 39.

calculate in theory. These heat capacities are connected by the thermodynamic relations

$$C_p - C_V = \frac{TV\alpha^2}{\gamma}, \tag{7.129}$$

where α is the bulk expansion coefficient and γ is the isothermal compressibility of the material.

The application of classical statistics to the problem of heat capacitance of crystalline solids in the harmonic approximation leads to the empirically known law of Dulong and Petit for monatomic crystals:

$$C_V = 3R = 6 \text{ cal/mol deg} \equiv C_0.$$

In the harmonic theory, $\alpha = 0$ and γ is independent of temperature. Therefore,

$$C_p = C_V.$$

The harmonic classical approximation can be used in the classical approximation in the theory of heat capacity in view of the satisfaction of two conditions.

(1) $T \gg T_D$, where T_D is the Debye temperature (see Table XVI).

(2) The anharmonic contributions to heat capacity are small.

Neither condition is met for inert-gas crystals (see the experimental data on the heat capacities of these crystals at normal pressure in Tables XVII–XX, where $C_1 = C/C_0$ is the ratio of the heat capacity $C = C_p$ to the heat capacity in accordance with the law of Dulong and Petit, $C_2 = C - C_D/C_D$ is the relatively small contribution to the heat capacity, calculated by the Debye formula, and $C_3 = C_0 - C_D/C_D$ is a quantity that describes the contribution of the quantum corrections to the heat capacity).

It is seen from Tables XVII–XX that, first, inert gases are strongly anharmonic and, second, the classical theory of heat capacitance cannot be applied to Ne and Ar, but can be used for Kr and Xe crystals at $T > T_D$.

Figure 40 shows the temperature dependences of C_2 and C_3 for a Kr crystal (see Table XIX).

Table XVI. Caloric properties of inert gases.[196]

Element	T_D (K)	C_0 (cal/g deg)
Ne	65	0.282
Ar	93	0.149
Kr	95	0.071
Xe	72	0.075

Table XVII. Heat capacity of Ne.[101]

T (K)	C_D (cal/g deg)	$C_1 = \frac{C_0}{C_0}$	$C_2 = \frac{C - C_D}{C_D}$	$C_3 = \frac{C_0 - C_D}{C_D}$
1	2	3	4	5
11	0.0866	0.29	0.05	2.64
12	0.1000	0.34	0.13	2.33
14	0.1282	0.43	0.05	1.44
16	0.1564	0.53	0.08	1.04
18	0.1850	0.62	0.12	0.82
20	0.2143	0.72	0.16	0.61
22	0.2495	0.84	0.25	0.49
23.5	0.2886	0.97	0.36	0.41

Table XVIII. Heat capacity of Ar.[104]

1	2	3	4	5
10	0.0198	0.13	0.03	12.33
15	0.0486	0.32	0.42	3.44
20	0.0749	0.50	0.33	1.66
25	0.0960	0.64	0.56	1.44
30	0.1118	0.74	0.16	0.56
35	0.1257	0.83	0.17	0.41
40	0.1348	0.90	0.17	0.29
45	0.1429	0.95	0.17	0.28
50	0.1502	1.00	0.19	0.19
55	0.1571	1.05	0.21	0.15
60	0.1633	1.09	0.22	0.12
65	0.1698	1.13	0.26	0.11
70	0.1778	1.19	0.31	0.09
75	0.1872	1.25	0.36	0.087
80	0.1985	1.32	0.42	0.075
83	0.2083	1.39	0.48	0.064

Table XIX. Heat capacity of Kr.[105,106]

1	2	3	4	5
10	0.169	0.24	−0.08	3
15	0.0334	0.47	−0.15	1.5
20	0.0455	0.62	−0.11	0.82
25	0.0539	0.75	−0.04	0.62
30	0.0596	0.83	−0.01	0.18
35	0.0638	0.89	0.01	0.14
40	0.0671	0.94	0.03	0.10
45	0.0694	0.97	0.05	0.09
50	0.0715	1.00	0.08	0.08
55	0.0734	1.03	0.08	0.06
60	0.0751	1.05	0.11	0.05
65	0.0772	1.08	0.14	0.05
70	0.0785	1.10	0.16	0.05
75	0.0802	1.12	0.18	0.05
80	0.0815	1.14	0.20	0.05
85	0.0833	1.16	0.21	0.04
90	0.0853	1.19	0.24	0.04
100	0.0906	1.27	0.32	0.04
110	0.0972	1.36	0.42	0.03
113	0.1022	1.43	0.49	0.03

Table XX. Heat capacity of Xe.[197]

1	2	3	4	5
10	0.015 23	0.33	0	2.03
15	0.024 98	0.55	0.37	1.5
20	0.031 60	0.69	0.21	0.75
25	0.036 17	0.79	0.16	0.47
30	0.030 0	0.85	0.10	0.29
40	0.043 3	0.95	0.11	0.17
60	0.046 9	1.03	0.11	0.07
80	0.049 4	1.08	0.13	0.04
100	0.051 8	1.13	0.18	0.04
120	0.054 8	1.20	0.24	0.03
140	0.058 6	1.28	0.32	0.03
158	0.547	1.42	0.45	0.02

The values of T_D listed in Table XVI (Ref. 190) correspond to $T=0$. When the condition $T>T_D$ is used, account must be taken of the temperature dependence of T_D. It must be borne in mind, however, that the experimental values of T_D have a rather appreciable scatter as functions of the temperature (see Fig. 41, which shows the values of T_D measured for Ar by different workers[191–193] at various temperatures).

To calculate the heat capacity of crystals by the correlation theory we begin with the expression obtained for the internal energy

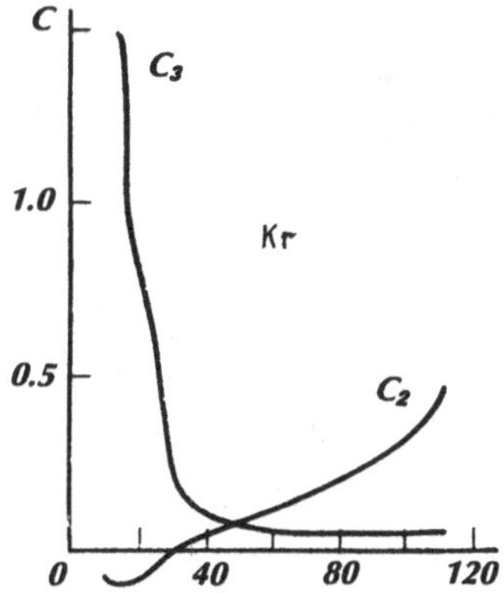

Figure 40.

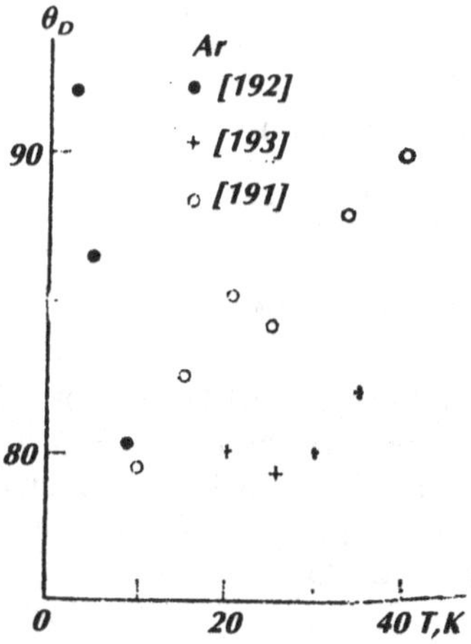

Figure 41.

$$
\begin{aligned}
E = {} & \frac{3}{2}N\theta + \frac{1}{2}\int \Phi(|q_1 - q_2|)\rho_1(q_1)\rho_1(q_2)dq_1\,dq_2 \\
& - \frac{\beta}{2}\left\{\int \Phi(|q_1 - q_2|)\rho_1(q_1)\rho_1(q_2)dq_1\,dq_2 \right. \\
& \left. - \frac{1}{2N}\left[\int \Phi(|q_1 - q_2|)\rho_1(q_1)\rho_1(q_2)dq_1\,dq_2\right]^2\right\}.
\end{aligned}
$$

Accurate to anharmonicities of fourth order we have

$$
\begin{aligned}
E = E_0 + \overline{\Phi}_4 - \beta\,(& \overline{\Phi_3\tilde{\Phi}_3} + \overline{\Phi_1\tilde{\Phi}_3} \\
& + \overline{\tilde{\Phi}_2^2} + \overline{\tilde{\Phi}_2\Phi_4} + \overline{\Phi_2\tilde{\Phi}_4} - \overline{\Phi_2}\;\overline{\tilde{\Phi}_4}),
\end{aligned}
\tag{7.130}
$$

where

$$E_0 = \frac{N}{2}\sum_k z_k \Phi(\nu_k a) + \overline{\Phi}_2, \tag{7.131}$$

where z_k is the number of neighbors on the kth sphere.[33]

We now express Eq. (7.130) in the form

$$E = E^0 + E_4 + E_4^{kk}, \tag{7.132}$$

where

$$E^0 = E_0 - \beta(\overline{\tilde{\Phi}_2^2} + \overline{\Phi_1\tilde{\Phi}_3}),$$

$$E_4 = \overline{\Phi}_4,$$

$$E_4^{kk} = -\beta(\overline{\Phi_1\tilde{\Phi}_3} + \overline{\tilde{\Phi}_2\Phi_4} + \overline{\Phi_2\tilde{\Phi}_4} - \overline{\Phi_2}\,\overline{\tilde{\Phi}_4}).$$

Consequently,

$$C_V = C_0 + C_4 + C_4^{kk}, \tag{7.133}$$

where

$$C_0 = \left(\frac{\partial E^0}{\partial T}\right)_V, \quad C_4 = \left(\frac{\partial E_4}{\partial T}\right)_V, \quad C_4^{kk} = \left(\frac{\partial E_4^{kk}}{\partial T}\right)_V.$$

Calculations yield the following results:

$$C_4 \approx \frac{2\theta}{\varepsilon}, \quad C_4^{kk} \approx -1.36\frac{\theta}{\varepsilon}, \tag{7.134}$$

$$C_4 + C_4^{kk} \approx 0.64\frac{\theta}{\varepsilon}. \tag{7.135}$$

We see thus that in the self-consistent-field approximation the anharmonic corrections overestimate the heat capacity. Allowance for the correlations decreases the contribution made to the heat capacity by the harmonic corrections (in the self-consistent approximation the contribution of the fourth-order anharmonicities to the heat capacity reaches 45%, as against 15% when the correlations are taken into account). Some deviation of C_0 from the Dulong and Petit law is due to allowance for the correlations in only the first approximation. This deviation can be eliminated by the method of partial summation of the harmonic terms (see Sec. 10).

47. Absolute stability limit of the crystalline state

An equation for the limit of the absolute stability of face-centered-cubic crystals was obtained in Ref. 198 in the self-consistent-field approximation. It was found that the plot of this limit is close to the melting curve. We obtain here the absolute-stability limit of the crystalline state by correlation theory. The limit of absolute stability to compression is defined by the equation

$$\left(\frac{\partial p}{\partial V}\right)_T = 0.$$

Element	Σ (K)	$\sigma \times 10^8$ cm
Ne	36.3	2.82
Ar	119.3	3.45
Kr	159	3.98
Xe	228	3.61

From Eqs. (7.126), (7.127), and (7.128) we obtain the equations for the crystal absolute-stability limit under the corresponding approximations:

$$p = \frac{a}{3}\left(\frac{\partial(vp_2)}{\partial a}\right)_T + \frac{a}{vN}\Big\{\overline{f_4} - \frac{1}{\theta}\big[\,\overline{f_1\tilde{\Phi}_3} + \overline{f_3\tilde{\Phi}_3} + \overline{f_4\tilde{\Phi}_2} + \overline{f_2\tilde{\Phi}_2} + \overline{f_2\tilde{\Phi}_4} - \overline{f_2}\,\overline{\tilde{\Phi}_4}\big]\Big\}, \tag{7.136}$$

$$p = \frac{a}{3}\left(\frac{\partial(vp)}{\partial a}\right)_T, \tag{7.137}$$

$$p = \frac{a}{3}\left(\frac{\partial(vp_2)}{\partial a}\right)_T + \frac{a}{vN}\left(\frac{\partial \overline{f_4}}{\partial a}\right). \tag{7.138}$$

Equation (7.136) must be considered here jointly with Eq. (7.121), Eq. (7.137) jointly with Eq. (7.122), and Eq. (7.138) jointly with Eq. (7.123). It must be recognized that for all face-centered crystals we have

$$v = \frac{\sqrt{2}}{2}\,a^3.$$

Following Ref. 198, we calculate the values of the temperature T_0 of the limit of the absolute stability of inert-gas crystals compared with Eq. (7.137) of the correlation theory, and compare them with the experimentally known melting temperatures of these crystals at various pressures (see Tables XXI and XXII). The parameters ε and σ for inert gases are given below.

We see that the absolute-stability temperature limit calculated by correlation theory for a flat crystal surface is quite close to the melting temperature.

48. Chain of Bogolyubov for the correlation-density matrices, its uncoupling, and its solution

We consider a system of N identical particles with the Hamiltonian

$$H = \sum_{1 \leqslant i \leqslant N} H(i) + \sum_{1 \leqslant i_1 < i_2 \leqslant N} \Phi(i_1, i_2), \tag{7.139}$$

where

Table XXI. Melting temperatures of Ne and Ar at various pressures.

Ne			Ar		
p (kg/cm) 1	T_0 (K) 2	T_m (K) 3	p (kg/cm) 4	T_0 (K) 5	T_m (K) 6
0	25.5	24.4	0	86.0	83.8
140	27.8	26.8	600	98.5	96.6
350	30.7	29.5	620	104.8	97.5
440	31.7	30.1	920	105.8	102.5
490	32.2	31.3	950	109.2	102.5
620	33.1	31.8	1160	112.4	107.2

Table XXII. Melting temperatures of Kr and Xe at various pressures.

Kr			Xe		
p (kg/cm) 1	T_0 (K) 2	T_m (K) 3	p (kg/cm) 4	T_0 (K) 5	T_m (K) 6
5.5	118.0	115.9	5.5	164.0	161.6
110.5	121.8	119.1	147.3	170.4	167.1
259.3	126.2	123.5	259.2	175.6	171.5
297.4	137.0	133.2	596.2	188.6	184.0
794.1	143.0	138.6	794.1	196.2	191.1
1129	152.8	147.6	984	203.3	198.8
1251	156.9	150.8	1255	219.2	207.2

$$H(i) = -\frac{\hbar^2}{2m}\nabla^2_{r_i} + u(r_i), \tag{7.140}$$

$$\Phi(i_1,i_2) = \Phi(|r_{i_1} - r_{i_2}|), \tag{7.141}$$

$\Phi(r)$ is the pair-interaction potential and $u(r)$ is the potential of the external field, which we assume hereafter to be zero. The chain of equilibrium Bogolyubov equations for quantum systems with pair interaction is of the form[2]

$$\begin{aligned}&[H_s(1,...,s),R_s(1,...,s)]\\ &\quad + \operatorname*{Tr}_{(s+1)}\left[\sum_{1\leq i\leq s}\Phi(i,s+1),R_{s+1}(1,...,s+1)\right] = 0,\end{aligned} \tag{7.142}$$

where

$$H_s(1,...,s) = \sum_{1\leq i\leq s} H(i) + \sum_{1\leq i_1<i_2\leq s}\Phi(i_1,i_2) \tag{7.143}$$

is the Hamiltonian of an isolated system of s particles; $R_s(1,...,s)$ are statistical operators defined in terms of the trace over the subspace of the total statistical operator ρ:

$$R_s(1,\ldots,s) = \frac{N!}{(N-s)!} \operatorname*{Tr}_{(s+1,\ldots,N)} \rho(1,\ldots,N) \tag{7.144}$$

and satisfy the normalization conditions

$$\operatorname*{Tr}_{1,\ldots,s} R_s(1,\ldots,s) = \frac{N!}{(N-s)!} \,. \tag{7.145}$$

We transform from operators of aggregates of particles $R_1(1)$, $R_2(1,2),\ldots,R_s(1,2,\ldots,s)$ to correlation operators $G_2(1,2)$, $G_3(1,2,3),\ldots,G_s(1,2,\ldots,s)$ with the aid of the relations[2]

$$\begin{aligned}
R_2(1,2) &= R_1(1)R_1(2) + G_2(1,2),\\
R_3(1,2,3) &= R_1(1)R_1(2)R_1(3) + R_1(1)G_2(2,3)\\
&\quad + R_1(2)G_2(1,3) + R_1(3)G_2(1,2) + G_3(1,2,3),\\
R_4(1,2,3,4) &= R_1(1)R_1(2)R_1(3)R_1(4)\\
&\quad + R_1(1)R_1(2)G_2(3,4) + R_1(1)R_1(3)G_2(2,4)\\
&\quad + R_1(1)R_1(4)G_2(2,3) + R_1(2)R_1(3)G_2(1,4)\\
&\quad + R_1(2)R_1(4)G_2(1,3) + R_1(3)R_1(4)G_2(1,2)\\
&\quad + R_1(1)G_3(2,3,4) + R_1(2)G_3(1,3,4)\\
&\quad + R_1(3)G_3(1,2,4) + R_1(4)G_3(1,2,3)\\
&\quad + G_2(1,2)G_2(3,4) + G_2(1,3)G_2(2,4)\\
&\quad + G_2(1,4)G_2(2,3) + G_4(1,2,3,4).
\end{aligned} \tag{7.146}$$

Substituting Eq. (7.146) in Eq. (7.142) and retaining, on grouping the terms, the correlation matrices only under the trace sign, we obtain the following system of equations for R_1, G_2, R_2, R_3, G_3, G_4:

$$\begin{aligned}
&[\widetilde{H}_1(1),R_1(1)] + \operatorname*{Tr}_2[\Phi(1,2),G_2(1,2)] = 0,\\
&[\widetilde{H}_2(1,2),R_2(1,2)] + \operatorname*{Tr}_3[\Phi(1,3) + \Phi(2,3),G_2(2,3)R_1(1)\\
&\qquad + R_1(2)G_2(1,3) + G_3(1,2,3)] = 0,\\
&[\widetilde{H}_3(1,2,3),R_3(1,2,3)] + \operatorname*{Tr}_4[\Phi(1,4) + \Phi(2,4)\\
&\qquad + \Phi(3,4),R_2(1,2)G_2(3,4) + R_2(2,3)G_2(1,4)\\
&\qquad + R_2(1,3)G_2(2,4) + R_1(1)G_3(2,3,4) + R_1(2)G_3(1,3,4)\\
&\qquad + R_1(3)G_3(1,2,4) + G_4(1,2,3,4)] = 0,
\end{aligned} \tag{7.147}$$

where

$$\begin{aligned}
\widetilde{H}_1(1) &= H_1(1) + \operatorname*{Tr}_2 \Phi(1,2)R_1(2),\\
H_2(1,2) &= H_2(1,2) + \operatorname*{Tr}_3[\Phi(1,3) + \Phi(2,3)]R_1(3),\\
\widetilde{H}_s(1,\ldots,s) &= H_s(1,\ldots,s) + \operatorname*{Tr}_{s+1}\left[\sum_{1\le i\le s} \Phi(i,s+1)R_1(s+1)\right].
\end{aligned} \tag{7.148}$$

Just as in the classical case, we exclude here the self-action of the particles. Expressions (7.14) impose the following conditions on the correlation density matrices:

$$\operatorname*{Tr}_{1,\ldots,s} G_s(1,\ldots,s) = 0. \tag{7.149}$$

To uncouple the chain of equations (7.147), just as for classical systems, we assume in the zeroth approximation that the traces of the correlation matrices (correlation traces) are small compared with the remaining terms of the equations (the second solution method).

We have thus from Eq. (7.147), in the zeroth approximation,

$$\begin{aligned}
[\tilde{H}_1(1),R_1(1)] &= 0,\\
[\tilde{H}_2(1,2),R_2(1,2)] &= 0,\\
[\tilde{H}_3(1,2,3),R_3(1,2,3)] &= 0,\\
[\tilde{H}_s(1,\ldots,s),R_s(1,\ldots,s)] &= 0.
\end{aligned} \tag{7.150}$$

The commutation relation for an operator of the Gibbs type

$$D = \frac{\exp[-\beta H]}{\operatorname*{Tr}_{(1,\ldots,N)} \exp[-\beta H]} \tag{7.151}$$

and the Hamiltonian H is, obviously, of the form

$$[H,D] = 0. \tag{7.152}$$

Since we are considering a statistical system, we are interested in a Gibbs-type solution of Eqs. (7.150). Comparing Eqs. (7.150), (7.151), and (7.152), we obtain

$$R_1(1) = \exp[-\beta\tilde{H}_1(1)]/Q_1, \tag{7.153}$$

$$R_2(1,2) = \exp[-\beta\tilde{H}_2(1,2)]/Q_2, \tag{7.154}$$

$$R_3(1,2,3) = \exp[-\beta\tilde{H}_3(1,2,3)]/Q_3, \tag{7.155}$$

$$R_s(1,\ldots,s) = \exp[-\beta\tilde{H}_s(1,\ldots,s)]/Q_s, \tag{7.156}$$

where

$$Q_\alpha = \operatorname*{Tr}_{1,\ldots,\alpha} \exp[-\beta\tilde{H}_\alpha(1,\ldots,\alpha)]/N^\alpha. \tag{7.157}$$

To find the thermodynamic properties of the system we must know also the unary and binary density matrices. It is therefore necessary to solve only the first two equations of Eq. (7.150), i.e., to stipulate knowledge of the analytic forms of the density matrices (7.153) and (7.154).

Taking into account the periodic structure of the crystal, we introduce statistical operators whose matrix elements differ from zero only in the vicinity of a lattice site with coordinate a_i:

$$R_1(q_1,q_2') = \sum_{1\cdot i_1\cdot N-1} R_1(q_1 - a_{i_1}, q_1' - a_{i_1}). \tag{7.158}$$

It is easily seen that expression (7.154) obtained for $R_2(1,2)$ satisfies the correlation-decay condition

$$R_2(1,2)\to R_1(1)R_1(2) \tag{7.159}$$

as $|q_1-q_2|\to\infty$.

We neglect exchange effects in our analysis. The accuracy of this procedure is quite high for an overwhelming majority of crystals, and is greatly beyond the experimental capabilities.[200]

Various methods of determining the coordinate representation for a unitary density, starting from the operator equation (7.150), are described in Ref. 58. Thus, the expression for the diagonal elements of the non-normalized density matrix takes in the harmonic approximation the form

$$R_1(q,q)=\frac{m^{3/4}}{(2\pi\hbar)^{3/2}}\left\{\prod_{\alpha=1}^{3}\lambda_\alpha^{1/4}\left[\sinh\left(\beta\hbar\sqrt{\frac{\lambda_\alpha}{m}}\right)\right]^{-1/2}\right\}$$
$$\times\exp\left[-\beta u(0)-\frac{m^{1/2}}{\hbar}\sum_{\alpha=1}^{3}\lambda_\alpha^{1/2}\tanh\left(\frac{\beta\hbar}{2}\sqrt{\frac{\lambda_\alpha}{m}}\right)q^{\alpha^2}\right], \tag{7.160}$$

where m is the particle mass, λ_α are the diagonal elements of the matrix

$$\left\|\left(\frac{\partial^2u(r)}{\partial r^\alpha\partial r^\beta}\right)_{r=a_i}\right\|, \tag{7.161}$$

and $n(r)$ is the self-consistent potential.

Using the Bogolyubov variational principle

$$F=F_0+\left\langle U_N-\sum_{i=1}^{N}u_i\right\rangle_0, \tag{7.162}$$

where $\langle\cdots\rangle$ denotes averaging over the self-consistent field, we obtain an expression for the crystal free energy per particle in the quasiharmonic approximation:

$$F/N=\beta^{-1}\sum_{\alpha=1}^{3}\ln\left[2\sinh\left(\frac{\beta\hbar}{2}\sqrt{\frac{\lambda_\alpha}{m}}\right)\right]+E, \tag{7.163}$$

where E is the static energy

$$E=\frac{1}{2N}\sum_{i\neq j}\Phi(|a_i-a_j|)=\frac{1}{2}\sum_{i=1}^{N-1}\Phi(|a_i|). \tag{7.164}$$

At high temperatures ($\beta\ll\sqrt{m/\lambda_\alpha}/\hbar$) this equation takes the classical form

$$F/N=\tfrac{1}{2}\,\theta\sum_\alpha\ln\lambda_\alpha-3\theta\ln(\theta\sqrt{m}/\hbar)+E. \tag{7.165}$$

Note that the quantities λ_α and E depend on the crystal lattice parameters, and through them on the temperature and pressure. We obtain now a coordinate representation for the binary density matrix in the quasiharmonic approximation. To this end we assume, when determining the positions of the two particles in lattice sites, that the first and second axes are parallel while the third is directed along the one joining the sites. By the change of variables

$$
\begin{aligned}
q_1 &= t_1 + t_2, \\
q_2 &= t_1 - t_2,
\end{aligned}
\tag{7.166}
$$

the operator expression for the binary density matrix becomes in the quasiharmonic approximation a product of two unitary density matrices whose coordinate representations are known.[58] In fact,

$$
R_2(1,2) = R_{11}(1)R_{12}(2),
$$

where

$$
R_{1i}(i) = \frac{\exp[-\beta H_{1i}]}{\mathrm{Tr}\exp[-\beta H_{1i}]},
$$

$$
\begin{aligned}
H_{1i} = {} & -\frac{\hbar^2}{4m}\nabla^2_{q_i} + m\omega^2 \sum_{j=1}^{2} t_i^{j2} + (-1)^i \Phi''(\nu_k a)(t_i^3)^2 \\
& + (-1)^i \frac{1}{\nu_k a}\Phi'(\nu_k a)[(t_i^1)^2 + (t_i^2)^2] \quad (i = 1,2).
\end{aligned}
$$

Returning to the old variables, we have

$$
R_2(q_1,q_2,q_1',q_2') = \prod_{1 \le i \le s} R_2(q_1^i,q_1'^i,q_2^i,q_2'^i),
\tag{7.167}
$$

$$
\begin{aligned}
R_2(q_1^i,q_1'^i,q_2^i,q_2'^i) = {} & \left[\frac{m^2\omega_{1i}\omega_{2i}}{\pi^2\hbar^2}\tanh\left(\frac{\hbar\omega_i}{2}\beta\right)\tanh\left(\frac{\hbar\omega_{2i}}{2}\beta\right)\right]^{1/2} \\
& \times\exp\left\{-\frac{\alpha_{1i}}{4}[(q_1^i + q_1'^i)^2 + (q_2^i + q_2'^i)^2]\right. \\
& -\frac{\alpha_{2i}}{4}[(q_1^i - q_1'^i)^2 + (q_2^i - q_2'^i)^2] \\
& \left.+\frac{\gamma_{1i}}{2}[(q_1^i + q_1'^i)(q_2^i + q_2'^i)] + \frac{\gamma_{2i}}{2}[(q_1^i - q_1'^i)(q_2^i - q_2'^i)]\right\},
\end{aligned}
\tag{7.168}
$$

where

$$\alpha_{1i} = \frac{m}{2\hbar}\left[\omega_{1i}\tanh\frac{\hbar\omega_{1i}}{2}\beta + \omega_{2i}\tanh\frac{\hbar\omega_{2i}}{2}\beta\right],$$

$$\alpha_{2i} = \frac{m}{2\hbar}\left[\omega_{2i}\coth\frac{\hbar\omega_{2i}}{2}\beta + \omega_{1i}\coth\frac{\hbar\omega_{1i}}{2}\beta\right],$$

$$\gamma_{1i} = \frac{m}{2\hbar}\left[\omega_{2i}\tanh\frac{\hbar\omega_{2i}}{2}\beta - \omega_{1i}\tanh\frac{\hbar\omega_{1i}}{2}\beta\right],$$

$$\gamma_{2i} = \frac{m}{2\hbar}\left[\omega_{2i}\coth\frac{\hbar\omega_{2i}}{2}\beta - \omega_{1i}\coth\frac{\hbar\omega_{1i}}{2}\beta\right],$$

$$\omega_{11} = \omega_{12} = \sqrt{\omega^2 - \frac{\Phi'(\nu_k a)}{m}}, \quad \omega_{13} = \sqrt{\omega^2 - \frac{\Phi''(\nu_k a)}{m}},$$

$$\omega_{21} = \omega_{22} = \sqrt{\omega^2 + \frac{\Phi'(\nu_k a)}{ma\nu_k}},$$

$$\omega_{23} = \sqrt{\omega^2 + \frac{\Phi''(\nu_k a)}{m}}.$$

At higher temperatures, when the perturbation-theory series converges rapidly enough and the anharmonic terms become significant, the problem of finding a coordinate representation for $R_2(1,2)$ can be reduced to determining the coordinate representation of $R_1(1)$ (Ref. 37). In fact, the density matrix satisfies the differential equation

$$\frac{\partial\rho}{\partial\beta} = -H\rho\,.$$

If the Hamiltonian H is close to the Hamiltonian H_0, it is convenient to write

$$H = H_0 + H_1,$$

$$\frac{\partial\rho_0}{\partial\beta} = -H_0\rho_0.$$

The problem is to obtain, knowing ρ_0, an approximate expression for the density matrix ρ. The latter is expected to be close to $\rho_0 = e^{-\beta H_0}$, with $e^{H_0\beta}\rho$ slowly varying with β:

$$\frac{\partial}{\partial\beta}(e^{H_0\beta}\rho) = H_0 e^{-H_0\beta}\rho + e^{H_0\beta}\frac{\partial\rho}{\partial\beta} = e^{H_0\beta}H_0\rho - e^{-H_0\beta}\rho = -e^{-H_0\beta}H_1\rho.$$

Integrating this equation from zero to β and recognizing that

$$e^{-H_0\beta}\rho = 1$$

at $\beta = 0$, we have

$$e^{H_0\beta}\rho(\beta) - 1 = -\int_0^\beta e^{-H_0\beta}H_1\rho(\beta')d\beta'\,.$$

Consequently,

$$\rho(\beta)=\rho_0(\beta)-\int_0^{\beta}\rho_0(\beta-\beta')H_1\rho(\beta')d\beta'.$$

The second term on the right-hand side of the equation is small if H_1 is small: this term is a correction to the approximate equation $\rho\approx\rho_0$. If the correction term is indeed small, we can use an approximate equation for $\rho(\beta^1)$. For example, putting approximately $\rho(\beta')=\rho_0(\beta')$, we obtain

$$\rho(\beta)=\rho_0(\beta)-\int_0^{\beta}\rho_0(\beta-\beta')H_1\rho_0(\beta')d\beta'.$$

Using this expression as the next approximation for $\rho(\beta')$, we find an even better approximation for ρ. Continuing this iteration, we arrive at the expression

$$\begin{aligned}\rho(\beta)=\rho_0(\beta)&-\int_0^{\beta}d\beta'\,[\rho_0(\beta-\beta')H_1\rho_0(\beta')]\\&+\int_0^{\beta}d\beta'\int_0^{\beta'}d\beta''\,[\rho_0(\beta-\beta')H_1\rho_0(\beta'-\beta'')H_1\rho_0(\beta'')]\\&-\int_0^{\beta}d\beta'\int_0^{\beta'}d\beta''\,[\cdots]+\cdots.\end{aligned}\tag{7.169}$$

For the unrenormalized density matrix

$$\tilde{R}_2(1,2)=\exp[-\beta\tilde{H}_2(1,2)]$$

we obtain then in accordance with Eq. (7.169),

$$\begin{aligned}\tilde{R}_2(1,2)=\tilde{R}_1(1)\tilde{R}_1(2)&-\int_0^{\beta}d\beta'\,[\tilde{R}_1(1,\beta-\beta')\\&\times R_1(2,\beta-\beta')\,\tilde{\Phi}(1,2)\tilde{R}_1(1,\beta')\tilde{R}_1(2,\beta')\,]\\&+\int_0^{\beta}d\beta'\int_0^{\beta'}d\beta''\,[\cdots]+\cdots.\end{aligned}\tag{7.170}$$

This yields readily an expansion for Q_2:

$$Q_2=Q_1^2\left[1-\frac{\beta}{2N}\operatorname*{Tr}_{1,2}\tilde{\Phi}(1,2)R_1(1)R_1(2)+\cdots\right].\tag{7.171}$$

The final expansion of the normalized binary density matrix is

$$\begin{aligned}R_2(1,2)&=R_1(1)R_1(2)\\&-\int_0^{\beta}d\beta'\,[R_1(1,\beta-\beta')R_1(2,\beta-\beta')\,\tilde{\Phi}(1,2)R_1(1,\beta')R_1(2,\beta')\,]\\&+R_1(1)R_1(2)\frac{\beta}{2N}\operatorname*{Tr}_{3,4}\tilde{\Phi}(3,4)R_1(3)R_1(4)+\cdots.\end{aligned}\tag{7.172}$$

Having thus derived expressions for $R_1(1)$ and $R_2(1,2)$, we can formulate the thermodynamics of the system.

49. Modified statistical variational principle for quantum systems

We introduce an approximating Hamiltonian H_0 in the form of an additive operator of the unitary type

$$H_0 = \sum_{1 \le i \le N} h_{0i} = \sum_{1 \le i \le N} \left\{ -\frac{\hbar^2}{2m} \nabla_i^2 + u_1(q_i) \right\}. \qquad (7.173)$$

As shown in Ref. 33, the maximum value F_{max} of the free energy of the system is determined in this case by the self-consistent-field approximation.

We express F_{min} in terms of the binary density matrix

$$F_{min} = F_0 + \tfrac{1}{2}\,\mathrm{Tr}[\Phi(1,2) - u_1(1) - u_1(2)]R_2(1,2). \qquad (7.174)$$

Here $R_2(1,2)$ is the exact expression for the binary density matrix. Determination of F_{min} to some given degree of accuracy reduces to finding $R_2(1,2)$ to the same accuracy.

Knowing the upper and lower limits F_{max} and F_{min} of the system free energy we obtain a better approximation of its true value by using for the energy the expression

$$F = \tfrac{1}{2}(F_{max} + F_{min}). \qquad (7.175)$$

Just as the inequality (2.77) represents the Bogolyubov variational principle, the equation (7.175) is a modified variational principle.

Let us show that this principle determines F with accuracy not worse than perturbation theory to second-order inclusive (the result is actually more accurate, since the variational principle is applicable also when the perturbation-theory series does not converge). To this end, we represent the Hamiltonian of the system in the form[27]

$$H_g = H_0 + g\,(H - H_0). \qquad (7.175')$$

The unperturbed Hamiltonian is of the form (7.173), while $H - H_0$ is regarded as a perturbation. The parameter g was introduced only to simplify the calculations.

It is easily seen that

$$H_g|_{g=1} = H$$

is the true Hamiltonian, and

$$H_g|_{g=0} = H_0$$

is the unperturbed Hamiltonian.

The partition function is

$$Q = \sum_n \{\psi_n, \exp[-\beta H_0 - \beta(H - H_0)]\psi_n\}.$$

The energy $H - H_0$ is proportional to N. It follows hence that Q cannot be expanded in powers of $H - H_0$. We can, however, expand in powers of $H - H_0$

the free energy which is proportional to $\ln Q$. Each order of the expansion should then be proportional to N, and all the expansion should converge.

We put

$$Q = Q(g)$$

and expand $\ln Q(g)$ in powers of g. In contrast to Ref. 27, we consider here expansion both about the point $g = 0$ and about the point $g = 1$. Introducing the notation

$$Q'(0) = \left.\frac{\partial Q(g)}{\partial g}\right|_{g=0}, \quad Q''(0) = \left.\frac{\partial^2 Q(g)}{\partial g^2}\right|_{g=0},$$
$$Q'(1) = \left.\frac{\partial Q(g)}{\partial g}\right|_{g=1}, \quad Q''(1) = \left.\frac{\partial^2 Q(g)}{\partial g^2}\right|_{g=1},$$

we obtain

$$\ln Q(1) = \ln Q(0) + \frac{Q'(0)}{Q(0)} + \frac{1}{2}\left\{\frac{Q''(g_0)}{Q(g_0)} - \left(\frac{Q'(g_0)}{Q(g_0)}\right)^2\right\}, \tag{7.176}$$

$$\ln Q(0) = \ln Q(1) - \frac{Q'(1)}{Q(1)} + \frac{1}{2}\left\{\frac{Q''(g_1)}{Q(g_1)} - \left(\frac{Q'(g_1)}{Q(g_1)}\right)^2\right\}, \tag{7.177}$$

where $0 \leqslant g_0 \leqslant 1$ and $0 \leqslant g_1 \leqslant 1$.

Since the function $\ln Q(g)$ is convex, we note that Eqs. (2.77) and (2.79) follow from Eqs. (7.176) and (7.177), respectively. If $\ln Q(1)$ is defined as the arithmetic mean of its values obtained from Eqs. (7.176) and (7.177), we have

$$\ln Q(1) = \ln Q(0) + \frac{1}{2}\left[\frac{Q'(0)}{Q(0)} + \frac{Q'(1)}{Q(1)}\right]$$
$$+ \frac{1}{4}\left\{\left(\frac{Q''(g_0)}{Q(g_0)}\right) - \left(\frac{Q'(g_0)}{Q(g_0)}\right)^2 - \left[\frac{Q''(g_1)}{Q(g_1)} - \left(\frac{Q'(g_1)}{Q(g_1)}\right)^2\right]\right\}. \tag{7.178}$$

Recognizing that the sign of the second derivative of $\ln Q(g_0)$ is preserved, we see that the second-order terms in Eq. (7.178) are smaller than those in Eq. (7.176). If they are expanded about the point $g = \frac{1}{2}$ we find that the second-order terms in Eq. (7.178) are actually of third order of smallness. It follows from Eqs. (7.174) and (7.175) that

$$F = F_{\rm SCF} + \tfrac{1}{4}\,\mathrm{Tr}[\Phi(1,2) - u_1(1) - u_1(2)]\,G_2(1,2), \tag{7.179}$$

where

$$F_{\rm SCF} = -\theta N \ln \mathop{\mathrm{Tr}}_{1} \tilde{R}_1(1) + \tfrac{1}{2} \mathop{\mathrm{Tr}}_{1,2} [\Phi(1,2) - u_1(1) - u_1(2)]R_1(1)R_1(2) \tag{7.180}$$

is the free energy calculated in the self-consistent-field approximation.

Note that in addition to Eq. (7.175) we can put

$$F = \alpha F_{\min} + (1-\alpha)F_{\max},$$

where

$$0 \leqslant \alpha \leqslant 1$$

is a certain parameter chosen to obtain best agreement between the theoretical and experimental data.

50. Free energy of a crystal in the quasiharmonic approximation with allowance for correlation

We proceed now to a direct calculation of the explicit form of the free energy of a crystal.

We choose $u_1(q)$ in the form

$$u_1(q) = u_1^0 + \tfrac{1}{2} m\omega^2 (q_1^{1^2} + q_1^{2^2} + q_1^{3^2}). \tag{7.181}$$

In the quasiharmonic approximation we have according to Eq. (7.180),

$$F_{\text{SCF}}/N = \beta^{-1} \sum_{\alpha=1}^{3} \ln \left[2 \sinh \left(\frac{\beta \hbar}{2} \sqrt{\frac{\lambda \alpha}{m}} \right) \right] + E. \tag{7.182}$$

In addition, we now have

$$F - F_{\text{SCF}} = \tfrac{1}{4} \operatorname*{Tr}_{1,2} \Phi(1,2)\, R_2(1,2) - \tfrac{1}{2} E^h, \tag{7.183}$$

where

$$E^h = \frac{3\hbar\omega}{2} + \frac{3\hbar\omega}{\exp(\hbar\omega/\theta) - 1}. \tag{7.184}$$

Furthermore, since $\Phi''(a) \gg \Phi'(a)/a$, we regard $\Phi'(a)/a \sim 0$ as corrections to the correlation terms; therefore,

$$\tfrac{1}{2} \operatorname{Tr} \Phi(1,2) R_2(1,2) = \sum_k p_k E_k^h, \tag{7.185}$$

where

$$p_k = \frac{z_k \Phi''(\nu_k a)}{m\omega^2 + \Phi''(\nu_k a)},$$

$$E_k^h = \frac{3\hbar\omega_k}{2} + \frac{3\hbar\omega_k}{\exp(\hbar\omega_k/\theta) - 1},$$

$$\omega_k = \sqrt{\omega^2 + \Phi''(\nu_k a)/m}.$$

The final expression for the crystal free energy in the quasiharmonic approximation, with the correlations taken into account, is

$$F = N \left\{ \beta^{-1} 3 \ln \left[2 \sinh \left(\frac{\beta \hbar}{2} \sqrt{\frac{\lambda \alpha}{m}} \right) \right] + E + \frac{1}{2} \left(\sum_{k=1}^{N-1} p_k E_k^h - E^h \right) \right\}. \tag{7.186}$$

For the crystal heat capacity

$$C_V = -k\theta\left(\frac{\partial^2 F}{\partial\theta^2}\right)_V,$$

we obtain from Eq. (7.183),

$$C_V = 3R\left\{\left(\frac{\hbar\omega}{\theta}\right)^2 \frac{e^{\hbar\omega/\theta}}{(e^{\hbar\omega/\theta}-1)^2} - \frac{1}{2}\left(C^h - \sum_{k=1}^{N-1} p_k C_k^h\right)\right\}, \qquad (7.187)$$

where

$$C^h = 2\left(\frac{\hbar\omega}{\theta}\right)^2 \frac{\exp(\hbar\omega/\theta)}{[\exp(\hbar\omega/\theta)-1]^2} - \left(\frac{\hbar\omega}{\theta}\right)^3 \frac{\exp(\hbar\omega/\theta)[\exp(\hbar\omega/\theta)-1]}{[\exp(\hbar\omega/\theta)-1]^3},$$

and C_k^h is obtained by replacing ω by ω_k in Eq. (7.186). If terms of order, $(1/a)\Phi'(a)$ are not neglected, we have for F,

$$F = N\left\{\beta^{-1}3\ln\left[2\sinh\left(\frac{\beta\hbar}{2}\sqrt{\frac{\lambda}{m}}\right)\right] + E + \frac{1}{2}\left[\sum_{k=1}^{N-1}(p_k E_k^h + t_k E_{1k}^h) - E^h\right]\right\}, \qquad (7.188)$$

where

$$t_k = \frac{2z_k \Phi'(\nu_k a)}{a\nu_k\left[m\omega^2 + \Phi'(\nu_k a)/\nu_k a\right]},$$

$$E_{1k}^h = E^h(\omega_{1k}),$$

$$\omega_{1k} = \sqrt{\omega^2 + \frac{\Phi'(\nu_k a)}{ma\nu_k}}.$$

The heat capacity is in this case

$$C_V = 3R\left\{\left(\frac{\hbar\omega}{\theta}\right)^2 \frac{\exp(\hbar\omega/\theta)}{[\exp(\hbar/\theta)-1]^2} - \frac{1}{2}\left[E^h - \sum_{k=1}^{N-1}(p_k C_k^h + t_k C_{1k}^h)\right]\right\}. \qquad (7.189)$$

We see from Eq. (7.187) that, when allowance is made for the correlations, the decrease of the crystal heat capacity at low temperature is slower than that given by the Einstein formula, in accord with the experimental data.

To estimate the contribution of the correlation terms to the crystal free energy, we introduce the quantity

$$\delta F = \frac{N\Sigma_{k=1}^{N-1}(p_k E_k^h + t_k E_1^h) - E^h}{F}. \qquad (7.190)$$

Numerical values of δF are listed in Table XXIII. The experimental data for the heat capacity C_V were taken from Ref. 183. Figures 42–44 show, in arbitrary units, plots of $C = C_V/3R$, where R is the universal gas constant (the circles mark the experimental data), the heat capacities C_V calculated in the

Table XXIII. Numerical values of δ*F*.

T (K)	Ne	Ar	Kr	Xe
4	8.91	1.21	0.60	0.37
6	8.91	1.21	0.60	0.37
8	8.91	1.21	0.60	0.37
10	8.91	1.21	0.60	0.37
15	8.91	1.21	0.60	0.37
20	8.89	1.20	0.50	0.36
25	8.81	1.19	0.56	0.33
30	8.64	1.17	0.52	0.30
35	8.34	1.13	0.47	0.26
40	7.92	1.07	0.40	0.20

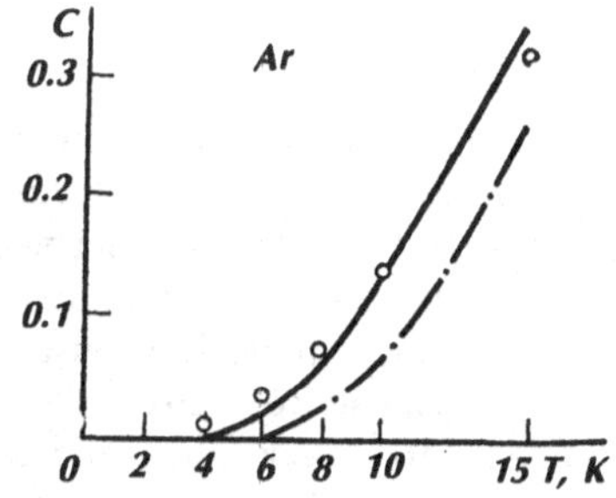

Figure 42.

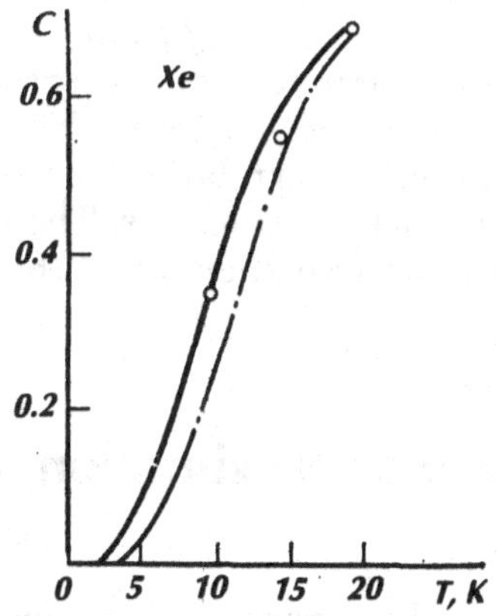

Figure 43.

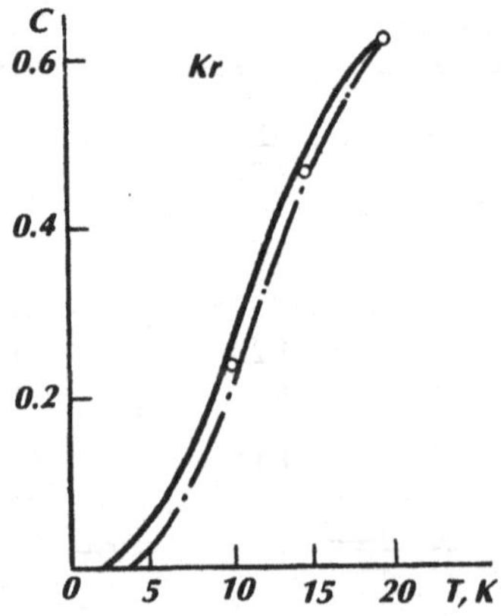

Figure 44.

self-consistent approximation are represented by dash–dot curves, and those obtained with allowance for correlation by continuous ones.

Numerical estimates show that allowance for the correlations improves the agreement between the experimental and theoretical curves at low temperatures for all substances. At sufficiently high temperatures (40, 30, and 20 K for Ar, Kr, and Xe, respectively), the heat-capacity plots in the self-consistent-field approximation agree with those obtained with allowance for the correlations. For Ar, the self-consistent approximation leads to good agreement with experiment from 20 K upwards (the error is 10%), allowance for the correlations leads to good agreement between theory and experiment to 9 K, and to approximate agreement all the way to 5 K. The Einstein temperature for Ne is higher than the melting point. A comparison with experiment is therefore possible only when account is taken for higher-order anharmonicities, which are signficiant near the melting point.

In the 4–20 K range, the contribution of the correlation terms to the free energy is ~9%, ~1.5%, ~0.6%, and ~0.4% for Ne, Ar, Kr, and Xe, respectively. Thus, the contribution of the correlations for Ne is of the same order as the contribution of three-particle interactions. The self-consistent approximation leads to fairly accurate free energies of the remaining considered substances.

51. First-order approximation of unary density matrix

We transform in the Bogolyubov operator equations (7.1) to the coordinate representation. The first equation is then

$$-\frac{\hbar^2}{2m}(\Delta_{q_1}-\Delta_{q_1'})D_1(q_1,q_1')+\int[(|q_1-q_2|)$$
$$-\Phi(|q_1-q_2|)]D_2(q_1,q_2,q_1',q_2)dq_2=0. \qquad (7.191)$$

If the binary density matrix is multiplicative, we obtain the known expression for the pair density matrix in the self-consistent-field approximation. We substitute Eq. (7.167) in Eq. (7.191) and recognize that the sum of the squares is invariant to rotations. We consider the factor

$$\mathscr{L}=\exp\left\{-\left[\frac{\alpha_1}{4}\left(1-\frac{\gamma_1^2}{\alpha_1^2}\right)-\frac{m\omega}{4\hbar}\tanh\frac{\hbar\omega}{2}\beta\right](q_1^3+q_1'^3)^2\right.$$
$$\left.-\left[\frac{\alpha_2}{4}-\frac{m\omega}{4\hbar}\coth\frac{\hbar\omega}{2}\beta\right](q_1^3-q_1'^3)^2\right\}. \qquad (7.192)$$

Then the first term in the argument of the exponential is quadratic in the ratio of the coupling constant $\Phi''(\nu_k a)$ to the elastic constant $m\omega^2$ of the self-consistent field. For face-centered crystals their ratio is of the order of $3.z_1$ and the respective square is of the order of $9/z^2$. A series expansion of $\mathscr{L}$ is therefore possible. In the harmonic approximation we obtain [recognizing that the coordinate-independent part of the self-consistent-field potential in Eq. (7.191) plays no role] the following equation for the frequencies:

$$\frac{m\omega^2}{2}=\frac{m}{2}\sum_{k\neq 1}\left\{\frac{z_k}{3}\Phi''(\nu_k a)\left(1-\frac{\omega_{23}\tanh(\hbar\omega_{23}/2)\beta-\omega_{13}\tanh(\hbar\omega_{13}/2)\beta}{\omega_{13}\tanh(\hbar\omega_{13}/2)\beta+\omega_{23}\tanh(\hbar\omega_{23}/2)\beta}\right)\right.$$
$$\left.+\frac{2z_k\Phi'(\nu_k a)}{3a\nu_k}\left(1-\frac{\omega_{21}\tanh(\hbar\,\omega_{21}/2)\beta-\omega_{11}\tanh(\hbar\omega_{11}/2)\beta}{\omega_{11}\tanh(\hbar\omega_{11}/2)\beta+\omega_{21}\tanh(\hbar\omega_{21}/2)\beta}\right)\right\}$$
$$\times\prod\left(\frac{2\omega_{1i}\omega_{2i}\tanh(\hbar\omega_{1i}/2)\beta\tanh(\hbar\omega_{2i}/2)\beta}{\omega\tanh(\hbar\omega/2)\beta\left[\omega_{1i}\tanh(\hbar\omega_{1i}/2)\beta+\omega_{2i}\tanh(\hbar\omega_{2i}/2)\beta\right]}\right)^{1/2}. \qquad (7.193)$$

In terms of new dimensionless variables

$$\omega_0=\left(\frac{z_1\Phi''(a)}{3m}\right)^{1/2},\quad x=\frac{\omega}{\omega_0},\quad \mu=\frac{\theta}{\hbar\omega_0},$$
$$l_{k1}=l_{k2}=\frac{\Phi'(V_k a)}{\omega_0^2 m V_k a},\quad p_{k1}=p_{k2}=\frac{z_k\Phi'(V_k a)}{z_1\Phi''(a)a},$$
$$l_{13}=\frac{\Phi''(V_k a)}{\omega_0^2 m},\quad p_{k3}=\frac{z_k\Phi''(V_k a)}{z_1\Phi''(a)},$$

we can rewrite Eq. (7.193) in the form

$$x^2 = \sum_{k=1} \left\{ \frac{2p_{k3}\sqrt{x^2 - l_{k3}} \tanh \frac{\sqrt{x^2 - l_{k3}}}{2\mu}}{\sqrt{x^2 - l_{k3}} \tanh \frac{x^2 - l_{k3}}{2\mu} + \sqrt{x^2 + l_k} \tanh \frac{\sqrt{x^2 + l_{k1}}}{2\mu}} \right\}$$

$$\times \prod_{i=1}^{3} \left(\frac{2\sqrt{x^4 - l_{ki}^2} \tanh \frac{\sqrt{x^2 - l_{ki}}}{2\mu} \tanh \frac{\sqrt{x^2 + l_{ki}}}{2\mu}}{xth\left(\frac{x}{2\mu}\right)\left(\sqrt{x^2 - l_{ki}} \tanh \frac{\sqrt{x^2 - l_{ki}}}{2\mu} + \sqrt{x^2 + l_{ki}} \tanh \frac{\sqrt{x^2 + l_k}}{2\mu}\right)} \right)^{1/2}. \tag{7.194}$$

From Eq. (7.194) we obtain $\omega = \omega(a,\theta)$ and, using the relation

$$\frac{\partial F}{\partial a} = -\frac{3pv}{a}, \tag{7.195}$$

from which we determine $a = a(p,\theta)$, we obtain $\omega = \omega(p,\theta)$.

We now examine the question of finding the first-approximation unary density matrix with account taken of anharmonicities of arbitrary order. We express the first Bogolyubov-chain equation in the form

$$[H_1(1),R_1(1)] + \operatorname*{Tr}_2\{\Phi(1,2)R_2(1,2)R_1^{-1}(1)R_1^{-1}(2)R_1(2)R_1(1)$$

$$- R_1(1)\operatorname*{Tr}_2 R_1(2)R_1^{-1}(1)R_1^{-1}(2)R_2(1,2)\Phi(1,2)\} = 0, \tag{7.196}$$

where

$$R_1^{-1}(1) = Q_1 \exp[\beta\tilde{H}_1(1)] .$$

To continue, we must analyze Eq. (7.196). We introduce for simplicity the notation

$$A = \tilde{H}_1(1) + \tilde{H}_1(2), \quad B = -[\tilde{H}_1(1) + \tilde{H}_1(2) + \Phi(1,2)] . \tag{7.197}$$

The commutator is

$$C = [A,B] = -[H_1(1) + H_1(2),\Phi(1,2)]. \tag{7.198}$$

Accordingly

$$[A,C] = \{H_1(1) + H_1(2),[H_1(1) + H_1(2),\Phi(1,2)]\}$$
$$- \{v(1),[H_1(1),\Phi(1,2)]\} - \{v(2),[H_1(2),\Phi(1,2)]\}, \tag{7.199}$$

where

$$v(1) = \operatorname*{Tr}_2 \Phi(1,2)R_1(2). \tag{7.200}$$

The different terms of Eq. (7.198) are of the order of

$$\{H_1(1)+H_1(2),[H_1(1)+H_1(2),\Phi_{12}]\}\sim\hbar^4,$$
$$\{v(1),[H_1(1),\Phi(1,2)]\}\sim\{v(2),[H_1(2),\Phi(1,2)]\}$$
$$\sim\hbar^2(\nabla_1\Phi)^2. \tag{7.201}$$

In the case of a crystal we have for the atom-localization region

$$|\nabla_1\Phi(1,2)|\ll|\nabla_1^2\Phi(1,2)|\sigma, \tag{7.202}$$

therefore,

$$[A,c]=O(c). \tag{7.203}$$

We have analogously

$$[B,c]=O(c). \tag{7.204}$$

We can therefore use in this case the Glauber theorem (the Becker–Hausdorff identity)

$$\exp[\lambda A]\exp[\lambda\beta\ ]=\exp[-\lambda(A+B)]\exp\left[\frac{\lambda^2}{2}[A,B]\right]. \tag{7.205}$$

In our case,

$$R_2(1,2)R_1^{-1}(1)R_1^{-1}(2)=\frac{Q_1^2}{Q_2}\exp[-\beta\Phi(1,2)]$$
$$\times\exp\left\{-\frac{\beta^2}{2}[H_1(1)+H_1(2),\Phi(1,2)]\right\}, \tag{7.206}$$
$$R_1^{-1}(1)R_1^{-1}(2)R_2(1,2)=\frac{Q_1^2}{Q_2}\exp[-\beta\Phi(1,2)]$$
$$\times\exp\left\{\frac{\beta^2}{2}[H_1(1)+H_1(2),\Phi(1,2)]\right\}, \tag{7.207}$$
$$R_1^{-1}(1)R_1^{-1}(2)R_2(1,2)=\frac{Q_1^2}{Q_2}\exp[-\beta\Phi(1,2)]$$
$$\times\exp\left\{\frac{\beta^2}{2}[H_1(1)+H_1(2),\Phi(1,2)]\right\}.$$

Neglecting the quantum corrections in the correlation terms, we can rewrite Eq. (7.196) in the form

$$[\tilde{\tilde{H}}_1(1),R_1(1)]=0, \tag{7.208}$$

where

$$\tilde{\tilde{H}}_1=H(1)+\operatorname*{Tr}_1 K(1,2)R_1(2),$$
$$K(1,2)=\Phi(1,2)\exp[-\beta\Phi(1,2)]Q_1^2/Q_2, \tag{7.209}$$

whence

$$R_1(1) = \frac{1}{Q_{11}} \exp[-\beta \tilde{\tilde{H}}_1(1)], \quad Q_{11} = \mathrm{Tr}\exp[-\beta \tilde{\tilde{H}}(1)]/N. \qquad (7.210)$$

We see thus that the particle motion is described by the effective potential $K(1,2)$.

The equation of motion of a particle in the effective-potential field takes the form

$$\Delta_1^2 \psi + \frac{2m}{\hbar}\left[E - \mathrm{Tr}_2\, K(1,2)R_1(2)\right]\psi = 0. \qquad (7.211)$$

References

[1]J. W. Gibbs, "Foundations of Statistical Mechanics," in *Collected Works of J. Willard Gibbs* (Longmans, New York, 1928).

[2]N. N. Bogolyubov, *Lectures on Quantum Statistics* (Macdonald, London, 1967).

[3]H. Goldstein, *Classical Mechanics* (Addison-Wesley, Reading, MA, 1950).

[4]M. A. Leontovich, *Statistical Physics* (in Russian) (Gostekhizdat, 1944).

[5]J. von Neumann, *Mathematical Foundations of Quantum Mechanics* (Princeton Univ. Press, Princeton, 1955).

[6]P. A. M. Dirac, *The Principles of Quantum Mechanics* (Oxford Univ. Press, Oxford, 1935).

[7]L. D. Landau and E. M. Lifshitz, *Quantum Mechanics, Nonrelativistic Theory* (Pergamon, New York, 1977).

[8]I. P. Bazarov, *Thermodynamics* (in Russian) (Vysshaya Shkola, Kiev, 1983).

[9]N. N. Bogolyubov, Fiz. Elem. Chast. At. Yad. **9**, 781 (1978) [Sov. J. Part. Nuclei **9**, 315 (1978)].

[10]N. N. Bogolyubov (Jr.), *Method of Investigating Model Hamiltonians* (in Russian) (Nauka, Moscow, 1974).

[11]N. N. Bogolyubov, "Certain Problems Connected with the Foundations of Statistical Mechanics" (in Russian), in *History and Methodology of Natural Sciences* (Moscow Univ. Press, Moscow, 1983).

[12]G. D. Birkhoff, Proc. Natl. Acad. Sci. U.S.A. **17**, 650 (1931).

[13]L. Boltzmann, *Vorlesungen über Gastheorie* (Barth, Leipzig, 1923).

[14]O. Penrose, Rep. Prog. Phys. **12**, No. 12 (1979).

[15]Ya. G. Sinai, Usp. Mat. Nauk **25**, No. 2, 141 (1970).

[16]D. Ruelle, *Statistical Mechanics, Rigorous Results* (Benjamin, Reading, MA, 1969).

[17]C. Croxton, *Liquid State Physics* (Cambridge Univ. Press, Cambridge, 1974).

[18]J. E. Mayer and M. G. Mayer, *Statistical Mechanics* (Wiley, New York, 1940).

[19]B. T. Geilikman, *Statistical Theory of Phase Transitions* (in Russian) (GITTL, 1954).

[20]H. D. Ursell, Proc. Cambridge Philos. Soc. **23**, 685 (1927).

[21]R. H. Fowler, *Statistical Mechanics* (Cambridge Univ. Press, Cambridge, 1936).

[22]J. O. Hirschfelder, C. F. Curtiss, and R. B. Bird, *Molecular Theory of Gases and Liquids* (Wiley, New York, 1954).

[23]F. H. Ree and W. G. Hoover, J. Chem. Phys. **40**, 939 (1964).

[24]E. M. Mason and T. H. Spurling, *The Virial Equation of State* (Pergamon, New York, 1969).

[25]J. Ram and Y. Singh, Mol. Phys. **16**, 539 (1973).

[26]J. G. Kirkwood, Phys. Rev. **44**, 31 (1933).

[27]G. Leibfried and W. Ludwig, Solid State Phys. **12**, 276 (1971).

[28]M. Born and E. Brody, Z. Phys. **6**, No. 2, 132 (1921).

[29]M. Born and E. Brody, Z. Phys. **11**, No. 6, 327 (1922).

[30]E. Schrödinger, Z. Phys. **11**, No. 3, 170 (1922).

[31]G. Leibfried, Handb. Phys. **7**, 104 (1955).

[32]S. Nakajima, S. Adv. Phys. **4**, 363 (1955).

[33]I. P. Bazarov, *Statistical Theory of the Crystalline State* (in Russian) (Moscow Univ. Press, Moscow, 1972).

[34]K. Westera and E. R. Cowley, Phys. Rev. B **11**, 4008 (1975).

[35]I. A. Kvasnikov, Author's abstract of candidate's dissertation, Moscow State Univ., 1958.

[36]S. V. Tyablikov, *Methods of Quantum Theory of Magnetism* (in Russian) (Nauka, Moscow, 1975).

[37]R. Feynman, *Statistical Mechanics* (Benjamin, Reading, MA, 1972).

[38]I. P. Bazarov and P. N. Nikolaev, *Crystal Correlation Theory* (in Russian) (Moscow Univ. Press, Moscow, 1981).

[39]I. P. Bazarov and P. N. Nikolaev, Zh. Fiz. Khim. **55**, 1405 (1981).

[40]S. V. Tyablikov and V. V. Tolmachev, Nauch. Dokl. Vyssh. Shkoly. Fiz.-Mat. Nauki, No. 1, 101 (1958).

[41]F. Wagner and H. Koppe, Z. Naturforsch. **20A**, 1553 (1965).

[42]A. A. Vedenov and A. I. Larkin, Zh. Eksp. Teor. Fiz. **36**, 1133 (1958) [Sov. Phys. JETP **9**, 836 (1958)].

[43]I. Z. Fisher, *Statistical Theory of Liquids* (in Russian) (Fizmatgiz, Moscow, 1961).

[44]J. K. Perkus and G. J. Yevick, Phys. Rev. **110**, 1 (1958).

[45]N. P. Kovalenko and I. Z. Fisher, Usp. Fiz. Mat. Nauk **108**, 300 (1972).

[46]E. Thiele, J. Chem. Phys. **39**, 474 (1963).

[47]M. S. Wertheim, Phys. Rev. Lett. **10**, 321 (1963).

[48]F. M. Kuni, *Statistical Physics and Thermodynamics* (in Russian) (Nauka, Moscow, 1981).

[49]R. Balescu, *Equilibrium and Nonequilibrium Statistical Mechanics* (Wiley, New York, 1975).

[50]A. A. Maradudin, E. W. Montroll, and G. H. Weiss, *Theory of Lattice Dynamics in the Harmonic Approximation* (Academic, New York, 1963).

[51]M. L. Klein and J. A. Venables, *Rare Gas Solids* (Academic, New York, 1976).

[52]I. P. Bazarov and P. N. Nikolaev, Teor. Mat. Fiz. **31**, 125 (1977).

[53]N. M. Plakida, in *Statistical Physics and Quantum Field Theory* (in Russian), edited by N. N. Bogolyubov (Nauka, Moscow, 1973), p. 205.

[54]B. N. Rolov, V. A. Ivin, and V. N. Kuzkov, *Statistics and Kinetics of Phase Transitions in Solids* (in Russian) (Riga, 1979).

[55]S. A. Shekatolina and L. N. Yakub, Ukr. Fiz. Zh. **21**, 535 (1976).

[56]I. P. Bazarov and P. N. Nikolaev, Vestn. Mosk. Univ. Ser. Fiz. Astronomiya **19**, No. 3, 59 (1978).

[57]I. P. Bazarov and P. N. Nikolaev, Teor. Mat. Fiz. **41**, 424 (1979).

[58]I. P. Bazarov, E. V. Gevorkyan, and V. V. Kotenok, *Statistical Theory of Polymer Transformations* (in Russian) (Moscow Univ. Press, Moscow, 1978).

[59]L. Tonks, Phys. Rev. **50**, 955 (1936).

[60]N. Metropolis, A. W. Rosenbluth, M. N. Rosenbluth, and A. H. Teller, J. Chem. Phys. **21**, 1087 (1953).

[61]R. H. Ree and W. G. Hoover, J. Chem. Phys. **40**, 639 (1964).

[62]*Padé Approximation and Its Applications, Proceedings of the 1979 Antwerp Conference* (Springer, Berlin, 1979).

[63]P. Heller and G. B. Bendek, Phys. Rev. Lett. **8**, 428 (1962).

[64]C. Domb and M. F. Sykes, J. Math. Phys. **2**, No. 1, 63 (1961).

[65]L. Boltzmann, Verslag. Gewone Vergrader. **7**, 484 (1899).

[66]H. Happerl, Ann. Phys. (Leipzig) **21**, 342 (1906).

[67]B. R. A. Nijboer, Phys. Rev. **85**, 777 (1952).

[68]B. T. Geilikman, Dokl. Akad. Nauk SSR **70**, No. 25 (1950).

[69]R. W. Zwanzig, J. Chem. Phys. **14**, 855 (1956).

[70]W. G. Hoover and A. G. de Rocco, J. Chem. Phys. **34**, 1059 (1964); W. G. Hoover and J. C. Poirier, *ibid.* **28**, 327 (1963); W. G. Hoover, *ibid.* **49**, 937 (1964).

[71]H. N. V. Temperly, Proc. Phys. Soc. London B **70**, Pt. 5, 536 (1957).

[72]H. B. Dwight, *Tables of Integrals* (Collier-MacMillan, 1961).

[73]T. Kihara, Rev. Mod. Phys. **25**, 831 (1953).

[74]A. E. Sherwood and E. A. Mason, Phys. Fluids **8**, 1577 (1965).

[75]W. L. Bruch, Phys. Fluids **10**, 2531 (1967); **11**, 1938 (1968).

[76]J. T. Vanderslice, Ind. Eng. Chem. **50**, 1033 (1958).

[77]J. Ambur and E. A. Masson, Phys. Fluids **1**, 370 (1958).

[78]D. Henderson and L. Oden, Phys. Fluids **9**, 1592 (1966).

[79]G. Mie, Ann. Phys. (Leipzig) **11**, 657 (1903).

[80]J. E. Lennard-Jones, Proc. R. Soc. London Ser. A **106**, 709 (1924).

[81]J. E. Lennard-Jones and A. E. Ingham, Proc. R. Soc. London Ser. A **107**, 636 (1925).

[82]J. A. Barker, P. J. Leonard, and A. Pompe, J. Chem. Phys. **44**, 4206 (1966).

[83]E. A. Sherwood and J. M. Prausnite, J. Chem. Phys. **41**, 429 (1964).

[84]V. Rabinovich, A. A. Vasserman, V. I. Nedostup, and V. S. Vekeler, *Thermophysical Properties of Neon, Argon, Krypton, and Xenon* (in Russian) (USSR Standards Publ. House, 1976).

[85]*Rare Gas Solids* (Pergamon, New York, 1977), Vol. 2.

[86]M. S. Green and J. V. Sengers, *Critical Phenomena*, U.S. National Bureau of Standards Misc. Publ. 273 (U.S. GPO, Washington, D.C., 1966).

[87]P. K. Suetin, *Classical Orthogonal Polynomials* (in Russian) (Nauka, Moscow, 1976).

[88]J. M. Richardson, A. B. Arons, and R. R. Halverson, J. Chem. Phys. **15**, 785 (1947).

[89]H. C. Anderson, J. D. Weeks, and D. Chandler, Phys. Rev. A **4**, 1597 (1971).

[90]J. D. Weeks and D. Chandler, J. Chem. Phys. **54**, 5237 (1971).

[91]L. Verlet and J. J. Weiss, Phys. Rev. A **5**, 939 (1972).

[92]C.-K. Ma, *Modern Theory of Critical Phenomena* (Benjamin-Cummings, Reading, MA, 1976).

[93]*Physics of Simple Liquids. Statistical Theory*, edited by H. N. V. Temperley, J. S. Rowlinson, and G. S. Rushbrook (North-Holland, Amsterdam, 1968).

[94]J. D. van der Waals and F. Kohnstamm, *Thermostatics* (Russian translation) (Mir, Leningrad, 1936).

[95]R. H. Brout, *Phase Transitions* (Benjamin, New York, 1965).

[96]L. P. Filippov, in *Equations of State of Gases and Liquids* (in Russian) (Nauka, Moscow, 1976), p. 46.

[97]M. P. Bukalovich and I. I. Novikov, *Thermodynamics* (in Russian) (Mashinostroenie, 1972).

[98]E. A. Guggenheim, J. Chem. Phys. **13**, 253 (1945).

[99]J. I. Frenkel, *Kinetic Theory of Gases* (Oxford Univ. Press, Oxford, 1946).

[100]J. G. Kirkwood and E. M. Boggs, J. Chem. Phys. **10**, 394 (1942).

[101]G. H. A. Cole, J. Chem. Phys. **36**, 1680 (1962).

[102]B. R. A. Nijboer and R. Fiescki, Physica (Utrecht) **19**, 545 (1953).

[103]J. G. Kirkwood, E. K. Mann, and B. J. Alder, J. Chem. Phys. **18**, 1040 (1950).

[104]T. Hill, *Statistical Mechanics* (McGraw–Hill, New York, 1956).

[105]J. G. Kirkwood and E. Monroe, J. Chem. Phys. **9**, 514 (1941).

[106]M. Klein, J. Chem. Phys. **39**, 1367 (1963); **39**, 1388 (1963).

[107]S. A. Rice and J. Yekner, J. Chem. Phys. **42**, 3559 (1965).

[108]A. S. Blokhin, *Use of Computation Methods for Problems in Hydrodynamics and Statistical Physics* (Moscow Univ. Press, Moscow, 1977).

[109]E. V. Arnshtein, Izv. Vyssh. Ucheb. Zaved. Fiz., No. 2, 92 (1959).

[110]G. A. Martynov, Zh. Eksp. Teor. Fiz. **45**, 656 (1963) [Sov. Phys. JETP **18**, 450 (1964)].

[111]N. E. Carnahan and K. E. Starling, J. Chem. Phys. **51**, 635 (1969).

[112]N. W. Asheroff and J. Lekner, Phys. Rev. **145**, 83 (1966).

[113]I. Z. Fisher, in *Equations of State in the Theory of Liquids* (in Russian) (Nauka, Moscow, 1975), pp. 27–45.

[114]I. R. Yukhnovskii and M. M. Golovkov, *Statistical Theory of Classical Equilibrium Systems* (in Russian) (Naukova Dumka, Kiev, 1980).

[115]J. A. Barker and D. Henderson, Rev. Mod. Phys. **48**, 548 (1976).

[116]N. H. March and M. Tosi, *Motion of Liquid Atoms* (Russian translation) (Metallurgiya, 1980).

[117]J. K. Perkus, Phys. Rev. Lett. **8**, 462 (1962).

[118]D. Levesque, Physica (Utrecht) **32**, 1985 (1966).

[119]M. S. Wertheim, J. Math. Phys. **8**, 927 (1967).

[120]D. Henderson, S. Kim, and L. Oden, Discuss. Faraday. Soc. **43**, 26 (1967).

[121]J. Yvon, Actual. Sci. Ind. **543** (1937).

[122]L. Verlet, Physica **30**, 95 (1964).

[123]P. N. Nikolaev, Vestn. Mosk. Univ. Ser. Fiz. Astronomiya **20** (No. 4), 81 (1979).

[124]I. P. Bazarov, Dokl. Akad. Nauk SSSR **170**, 312 (1966) [Sov. Phys. Dokl. **11**, 799 (1967)].

[125]A. A. Vlasov, *Many-Particle Theory* (in Russian) (M-L, 1950).

[126]A. A. Vlasov, Vestn. Mosk. Univ. Ser. Fiz. Astronomiya Nos. 3 and 4, 63 (1946).

[127]H. J. Raveche, J. Math. Phys. **17**, 1949 (1976).

[128]H. J. Raveche and C. A. Stuart, J. Chem. Phys. **63**, 1099 (1975).

[129]W. Kunkin and H. Y. Frisch, J. Chem. Phys. **50**, 1817 (1969).

[130]H. J. Raveche and C. A. Stuart, J. Chem. Phys. **65**, 2305 (1976).

[131]S. V. Tyablikov, Zh. Eksp. Teor. Fiz. **17**, 386 (1947).

[132]E. A. Arnshtein, Author's abstract of candidate's dissertation, Moscow State Polytechnic Institute, 1958.

[133]I. Z. Fisher, Usp. Fiz. Nauk **76**, 499 (1962) [Sov. Phys. Usp. **5**, 239 (1962)].

[134]I. Z. Fisher and B. L. Kopelovich, Dokl. Akad. Nauk SSR **133**, 81 (1960) [Sov. Phys. Dokl. **5**, 761 (1961)].

[135]G. H. A. Cole, Adv. Phys. **8**, 225 (1959).

[136]G. H. A. Cole, J. Chem. Phys. **28**, 912 (1958).

[137]R. Abe, Prog. Theor. Phys. **19**, 57 (1958).

[138]R. Abe, Prog. Theor. Phys. **19**, 407 (1958).

[139]*Physics of Simple Liquids, Experimental Research*, edited by V. H. N. Temperley *et al.* (North-Holland, Amsterdam, 1973).

[140]W. C. McMillan, Phys. Rev. **138**, A442 (1965).

[141]A. N. Tihonov and A. A. Samarskii, *Equations of Mathematical Physics* (Pergamon, New York, 1964).

[142]A. N. Lagar'kov and V. M. Sergeev, Usp. Fiz. Nauk **125**, 409 (1978) [Sov. Phys. Usp. **21**, 566 (1978)].

[143]S. A. Rice and A. R. Allnatt.

[144]I. Z. Fisher, Zh. Eksp. Teor. Fiz. **61**, 1647 (1971) [Sov. Phys. JETP **34**, 878 (1972)].

[145]S. A. Rice and A. R. Allnatt, J. Chem. Phys. **34**, 2144 (1961).

[146]A. R. Allnatt and S. A. Rice, J. Chem. Phys. **34**, 2156 (1961).

[147]T. Einwohner and B. J. Alder, J. Chem. Phys. **49**, 1458 (1968).

[148]W. W. Wood, J. Chem. Phys. **48**, 415 (1968).

[149]B. J. Alder and T. E. Wainwright, Phys. Rev. **127**, 359 (1962).

[150]B. Borstnik and A. Azman, Chem. Phys. Lett. **12**, 620 (1972).

[151]V. G. Bandakov, A. A. Galashev, and V. P. Skripov, Fiz. Nizk. Temp. **2**, 957 (1976).

[152]W. G. Hoover and F. H. Ree, J. Chem. Phys. **47**, 4873 (1967).

[153]W. G. Hoover and F. H. Ree, J. Chem. Phys. **49**, 3609 (1968).

[154]F. A. Lindemann, Z. Phys. **11**, 609 (1910).

[155]M. Born, Proc. Cambridge Philos. Soc. **36**, 160 (1940).

[156]Ph. F. Chagard, *The Anharmonic Crystal* (New York, 1967).

[157]N. M. Plakida and T. Siklos, "Theory of Anharmonic Crystals. I. General Analysis," JINR, Theoretical Physics Laboratory, Dubna Report, 1968.

[158]L. D. Landau and E. M. Lifshitz, *Statistical Physics*, Pt. I (Pergamon, New York, 1980).

[159]F. C. Andrews and J. M. Benson, Phys. Lett. **20**, 16 (1966).

[160]A. Rotenberg, J. Chem. Phys. **43**, 1198 (1965).

[161]P. A. Nelson, "Molecular Dynamics Study of Square-Well Fluid," thesis, Princeton University, 1966.

[162]B. J. Alder, J. Chem. Phys. **31**, 1666 (1959).

[163]L. Verlet, Phys. Rev. **159**, 98 (1967).

[164]A. Rotenberg, J. Chem. Phys. **42**, 1126 (1965).

[165]L. Verlet, Phys. Rev. **165**, 201 (1968).

[166]A. M. Evseev and M. Ya. Frenkel', in *Equations of State of Gases and Liquids* (in Russian) (Nauka, Moscow, 1975), pp. 106–116.

[167]B. J. Alder and T. E. Wainwright, Phys. Rev. Lett. **18**, 989 (1967).

[168]N. W. Ashcroft and N. D. Mermin, *Solid State Physics* (Holt, Reinhart and Winston, New York, 1974), Vol. 1.

[169]N. W. Ashcroft and N. D. Mermin, *Solid State Physics* (Holt, Reinhart and Winston, New York, 1974), Vol. 2.

[170]R. W. Myckoff, *Crystal Structures* (Interscience, New York, 1963).

[171]L. Pauling, *The Nature of the Chemical Bond* (Cornell Univ. Press, Ithaca, NY, 1940).

[172]B. M. Axilrod and E. Teller, J. Chem. Phys. **11**, 299 (1943).

[173]W. L. Bade, J. Chem. Phys. **27**, 1280 (1958).

[174]R. M. Axilrod, J. Chem. Phys. **17**, 1349 (1949); **19**, 719 (1951).

[175]*Quantum Crystals* (collection of translations) (in Russian) (Mir, Moscow, 1975).

[176]D. C. Wallace and J. L. Patrick, Phys. Rev. **137**, A152 (1965).

[177]J. E. Johes and A. E. Ingham, Proc. R. Soc. London Ser. A **107**, 636 (1925).

[178]M. L. Klein, G. K. Horton, and J. L. Feldman, Phys. Rev. **184**, 968 (1968).

[179]D. N. Batchelder, Phys. Rev. **162**, 767 (1967).

[180]E. R. Dobbs and G. O. Jones, Rep. Prog. Phys. **20**, 516 (1957).

[181]J. C. Slater, *Insulators, Semiconductors, and Metals* (McGraw–Hill, New York, 1967).

[182]A. Einstein, *Planck Radiation Theory and Specific-Heat Theory* (Russ. translation), *Collected Works, Vol. III* (Nauka, Moscow, 1966), p. 134.

[183]*Table of Physical Quantities* (in Russian) (Atomizdat, Moscow, 1976).

[184]M. Born and K. Huang, *Dynamical Theory of Crystal Lattices* (Oxford Univ. Press, Oxford, 1954).

[185]A. P. Prudnikov, Yu. A. Brichkov, and O. I. Marichev, *Integrals and Series* (in Russian) (Nauka, Moscow, 1981).

[186]N. M. Plakida and T. Siklos, JINR Dubna Report No. R4-3449, 1967.

[187]N. M. Plakida, Teor. Mat. Fiz. **12**, No. 1, 135 (1972).

[188]N. M. Plakida, Fiz. Tverd. Tela (Leningrad) **14**, 2841 (1972) [Sov. Phys. Solid State **14**, 2462 (1973)].

[189]N. M. Plakida, "Self-Consistent Dynamic Theory of Anharmonic Crystals," Author's abstract of doctoral dissertation, JINR, Dubna, 1973.

[190]R. Guyer, Solid State Phys. **23**, 413 (1969).

[191]K. Clusius, Z. Phys. Chem. **4**, 1 (1929); **31**, 459 (1936).

[192]R. W. Hill, thesis, Oxford University, 1954.

[193]B. F. Figgins, Ph.D. thesis, Queen Mary College, 1955.

[194]P. Flubacher *et al.*, Proc. Phys. Soc. **78**, 1449 (1961).

[195]R. Beaumont *et al.*, Proc. Phys. Soc. **78**, 1462 (1961).

[196]K. Clusius *et al.*, Ann. Phys. (Leipzig) **33**, 642 (1938); Z. Phys. Chem. **34**, 375 (1936).

[197]K. Clusius and L. Riccoboni, Z. Phys. Chem. **38**, 81 (1938).

[198]I. P. Bazarov, Fiz. Tverd. Tela (Leningrad) **11**, 840 (1969) [Sov. Phys. Solid State **11**, 692 (1969)].

[199]I. de Boer, Proc. R. Soc. London Ser. A **215**, 4 (1952).

[200]S. V. Tyablikov, Zh. Eksp. Teor. Fiz. **18**, 368 (1948).

[201]I. P. Bazarov and P. N. Nikolaev, Vestn. Mosk. Univ. Ser. Fiz. Astronomiya **24**, No. 6, 101 (1983).

[202]I. P. Bazarov and P. N. Nikolaev, Vestn. Mosk. Univ. Ser. Fiz. Astronomiya **25**, No. 2 (1984).

[203]I. P. Bazarov and P. N. Nikolaev, Teor. Mat. Fiz. **60**, No. 1 (1984).

[204]I. P. Bazarov and P. N. Nikolaev, Dokl. Akad. Nauka SSSR **296**, 321 (1987).

[205]P. N. Nikolaev, *Statistical Thermodynamics in the Correlation Expansion Method*, Moscow, 1988, pp. 202–204.

Printed in the United States
By Bookmasters